T0310121

MICROWAVE AMPLIFIER AND ACTIVE CIRCUIT DESIGN USING THE REAL FREQUENCY TECHNIQUE

Pierre Jarry and Jacques N. Beneat

WILEY

Published by John Wiley & Sons, Inc., Hoboken, New Jersey
Published simultaneously in Canada

For MATLAB® and Simulink® product information, or information on other related products, please contact:
The MathWorks, Inc.
3 Apple Hill Drive
Natick, MA 01760-2098 USA
Tel 508-647-7000
Fax: 508-647-7001
E-mail: info@mathworks.com
Web: www.mathworks.com

For general information on our other products and services or for technical support, please contact our Customer Care Department within the United States at (800) 762-2974, outside the United States at (317) 572-3993 or fax (317) 572-4002.

Wiley also publishes its books in a variety of electronic formats. Some content that appears in print may not be available in electronic formats. For more information about Wiley products, visit our web site at www.wiley.com.

Library of Congress Cataloging-in-Publication Data

Names: Jarry, Pierre, 1946– author. | Beneat, Jacques, 1964–
Title: Microwave amplifier and active circuit design using the real frequency technique / Pierre Jarry and Jacques N. Beneat.
Description: Hoboken : John Wiley & Sons, Inc., 2016. | Includes bibliographical references and index.
Identifiers: LCCN 2015040505 | ISBN 9781119073208 (hardback)
Subjects: LCSH: Microwave amplifiers–Design and construction. | Electric filters, Active–Design and construction. |
BISAC: TECHNOLOGY & ENGINEERING / Microwaves.
Classification: LCC TK7871.2 .J37 2016 | DDC 621.381/325–dc23
LC record available at http://lccn.loc.gov/2015040505

Set in 10/12pt Times by SPi Global, Pondicherry, India

Printed in the United States of America

10 9 8 7 6 5 4 3 2 1

Contents

Foreword

It has been a privilege for me to read through the manuscript of the *Microwave Amplifier and Active Circuit Design Using the Real Frequency Technique*. I found that the authors have been very thorough in putting together this outstanding book containing a unique blend of theory and practice. I sensed that the authors are very passionate about the design of microwave circuits, which is also evidenced from their other recent books.

The book provides an extensive use of an impedance matching methodology, known as the real frequency technique (RFT), in numerous applications. The topics treated include the RFT itself, design of a wide variety of multistage amplifiers, active filters, passive equalizers for radar pulse shaping, and antenna impedance matching applications. All topics are self-contained and also include practical aspects. Extensive analysis and optimization methods for these topics are discussed. The design techniques are well explained by means of solved examples.

The book is divided into nine chapters covering the basics of amplifiers, an overview of RFT, multistage distributed amplifiers, the use of RFT to design trans-impedance microwave amplifiers, the optimization of equalizers employing lossy distributed networks, the use of RFT to design multistage power amplifiers, the design of multistage active filters, the design of equalizers for radar pulse shaping, and antenna impedance matching. To solve impedance matching-related design problems using RFT from specifications to realization of the end product, the book provides a unique integration of analysis/optimization aspects. I found that the book is well balanced and treats the material in depth. With emphasis on theory, design, and practical aspects applied to numerous day-to-day applications, the book is suitable for graduate students, teachers, and design engineers.

Congratulations Profs. Jarry and Beneat on this excellent book that I am confident will be very well received in the RF and microwave community for many years to come.

Inder J. Bahl
Roanoke, VA
November 2015

Preface

Microwave and radio-frequency (RF) amplifiers play an important role in communication systems, and due to the proliferation of radar, satellite, and mobile wireless systems, there is a need for design methods that can satisfy the ever-increasing demand for accuracy, reliability, and fast development times. This book provides an original design technique for a wide variety of multistage microwave amplifiers and active filters, and passive equalizers for radar pulse shaping and antenna return loss applications. This technique is referred to as the real frequency technique (RFT). It has grown out of the authors own research and teaching and as such has a unity of methodology and style, essential for a smooth reading.

The book is intended for researchers and RF and microwave engineers but is also suitable for an advanced graduate course in the subject area. Furthermore, it is possible to choose material from the book to supplement traditional courses in microwave amplifier design.

Each chapter provides complete representation and characterization of the multistage or passive equalizer structure as well as the design methodology. We hope that this will provide the researcher with a set of approaches that he/she could use for current and future microwave amplifier designs. We also emphasize the practical nature of the subject by summarizing the design steps and giving numerous examples of amplifier realizations and measured responses so that RF and microwave engineers can have an appreciation of each amplifier in view of their needs. This approach, we believe, has produced a coherent, practical, and real-life treatment of the subject. The book is therefore theoretical but also experimental with over 18 microwave amplifier realizations.

The book is divided into nine chapters. In Chapter 1 recalls fundamental equations and definitions useful for understanding the design of the microwave amplifiers presented here.

Chapter 2 provides a complete description of the RFT as it is first used to design multistage lumped amplifiers. The chapter starts with the multistage amplifier representation made of field-effect transistors (FETs) and equalizers defined using their scattering matrices. In this introductory chapter, the equalizers are assumed to be realizable as lumped ladder structures with transmission zeroes at 0 or infinity. The equalizers are optimized to improve the overall transducer gain and voltage standing wave ratio (VSWR). The complete expressions for multistage transducer gain and VSWR are provided and further explained in Appendix A. The RFT uses a progressive optimization of the equalizers, leading to a small number of parameters to optimize simultaneously. This is a great advantage of the RFT over typical design techniques that require simultaneous optimization of all unknown parameters.

The RFT uses a practical implementation of the Levenberg–Marquardt optimization method and is summarized in Chapter 2 and detailed in Appendix B. The optimization is performed over hundreds of frequency data points and is numerical in nature so that measured scattering parameters of the transistors can be used. The chapter then presents a complete step-by-step example of the design of a four-stage amplifier. Intermediary and final results are provided. A first extension of the technique is shown. It consists in using a transistor feedback circuit instead of the standalone transistor to increase the useful bandwidth of the amplifier. The chapter concludes with reporting two realizations designed using this technique in the range of 0–6.7 GHz, and a realization intended for 45 MHz to 65 GHz operation.

Chapter 3 extends the RFT to the case multistage distributed amplifiers. The equalizers are made of quarter-wave transmission lines. A new variable and formalism are introduced that are better suited for the distributed medium. The modified RFT is able to optimize and synthesize the transmission lines for transducer gain, VSWR, and also for multistage noise figure. This chapter shows how the effects of the bias circuit can be combined with the scattering parameters of the transistor to present a more accurate representation of the structure. A method is provided to solve the realization problem of having too high or too low characteristic impedance requirements. The chapter concludes with realizations of a 1.15–1.5 GHz, a 2–8 GHz, and a 5.925–6.425 GHz amplifier.

Chapter 4 presents the modifications to the RFT to design trans-impedance microwave amplifiers that are used for example in the case of photodiodes acting as high impedance current sources. It must provide a flat gain for a load charge impedance of 50 Ω. Contrary to the previous cases, the RFT performs progressive optimization of the equalizers from load to source (i.e., right to left). It uses admittance and impedance matrices rather than solely relying on scattering matrix representations. A technique based on peaking inductor is described, and it is sued to reduce the noise at the input of the amplifier, critical for the overall noise figure. Results for lumped and distributed examples from 3 to 7 GHz are given.

In Chapter 5, a method is presented for optimizing equalizers made of a lossy distributed network. Compared to the previous RFT for broadband multistage amplifiers, parallel resistors at the gate and drain of the transistors become part of the RFT optimization process. The chapter sets the foundations for this new medium and topology. An image network technique is used for defining the properties of the equalizers as well as the synthesis technique needed for transmission lines with lossy junctions. The chapter presents two hybrid realizations of broadband lossy distributed amplifiers. The first is optimized for gain and VSWR from 0.1 to 5 GHz and the second from 0.1 to 9 GHz.

In Chapter 6, it is shown how the RFT can be used in the case of multistage power amplifiers. It is shown how added power is optimized in addition to gain and VSWR. The technique requires interpolating large-signal scattering parameters in two dimensions: frequency and power. The optimization is done from load to source as was the case in Chapter 4. The technique is used in the realization of a single-stage 2.25 GHz power amplifier and in the case of a three-stage 2.245 GHz power amplifier. The chapter also provides a method and examples for the design of linear multistage power amplifiers using arborescent structures.

Chapter 7 describes how to use the RFT to design multistage active filters. In this chapter, the transducer gain and VSWR are optimized in the passband, while finite transmission zeros can be placed in the stopband to provide the desired selectivity of the filter. The technique also optimizes the group delay in the passband, an important feature for radar and digital communications. Three active filter examples are presented. The first is a low-pass case with two transmission zeroes using two transistors and three equalizers operating in the 0.1–5 GHz band. The second is a band-pass

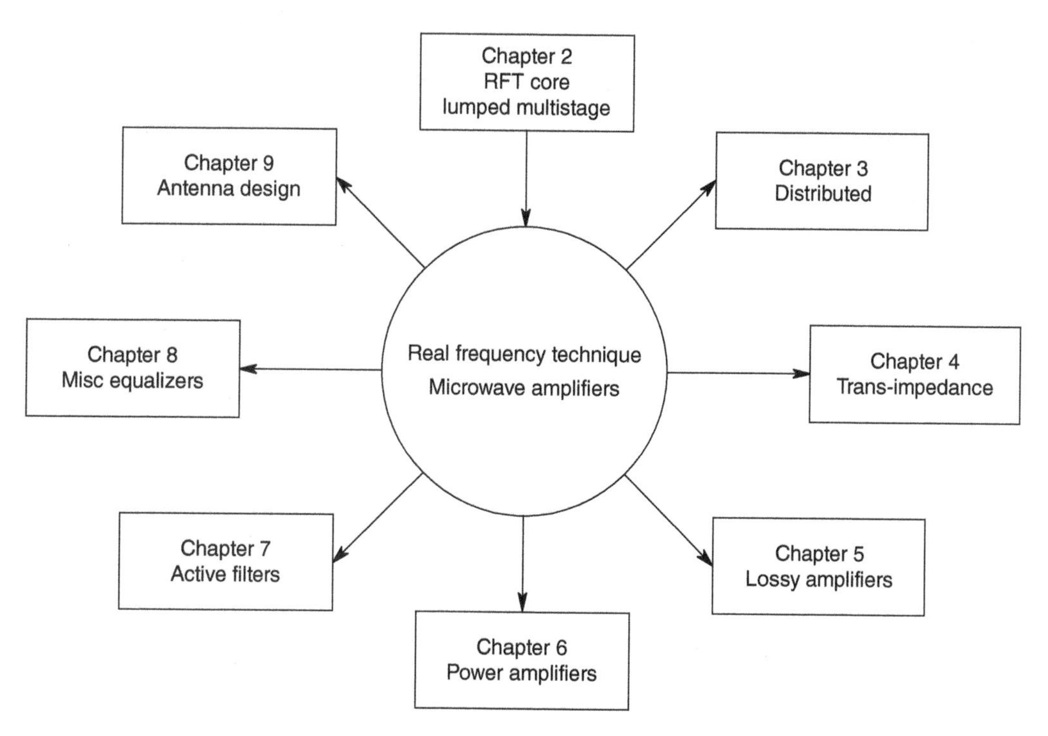

Figure 0.1 Summary of the organization of the book.

case with four transmission zeroes using one transistor and two equalizers with a 3–7 GHz pass band. The third is a band-pass case with two transmission zeroes with a 7.7–8.1 GHz pass band. This last case was realized in monolithic microwave integrated circuit (MMIC) technology.

Chapter 8 shows the flexibility of the RFT to solve a variety of microwave circuit design problems. In this chapter, the RFT is modified to optimize arbitrary responses using a structure made solely of passive equalizer. In this case, a transistor is shown to act as a "dummy passive device" when providing well-chosen scattering parameter data to represent it. Therefore with very little modification, the optimization capabilities of the traditional RFT can be used for passive structures. This chapter shows how the RFT no longer optimizes a constant transducer gain but rather optimizes a specific forward transmission curve for radar applications provided as a set of data points in a frequency band.

Two examples are provided. The first equalizer has an order six and used no transmission zeroes and the second equalizer has an order six and used two transmission zeroes to improve the optimization results in the stopband.

Finally, Chapter 9 describes a possible method for the synthesis of microwave antennas. The RFT is used to optimize the matching network placed between antenna and receiving/transmitting circuitry. The optimization focuses on improving the return loss in the receiving and transmitting bands.

The organization of the book is summarized in Figure 0.1. After a review chapter with important amplifier and microwave notations and equations, the central element is the RTF. The RFT is introduced in Chapter 2 for the design of multistage lumped amplifiers and is then modified to solve numerous problems in microwave amplifier design (Chapters 3–7), and adapted to solve more general microwave circuit design problems (Chapters 8 and 9).

It is believed that this book will provide the reader with an appreciation of the RFT as a tool that can be applied successfully in many disciplines.

We would like to acknowledge the contributions of our past and present research students whose collaboration has resulted in much of the material in the book. In particular, we would like to mention Professor Eric Kerherve; Associate Professors E.H. El Hendaoui, P. M. Martin, and A. Perennec; and Engineers L. Courcelle, M. Hazouard, M. Lecouve, and A. Olomo.

The book is based on the authors' research under the sponsorship of France Telecommunications Research & Development (CNET), National Center of Spatial Studies (CNES), ALCATEL SPACE, PHILIPS OMMIC, Commissariat Energie Atomique (CEA), PLESSEY (GB).

The work resulted in approximately nine contracts with different agencies and companies.

Pierre Jarry
Jacques N. Beneat

Acknowledgments

The authors are deeply indebted to Dr Inder J. Bahl (USA) editor of the International Journal of RF and Microwave Computer-Aided Engineering from Wiley. This book could not have been written without his help, and he is acknowledged with gratitude.

Their sincere appreciation extends to the Publisher of Wiley-Interscience and all the staff at Wiley involved in this project for their professionalism and outstanding efforts.

Pierre Jarry thanks his colleagues at the University Bordeaux Sciences, including Professor Eric Kerherve and Assistant Professors Nathalie Deltimple and Jean-Marie Pham. He thanks as well Professor Yves Garault who introduced Microwave and Telecommunications at the University of Limoges. Finally, he expresses his deep appreciation to his wife Roselyne and to his son Jean-Pierre for their tolerance and support.

Jacques Beneat is very grateful to Norwich University, a place conducive to trying and succeeding in new endeavors. He particularly thanks the Senior Vice President for Academic Affairs, Dr. Guiyou Huang, and the Director of the David Crawford School of Engineering, Professor Steve Fitzhugh, for their encouragements in such a difficult enterprise.

Pierre Jarry
Jacques N. Beneat

1

Microwave Amplifier Fundamentals

Microwave Amplifier and Active Circuit Design Using the Real Frequency Technique, First Edition.
Pierre Jarry and Jacques N. Beneat.

1.1 Introduction

In many high-speed applications, there is a need for microwave amplifier circuits. For example, satellite communications can be used when radio signals are blocked between two terrestrial transceiver stations as shown in Figure 1.1. The satellite then acts as a repeater, and the signal being repeated must be amplified before being sent back.

Important amplifier characteristics are center frequency and span of the pass band, gain, stability, input and output matching to the rest of the communication system, and noise figure [1].

At microwave frequencies, a common amplification component that has minimum noise is a field effect transistor (FET) as shown in Figure 1.2.

In this book, the FET will be typically modeled as a two-port network, where the input is on the gate and the output is on the drain. The source is mainly used for biasing of the transistor.

1.2 Scattering Parameters and Signal Flow Graphs

At high frequencies, voltages and currents are difficult to measure directly. However, scattering parameters determined from incident and reflected waves can be measured with resistive terminations. The scattering matrix of a two-port system provides relations between input and output reflected waves b_1 and b_2 and input and output incident waves a_1 and a_2 when the structure is

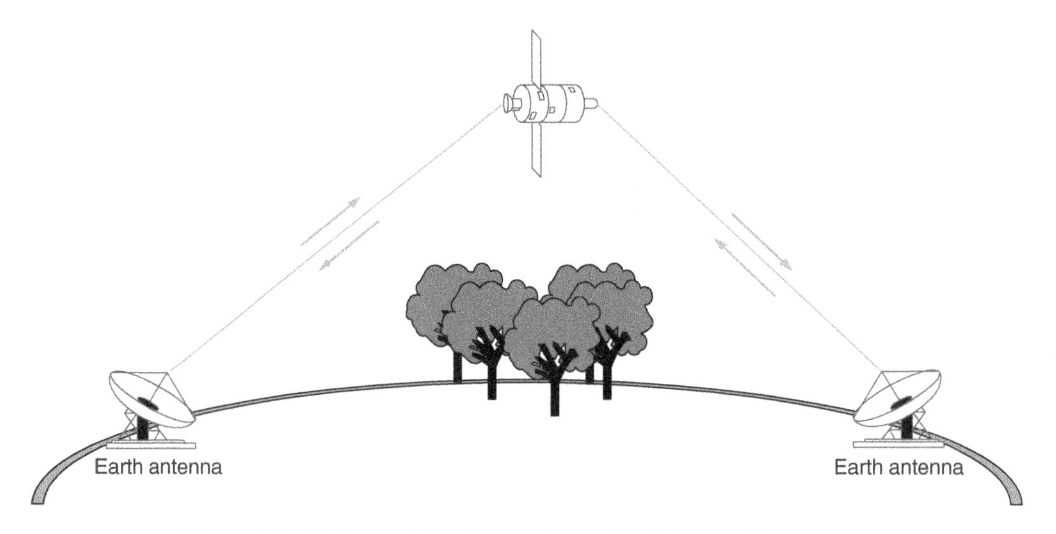

Figure 1.1 High-speed signals must be amplified in a satellite repeater.

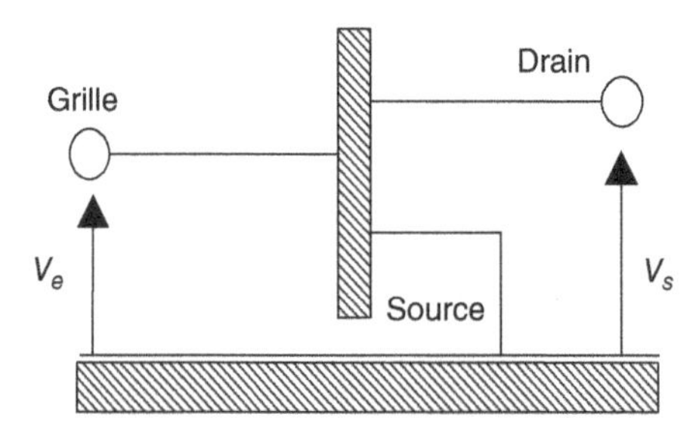

Figure 1.2 A FET modeled as a two-port network.

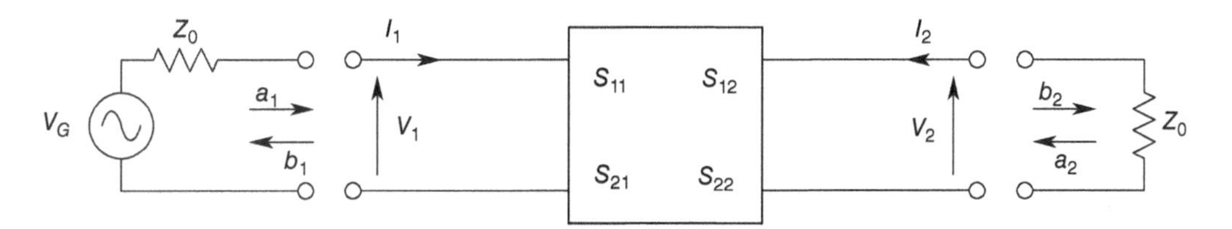

Figure 1.3 Scattering matrix of a two-port system terminated on characteristic impedance Z_0.

terminated on its characteristic impedance Z_0 as shown in Figure 1.3. Typically the reference source and load Z_0 used in commercial network analyzers is 50 Ω.

In the case of a two-port system, the equations relating incident and reflected waves and the scattering parameters are given by

$$\begin{aligned} b_1 &= S_{11}a_1 + S_{12}a_2 \\ b_2 &= S_{21}a_1 + S_{22}a_2 \end{aligned} \tag{1.1}$$

$$\begin{pmatrix} b_1 \\ b_2 \end{pmatrix} = \begin{pmatrix} S_{11} & S_{12} \\ S_{21} & S_{22} \end{pmatrix} \begin{pmatrix} a_1 \\ a_2 \end{pmatrix} \tag{1.2}$$

The incident and reflected waves are related to the voltages and currents in Figure 1.3.

$$a_1 = \frac{V_1 + Z_0 I_1}{2\sqrt{Z_0}} \tag{1.3}$$

$$a_2 = \frac{V_2 + Z_0 I_2}{2\sqrt{Z_0}} \tag{1.4}$$

$$b_1 = \frac{V_1 - Z_0 I_1}{2\sqrt{Z_0}} \tag{1.5}$$

$$b_2 = \frac{V_2 - Z_0 I_2}{2\sqrt{Z_0}} \tag{1.6}$$

The parameter S_{11} is the input reflection coefficient and is the ratio of input reflected wave over input incident wave when the output incident wave is equal to zero. The output incident wave a_2 is equal to zero when the output of the system is connected to the characteristic impedance Z_0:

$$S_{11} = \left.\frac{b_1}{a_1}\right|_{a_2=0} \tag{1.7}$$

The parameter S_{21} is the forward transmission coefficient and is the ratio of the output reflected wave over the input incident wave when the output incident wave is equal to zero:

$$S_{21} = \left.\frac{b_2}{a_1}\right|_{a_2=0} \tag{1.8}$$

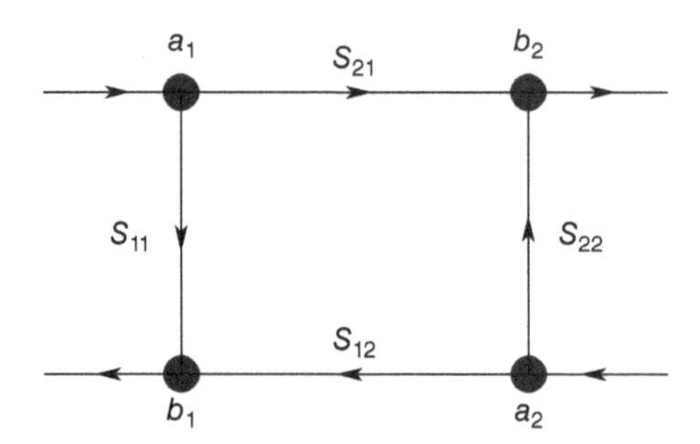

Figure 1.4 Signal flow graph representation of a two-port network.

The parameter S_{22} is the output reflection coefficient and is the ratio of the output reflected wave over the output incident wave when the input incident wave is equal to zero. The input incident wave a_1 is equal to zero when the input of the system is connected to the characteristic impedance Z_0:

$$S_{22} = \left.\frac{b_2}{a_2}\right|_{a_1=0} \tag{1.9}$$

The parameter S_{12} is the reverse transmission coefficient and is the ratio of the input reflected wave over the output incident wave when the input incident wave is equal to zero:

$$S_{12} = \left.\frac{b_1}{a_2}\right|_{a_1=0} \tag{1.10}$$

The two-port network and scattering parameters can be modeled using the signal flow graph representation of Figure 1.4.

A useful tool when defining system gains using signal flow graphs is the Mason gain formula [2]. It provides the gain T of a system between a source node and an output node:

$$T = \frac{\sum T_k \Delta_k}{\Delta} \tag{1.11}$$

with

$$\Delta = 1 - \sum L_i + \sum L_i L_j - \sum L_i L_j L_k \cdots$$

where

T_k is the gain of the kth forward path between the source node and the output node

$\sum L_i$ is the sum of all individual loop gains

$\sum L_i L_j$ is the sum of two loop gain products of any two nontouching loops

$\sum L_i L_j L_k$ is the sum of three loop gain products of any three nontouching loops

Δ_k is the part of Δ that does not touch the kth forward path

1.3 Reflection Coefficients

As shown in Figure 1.5, the input reflection coefficient when the output is connected to characteristic impedance Z_0 can be expressed in terms of the input impedance $Z_{IN} = V_1/I_1$ by replacing a_1 and b_1 by their expressions in terms of voltages and currents:

$$S_{11} = \left.\frac{b_1}{a_1}\right|_{a_2=0} = \frac{\dfrac{V_1 - Z_0 I_1}{2\sqrt{Z_0}}}{\dfrac{V_1 + Z_0 I_1}{2\sqrt{Z_0}}} = \frac{Z_{IN} - Z_0}{Z_{IN} + Z_0}$$

$$S_{11} = \frac{Z_{IN} - Z_0}{Z_{IN} + Z_0} \tag{1.12}$$

The input impedance can be expressed in terms of the input reflection coefficient by

$$Z_{IN} = Z_0 \frac{1 + S_{11}}{1 - S_{11}} \tag{1.13}$$

Figure 1.6 defines additional reflection coefficients when the two-port is terminated on arbitrary loads Z_G and Z_L.

Reflection coefficient of the source:

$$\rho_G = \frac{a_1}{b_1} = \frac{Z_G - Z_0}{Z_G + Z_0} \tag{1.14}$$

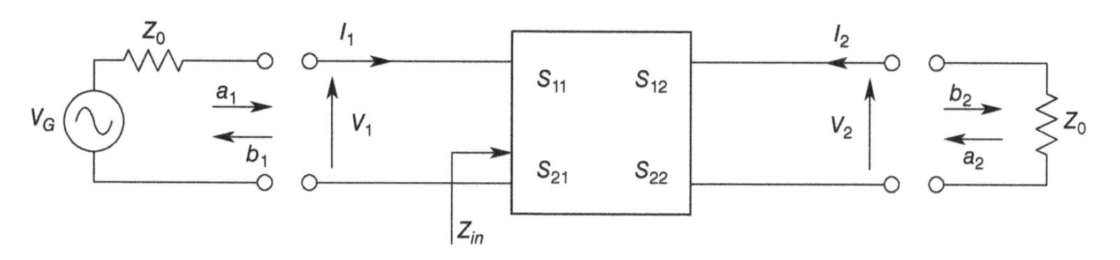

Figure 1.5 Input reflection coefficient and input impedance.

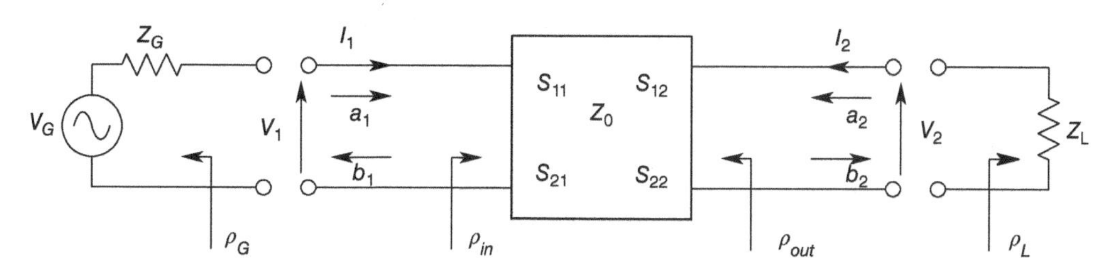

Figure 1.6 Reflection coefficients of a two-port when terminated on arbitrary loads.

Reflection coefficient of the load:

$$\rho_L = \frac{a_2}{b_2} = \frac{Z_L - Z_0}{Z_L + Z_0} \tag{1.15}$$

Input reflection coefficient of the two-port when output loaded on ρ_L:

$$\rho_{in} = \frac{b_1}{a_1} = S_{11} + \frac{S_{12}S_{21}\rho_L}{1 - S_{22}\rho_L} \tag{1.16}$$

Output reflection coefficient of the two-port when input loaded on ρ_G:

$$\rho_{out} = \frac{b_2}{a_2} = S_{22} + \frac{S_{12}S_{21}\rho_G}{1 - S_{11}\rho_G} \tag{1.17}$$

For example, the expression of the input reflection coefficient when loaded on ρ_L is obtained by first using the general scattering parameter definition of the two-port:

$$b_1 = S_{11}a_1 + S_{12}a_2$$
$$b_2 = S_{21}a_1 + S_{22}a_2$$

Then, using the relation between incident and reflected waves $a_2 = \rho_L b_2$ gives

$$b_1 = S_{11}a_1 + S_{12}\rho_L b_2$$
$$b_2 = S_{21}a_1 + S_{22}\rho_L b_2$$

then from the second equation, $b_2(1 - S_{22}\rho_L) = S_{21}a_1$ and $b_2 = \dfrac{S_{21}}{(1 - S_{22}\rho_L)}a_1$, so that

$$b_1 = S_{11}a_1 + S_{12}\rho_L \frac{S_{21}}{1 - S_{22}\rho_L}a_1 = \left(S_{11} + \frac{S_{21}S_{12}\rho_L}{1 - S_{22}\rho_L}\right)a_1$$

and

$$\rho_{in} = \frac{b_1}{a_1} = S_{11} + \frac{S_{12}S_{21}\rho_L}{1 - S_{22}\rho_L}$$

The voltage standing wave ratio (VSWR) is given in terms of a reflection coefficient ρ by

$$\text{VSWR} = \frac{1 + |\rho|}{1 - |\rho|} \tag{1.18}$$

The input VSWR for the two-port in Figure 1.6 is therefore

$$\text{VSWR}_{IN} = \frac{1 + |\rho_{in}|}{1 - |\rho_{in}|} \tag{1.19}$$

The output VSWR for the two-port in Figure 1.6 is therefore

$$\text{VSWR}_{OUT} = \frac{1+|\rho_{out}|}{1-|\rho_{out}|} \tag{1.20}$$

1.4 Gain Expressions

Figure 1.7 shows the different reflection coefficients used to define various power gains.

The transducer power gain can be computed using the signal flow graph and the Mason gain formula as shown in Figure 1.8.

There is one forward path from node b_G to node b_2. The path gain of this path is $T_1 = 1 \times S_{21} = S_{21}$.

There are three individual loops: $\rho_G S_{11}$, $\rho_L S_{22}$, and $\rho_G S_{21} \rho_L S_{12}$.

This gives

$$\Delta = 1 - \sum L_i + \sum L_i L_j - \sum L_i L_j L_k \cdots = 1 - [S_{11}\rho_G + S_{22}\rho_L + S_{12}S_{21}\rho_G\rho_L] + [S_{11}\rho_G S_{22}\rho_L] - 0$$

and

$$T = \frac{\sum T_k \Delta_k}{\Delta} = \frac{T_1 \times (1-0)}{1 - S_{11}\rho_G - S_{22}\rho_L - S_{12}S_{21}\rho_G\rho_L + S_{11}\rho_G S_{22}\rho_L}$$

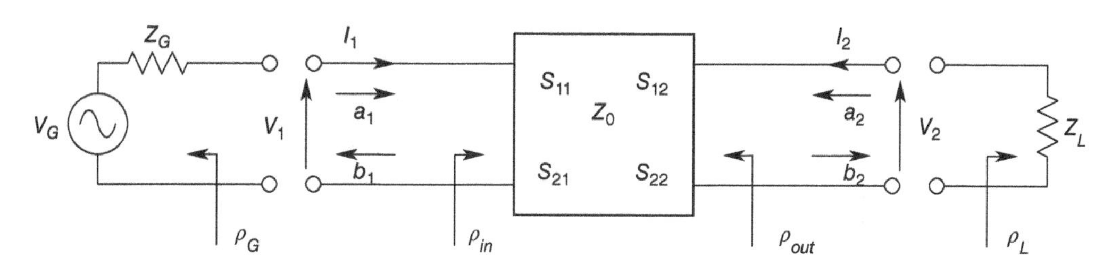

Figure 1.7 Gain definitions.

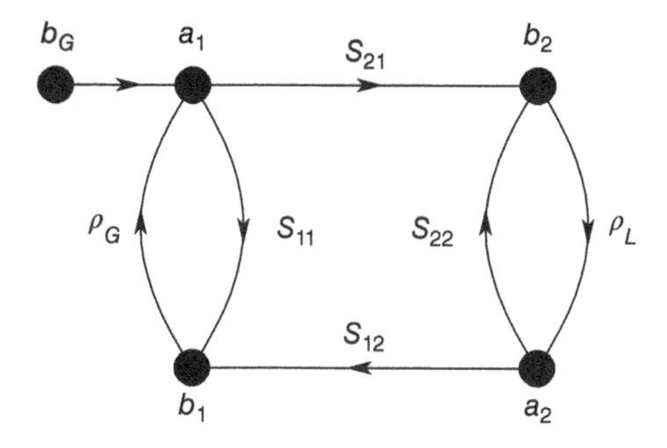

Figure 1.8 Signal flow graph representation for defining the gain.

$$T=\frac{S_{21}}{1-S_{11}\rho_G-S_{22}\rho_L+S_{11}S_{22}\rho_G\rho_L-S_{12}S_{21}\rho_G\rho_L}$$

The transducer power gain is defined as

$$G_T=\frac{\text{power delivered to the load}}{\text{maximum available power from the source}}$$

so that

$$G_T=\frac{|b_2|^2\left(1-|\rho_L|^2\right)}{\dfrac{|b_G|^2}{\left(1-|\rho_G|^2\right)}}=|T|^2\left(1-|\rho_G|^2\right)\left(1-|\rho_L|^2\right)$$

and

$$G_T=\frac{\left(1-|\rho_G|^2\right)\left(1-|\rho_L|^2\right)|S_{21}|^2}{|1-S_{11}\rho_G-S_{22}\rho_L+S_{11}S_{22}\rho_G\rho_L-S_{12}S_{21}\rho_G\rho_L|^2} \tag{1.21}$$

Note that G_T can also be written as

$$G_T=\frac{\left(1-|\rho_G|^2\right)\left(1-|\rho_L|^2\right)|S_{21}|^2}{|(1-S_{11}\rho_G)(1-S_{22}\rho_L)-S_{12}S_{21}\rho_G\rho_L|^2} \tag{1.22}$$

or as

$$G_T=\frac{\left(1-|\rho_G|^2\right)\left(1-|\rho_L|^2\right)|S_{21}|^2}{|1-\rho_G\rho_{in}|^2|1-S_{22}\rho_L|^2} \tag{1.23}$$

or as

$$G_T=\frac{\left(1-|\rho_G|^2\right)\left(1-|\rho_L|^2\right)|S_{21}|^2}{|1-S_{11}\rho_G|^2|1-\rho_L\rho_{out}|^2} \tag{1.24}$$

Note that when $S_{12}=0$, the transducer power gain reduces to the unilateral transducer power gain G_{TU} given by [3]

$$G_{TU}=G_G G_0 G_L$$

where

$$G_G=\frac{1-|\rho_G|^2}{|1-S_{11}\rho_G|^2},\quad G_0=|S_{21}|^2\ \text{ and }\ G_L=\frac{1-|\rho_L|^2}{|1-S_{22}\rho_L|^2}$$

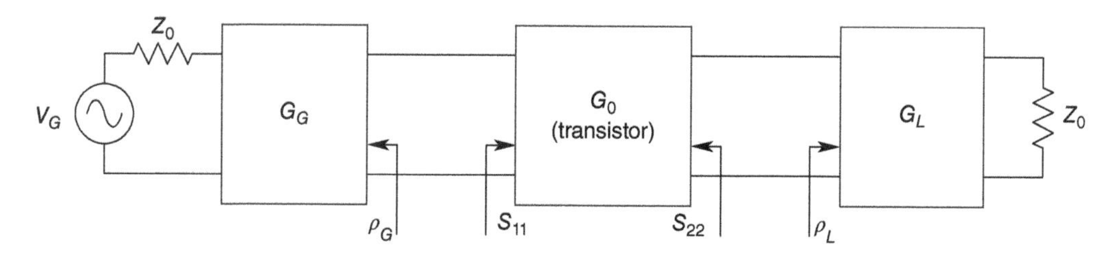

Figure 1.9 Representation in the case of a unilateral amplifier ($S_{12}=0$).

In this case, G_G represents the losses in the source, G_0 is the intrinsic gain, and G_L represents the losses in the load and can be modeled as in Figure 1.9.

The maximum unilateral gain occurs when there is perfect matching of source and load impedances. For maximum unilateral gain, one would match the source to the input of the transistor by making $\rho_G = S_{11}^*$ and match the load to the output of the transistor by making $\rho_L = S_{22}^*$.

The maximum unilateral gain $G_{TU\ MAX}$ is then given by

$$G_{TU\ MAX} = \frac{1}{1-|S_{11}|^2}|S_{21}|^2\frac{1}{1-|S_{22}|^2} \tag{1.25}$$

The available power gain is defined as

$$G_A = \frac{\text{maximum power the amplifier can deliver to the load}}{\text{maximum available power from the source}}$$

and it is given by

$$G_A = \frac{\left(1-|\rho_G|^2\right)|S_{21}|^2}{|1-S_{11}\rho_G|^2\left(1-|\rho_{out}|^2\right)} \tag{1.26}$$

Note that when the input is perfectly matched then $Z_G = Z_0$ and $\rho_G = 0$ and

$$\rho_{out} = S_{22} + \frac{S_{12}S_{21}\rho_G}{1-S_{11}\rho_G} = S_{22}$$

so that the available power gain becomes

$$G_A = \frac{|S_{21}|^2}{1-|S_{22}|^2} \tag{1.27}$$

1.5 Stability

The stability of the amplifier depends on the scattering parameters of the transistor but also on the matching networks and terminations [4].

For the two-port shown in Figure 1.10, ρ_{in} is the input reflection coefficient of the transistor when output loaded on ρ_L and ρ_{out} is the output reflection coefficient of the transistor when input

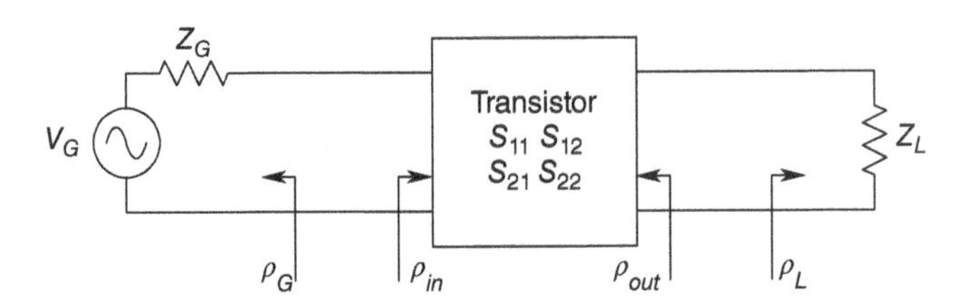

Figure 1.10 Stability of a two-port.

loaded on ρ_G. The system is said to be unconditionally stable if the amplitude of ρ_{in} and ρ_{out} are less than unity for all the real parts of load impedance Z_L and source impedance Z_G:

$$\forall Z_L(\text{with}\,\text{Re}\{Z_L\}>0);\ |\rho_{in}|^2<1$$
$$\forall Z_G(\text{with}\,\text{Re}\{Z_G\}>0);\ |\rho_{out}|^2<1$$

It can be shown that for unconditional stability one must satisfy three conditions:

$$\begin{cases} K>1 \\ B_1>0 \\ B_2>0 \end{cases} \tag{1.28}$$

where

$$K=\frac{1+|\Delta|^2-|S_{11}|^2-|S_{22}|^2}{2|S_{12}S_{21}|} \tag{1.29}$$

$$B_1=1-|S_{22}|^2-|S_{12}S_{21}| \tag{1.30}$$

$$B_2=1-|S_{11}|^2-|S_{12}S_{21}| \tag{1.31}$$

It is seen that these conditions only depend on the scattering parameters of the transistor. When these three conditions are met the amplifier can be connected to the loads without risk of becoming unstable and producing oscillations.

1.6 Noise

Figure 1.11 shows an active two-port between input impedance Z_G and load impedance Z_G.

The noise in the amplifier can be characterized by the noise figure F defined by

$$F=F_{\min}+\frac{R_n}{G_G}|Y_G-Y_{G\min}|^2 \tag{1.32}$$

where

$F_{\min}$ is the minimum noise figure obtained when $Y_G=Y_{G\min}$
R_n is the equivalent noise resistance of the active device

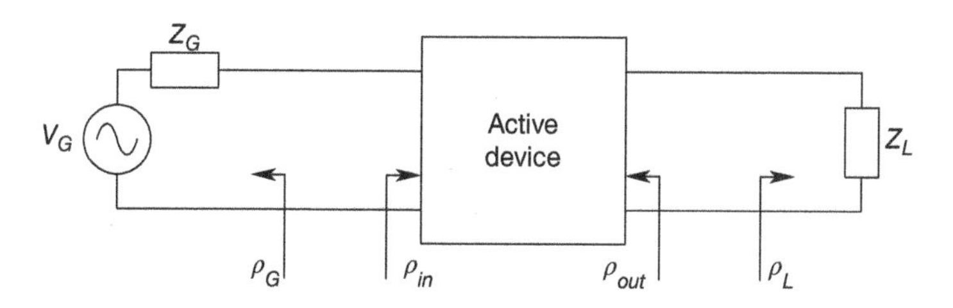

Figure 1.11 Active device and source and load impedances.

$Y_{G\min}$ is the source admittance that makes the noise figure minimum
Y_G is the source admittance such that $Y_G = G_G + jB_G$

The Section **1.7** consists of the rewritten noise figure formula in terms of reflection coefficients rather than in terms of admittances.

Referring to Figure 1.11, the reflection coefficient ρ_G from the source admittance is given by

$$\rho_G = \frac{Y_0 - Y_G}{Y_0 + Y_G} \tag{1.33}$$

where Y_0 is the characteristic admittance used.

This gives the source admittance in terms of the source reflection coefficient such that

$$Y_G = \frac{(1-\rho_G)}{(1+\rho_G)} Y_0 \tag{1.34}$$

In (1.32), taking $Y_G = Y_{G\min}$ makes the noise figure become minimum. This translates to a reflection coefficient $\rho_{G\min}$, where the noise figure is at a minimum such that

$$\rho_{G\min} = \frac{Y_0 - Y_{G\min}}{Y_0 + Y_{G\min}} \tag{1.35}$$

$$Y_{G\min} = \frac{(1-\rho_{G\min})}{(1+\rho_{G\min})} Y_0 \tag{1.36}$$

Then, replacing $Y_{G\min}$ and Y_G by their expressions in terms of ρ_G and $\rho_{G\min}$ gives us

$$|Y_G - Y_{G\min}|^2 = Y_0{}^2 \left| \frac{(1-\rho_G)}{(1+\rho_G)} - \frac{(1-\rho_{G\min})}{(1+\rho_{G\min})} \right|^2$$
$$= Y_0{}^2 \left| \frac{1-\rho_G+\rho_{G\min}-\rho_G\rho_{G\min}-(1-\rho_{G\min}+\rho_G-\rho_G\rho_{G\min})}{(1+\rho_G)(1+\rho_{G\min})} \right|^2$$

and

$$|Y_G - Y_{G\min}|^2 = Y_0{}^2 \left| \frac{2(\rho_{G\min}-\rho_G)}{(1+\rho_G)(1+\rho_{G\min})} \right|^2 = 4Y_0{}^2 \frac{|\rho_{G\min}-\rho_G|^2}{|1+\rho_G|^2 |1+\rho_{G\min}|^2}$$

So that the noise figure can first be expressed as

$$F = F_{\min} + \frac{4R_n}{G_G} {Y_0}^2 \frac{|\rho_G - \rho_{G\min}|^2}{|1+\rho_G|^2 |1+\rho_{G\min}|^2}$$

Then, one expresses G_G as the real part of the source admittance in terms of reflection coefficients:

$$G_G = \mathrm{Re}\{Y_G\} = \frac{Y_G + Y_G^*}{2} = \frac{Y_0}{2}\left[\frac{(1-\rho_G)}{(1+\rho_G)} + \frac{(1-\rho_G^*)}{(1+\rho_G^*)}\right]$$
$$= \frac{Y_0}{2}\left[\frac{1-\rho_G+\rho_G^* - |\rho_G|^2 + 1 + \rho_G - \rho_G^* - |\rho_G|^2}{|1+\rho_G|^2}\right]$$

and

$$G_G = Y_0 \frac{1-|\rho_G|^2}{|1+\rho_G|^2}$$

so that

$$F = F_{\min} + 4R_n {Y_0}^2 \frac{1}{Y_0} \frac{|1+\rho_G|^2}{1-|\rho_G|^2} \frac{|\rho_G - \rho_{G\min}|^2}{|1+\rho_G|^2 |1+\rho_{G\min}|^2}$$

and the noise figure is given in terms of the source reflection parameter ρ_G and the optimum source reflection parameter $\rho_{G\min}$ by

$$F = F_{\min} + 4R_n Y_0 \frac{|\rho_G - \rho_{G\min}|^2}{\left(1-|\rho_G|^2\right)|1+\rho_{G\min}|^2} \tag{1.37}$$

Typically, the manufacturer provides the three parameters $\rho_{G\min}$, $F_{\min}$, and $r_n = R_n/R_0$, the normalized equivalent noise resistance. Note that the reflection coefficient $\rho_{G\min}$ is complex and is often given as magnitude and phase. Note that these parameters do change with frequency so they are provided in table form.

Next, we provide the noise figure corresponding to a cascade of active devices. In Figure 1.12, a first active device is characterized by a gain G_1 and noise figure F_1, and a second active device is characterized by a gain G_2 and noise figure F_2.

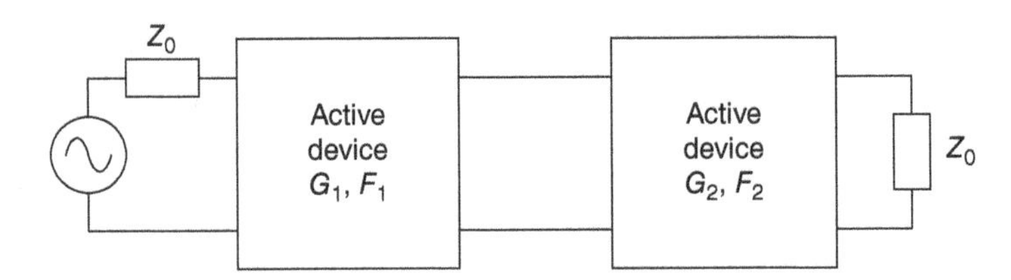

Figure 1.12 Noise figure of the cascade of two active devices.

It can be shown [5] that the noise figure F'_2 corresponding to the cascade of the two systems is given by

$$F'_2 = F_1 + \frac{1}{G_1}(F_2 - 1) \tag{1.38}$$

and the gain G'_2 of the cascaded system is given by

$$G'_2 = G_1 G_2 \tag{1.39}$$

The system is then placed in cascade with a third active device characterized by a gain G_3 and noise figure F_3 as shown in Figure 1.13.

Then, the noise figure F'_3 corresponding to the cascade of the three systems is given by

$$F'_3 = F'_2 + \frac{1}{G'_2}(F_3 - 1)$$

and the gain G'_3 of the cascaded system is given by

$$G'_3 = G'_2 G_3$$

We repeat this approach for the case of a cascade of k active devices as shown in Figure 1.14.

In this case, the noise figure F'_k corresponding to the cascade of the k systems is given by

$$F'_k = F'_{k-1} + \frac{1}{G'_{k-1}}(F_k - 1) \tag{1.40}$$

and the gain G'_k of the cascaded system is given by

$$G'_k = G'_{k-1} G_k \tag{1.41}$$

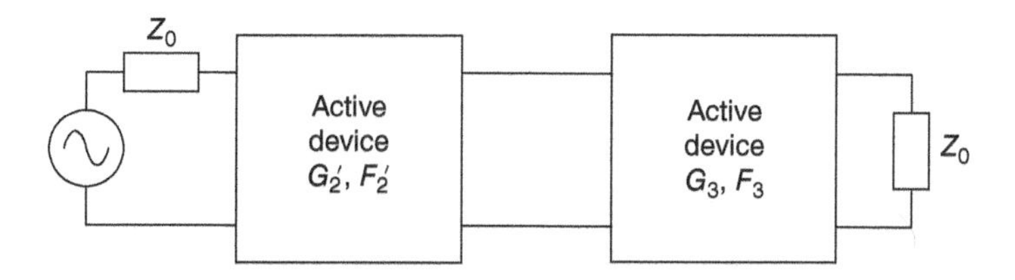

Figure 1.13 Noise figure of the cascade of three active devices.

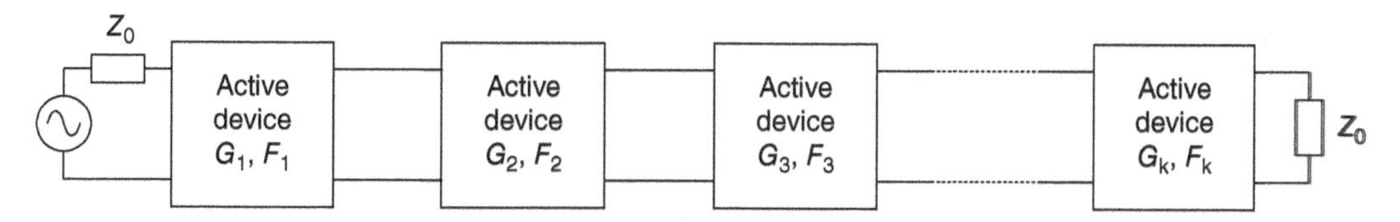

Figure 1.14 Noise figure of the cascade of k active devices.

where $k \geq 2$ with $F'_1 = F_1$ and $G'_1 = G_1$.

Note that this can also be written as

$$F'_k = F_1 + \frac{1}{G_1}(F_2 - 1) + \frac{1}{G_1 G_2}(F_3 - 1) + \cdots + \frac{1}{G_1 G_2 \cdots G_{k-1}}(F_k - 1) \tag{1.42}$$

1.7 *ABCD* Matrix

The *ABCD* matrix of a two-port is defined using the voltages and currents shown in Figure 1.15 [6].

The *ABCD* matrix shows the relation between input and output voltages and currents. It is given by

$$\begin{pmatrix} V_1 \\ I_1 \end{pmatrix} = \begin{pmatrix} A & B \\ C & D \end{pmatrix} \begin{pmatrix} V_2 \\ -I_2 \end{pmatrix} \tag{1.43}$$

What follows are the *ABCD* matrices of several common elements and configurations found in microwave structures.

1.7.1 ABCD *Matrix of a Series Impedance*

Figure 1.16 shows the case of a two-port network made of impedance placed in series.

The equations of this system are

$$\begin{aligned} I_1 &= (-I_2) \\ V_1 &= V_2 + Z(-I_2) \end{aligned} \quad \text{or} \quad \begin{pmatrix} V_1 \\ I_1 \end{pmatrix} = \begin{pmatrix} 1 & Z \\ 0 & 1 \end{pmatrix} \begin{pmatrix} V_2 \\ (-I_2) \end{pmatrix}$$

Therefore, the *ABCD* matrix representing a two-port system made of impedance connected in series is given by

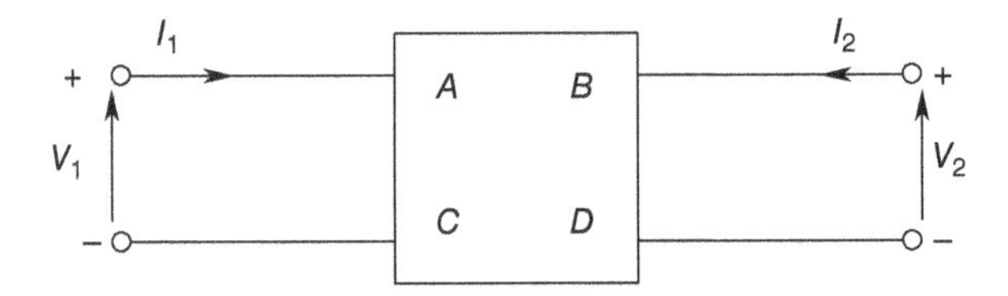

Figure 1.15 Notations for defining the *ABCD* matrix of a two-port system.

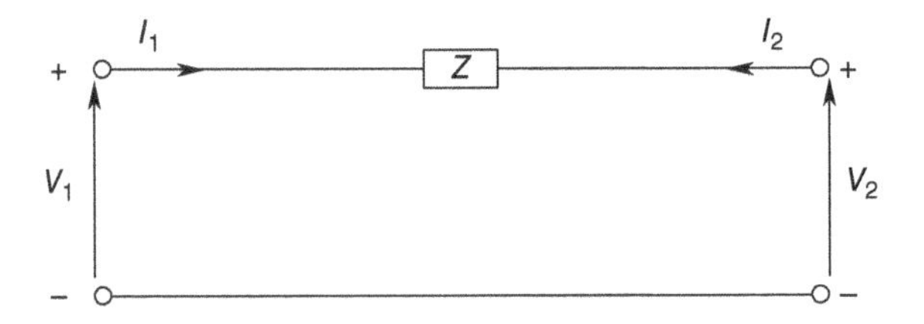

Figure 1.16 Impedance placed in series.

$$\begin{pmatrix} A & B \\ C & D \end{pmatrix} = \begin{pmatrix} 1 & Z \\ 0 & 1 \end{pmatrix} \tag{1.44}$$

1.7.2 ABCD *Matrix of a Parallel Admittance*

Figure 1.17 shows the case of a two-port network made of admittance placed in parallel.

The equations of this system are given by

$$\begin{aligned} V_1 &= V_2 \\ YV_2 &= I_1 + I_2 \end{aligned} \quad \text{or} \quad \begin{pmatrix} V_1 \\ I_1 \end{pmatrix} = \begin{pmatrix} 1 & 0 \\ Y & 1 \end{pmatrix} \begin{pmatrix} V_2 \\ (-I_2) \end{pmatrix}$$

Therefore, the *ABCD* matrix representing a two-port system made of admittance connected in parallel is given by

$$\begin{pmatrix} A & B \\ C & D \end{pmatrix} = \begin{pmatrix} 1 & 0 \\ Y & 1 \end{pmatrix} \tag{1.45}$$

1.7.3 *Input Impedance of Impedance Loaded Two-Port*

When a two-port is connected to load Z_L as shown in Figure 1.18, the output voltage V_2 and output current I_2 are such that $V_2 = Z_L(-I_2)$.

The equations of the system are given by

$$\begin{cases} V_1 = AV_2 + B(-I_2) = AZ_L(-I_2) + B(-I_2) \\ I_1 = CV_2 + D(-I_2) = CZ_L(-I_2) + D(-I_2) \end{cases}$$

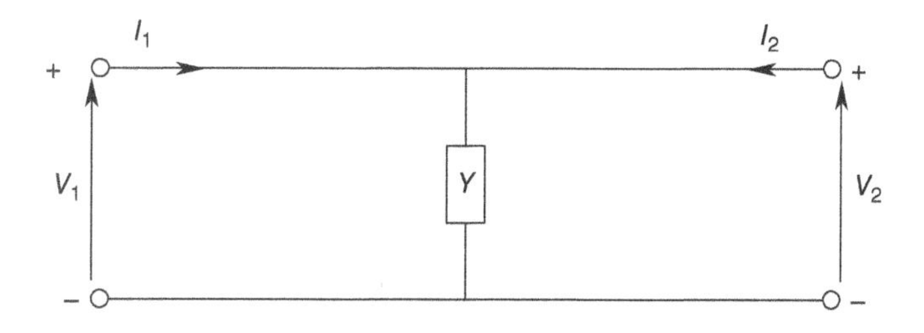

Figure 1.17 Admittance placed in parallel.

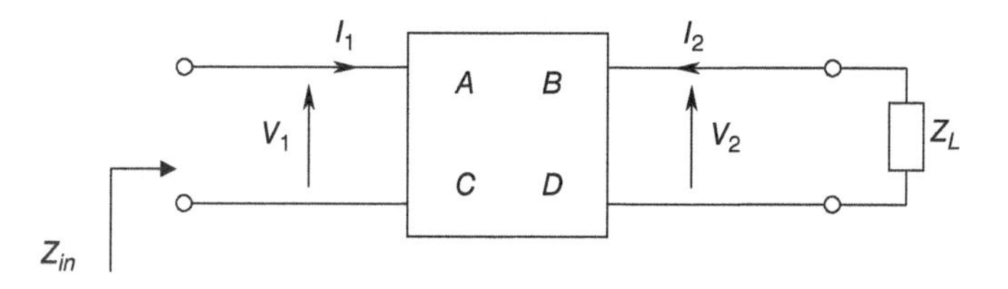

Figure 1.18 Two-port connected to load impedance Z_L.

from which it is straightforward to extract the input impedance of the two-port network:

$$Z_{IN} = \left.\frac{V_1}{I_1}\right|_{Z_L} = \frac{AZ_L + B}{CZ_L + D} \tag{1.46}$$

1.7.4 *Input Admittance of Admittance Loaded Two-Port*

When a two-port is connected to load admittance Y_L as shown in Figure 1.19, the output voltage V_2 and output current I_2 are such that $-I_2 = Y_L V_2$.

The equations of the system are given by

$$\begin{cases} V_1 = AV_2 + B(-I_2) = AV_2 + BY_L V_2 \\ I_1 = CV_2 + D(-I_2) = CV_2 + DY_L V_2 \end{cases}$$

from which it is straightforward to extract the input admittance of the two-port network:

$$Y_{IN} = \left.\frac{I_1}{V_1}\right|_{Y_L} = \frac{C + DY_L}{A + BY_L} \tag{1.47}$$

1.7.5 ABCD *Matrix of the Cascade of Two Systems*

One of the main advantages of the *ABCD* representation is that the *ABCD* matrix of the cascade of two systems as shown in Figure 1.20 is equal to the multiplication of the individual *ABCD* matrices.

The equations of this system are given by

$$\begin{cases} V_1 = A_1 V_2 + B_1(-I_2) \\ I_1 = C_1 V_2 + D_1(-I_2) \end{cases} \text{and} \begin{cases} V_1' = A_2 V_2' + B_2(-I_2') \\ I_1' = C_2 V_2' + D_2(-I_2') \end{cases}$$

In Figure 1.19, $V_2 = V_1'$ and $(-I_2) = I_1'$ so that the equations of the overall system are

$$\begin{cases} V_1 = A_1\left(A_2 V_2' + B_2(-I_2')\right) + B_1\left(C_2 V_2' + D_2(-I_2')\right) = (A_1 A_2 + B_1 C_2)V_2' + (A_1 B_2 + B_1 D_2)(-I_2') \\ I_1 = C_1\left(A_2 V_2' + B_2(-I_2')\right) + D_1\left(C_2 V_2' + D_2(-I_2')\right) = (C_1 A_2 + D_1 C_2)V_2' + (C_1 B_2 + D_1 D_2)(-I_2') \end{cases}$$

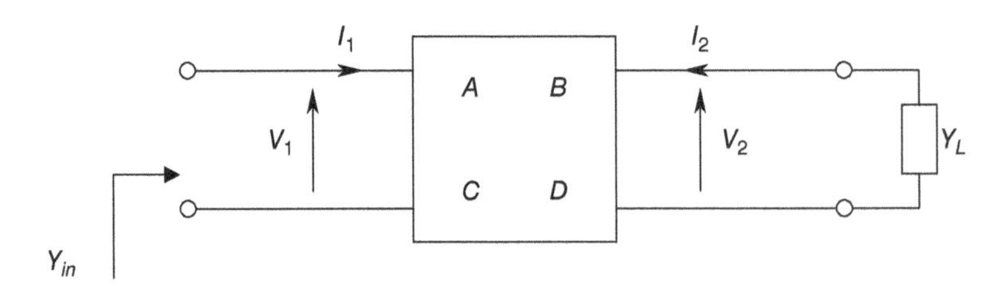

Figure 1.19 Two-port connected to load admittance Y_L.

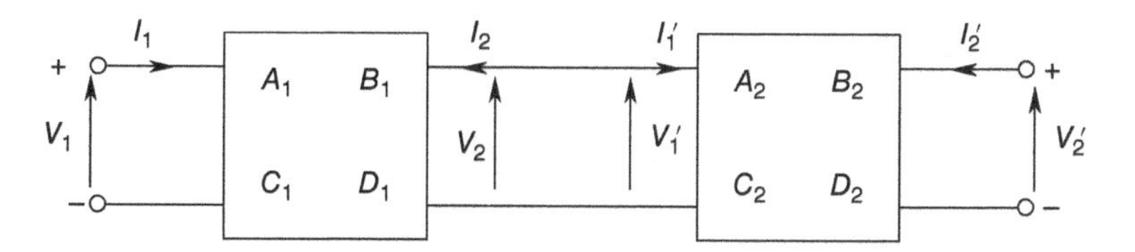

Figure 1.20 Cascade of two systems.

If we multiply the *ABCD* matrices representing the two systems, we get

$$\begin{pmatrix} A_1 & B_1 \\ C_1 & D_1 \end{pmatrix}\begin{pmatrix} A_2 & B_2 \\ C_2 & D_2 \end{pmatrix} = \begin{pmatrix} A_1A_2+B_1C_2 & A_1B_2+B_1D_2 \\ C_1A_2+D_1C_2 & C_1B_2+D_1D_2 \end{pmatrix}$$

Since both techniques provide the same answers, the *ABCD* matrix of the cascade of two systems will be given by the multiplication of their *ABCD* matrices:

$$\begin{pmatrix} A & B \\ C & D \end{pmatrix} = \begin{pmatrix} A_1 & B_1 \\ C_1 & D_1 \end{pmatrix}\begin{pmatrix} A_2 & B_2 \\ C_2 & D_2 \end{pmatrix} = \begin{pmatrix} A_1A_2+B_1C_2 & A_1B_2+B_1D_2 \\ C_1A_2+D_1C_2 & C_1B_2+D_1D_2 \end{pmatrix} \tag{1.48}$$

1.7.6 ABCD *Matrix of the Parallel Connection of Two Systems*

In this case, the two-port networks are connected in parallel as shown in Figure 1.21.

The *ABCD* matrix representing two systems connected in parallel is given by

$$\begin{pmatrix} A & B \\ C & D \end{pmatrix} = \begin{pmatrix} \dfrac{A_1B_2+B_1A_2}{B_1+B_2} & \dfrac{B_1B_2}{B_1+B_2} \\ C_1+C_2+\dfrac{(A_1-A_2)(D_2-D_1)}{B_1+B_2} & \dfrac{D_1B_2+B_1D_2}{B_1+B_2} \end{pmatrix} \tag{1.49}$$

1.7.7 ABCD *Matrix of the Series Connection of Two Systems*

In this case, the two-port networks are connected in series as shown in Figure 1.22.

The *ABCD* matrix representing two systems connected in series is given by

$$\begin{pmatrix} A & B \\ C & D \end{pmatrix} = \begin{pmatrix} \dfrac{A_1C_2+C_1A_2}{C_1+C_2} & B_1+B_2+\dfrac{(A_1-A_2)(D_2-D_1)}{C_1+C_2} \\ \dfrac{C_1C_2}{C_1+C_2} & \dfrac{D_1C_2+C_1D_2}{C_1+C_2} \end{pmatrix} \tag{1.50}$$

1.7.8 ABCD *Matrix of Admittance Loaded Two-Port Connected in Parallel*

In this case, a two-port loaded on admittance Y_{L1} is connected in parallel to form the two-port of network as shown in Figure 1.23.

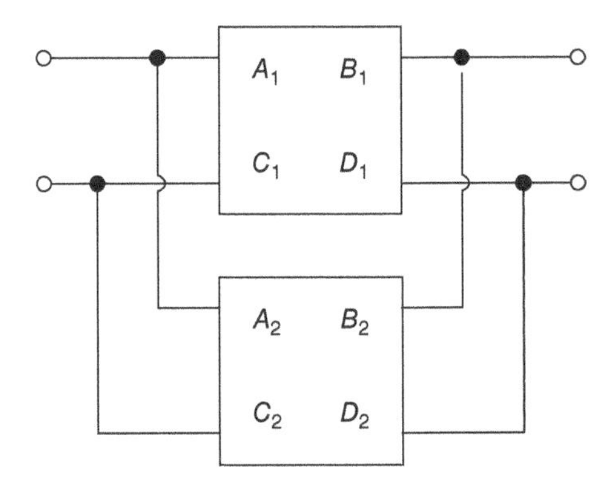

Figure 1.21 Two systems connected in parallel.

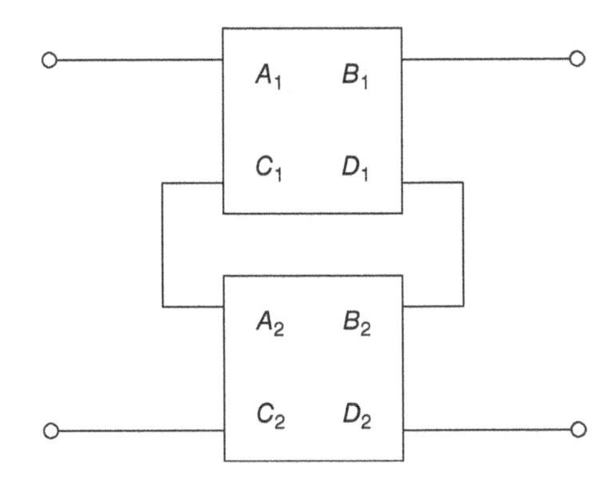

Figure 1.22 Two systems connected in series.

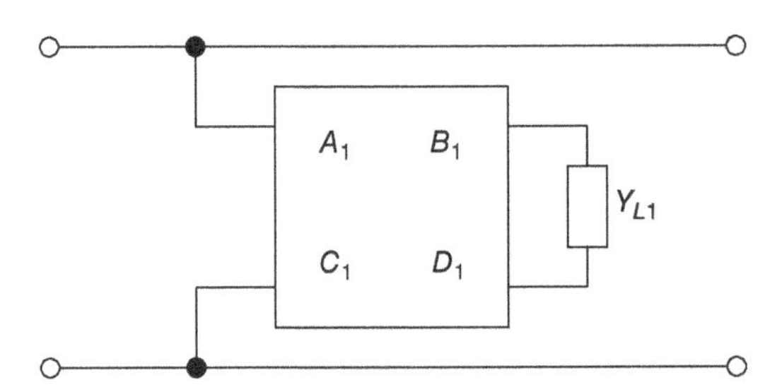

Figure 1.23 Two-port system made of admittance loaded two-port connected in parallel.

The *ABCD* matrix is that of an admittance in parallel and where the admittance is equal to the input admittance of the two-port loaded on Y_{L1}. It is given by

$$\begin{pmatrix} A & B \\ C & D \end{pmatrix} = \begin{pmatrix} 1 & 0 \\ \dfrac{C_1 + D_1 Y_{L1}}{A_1 + B_1 Y_{L1}} & 1 \end{pmatrix} \tag{1.51}$$

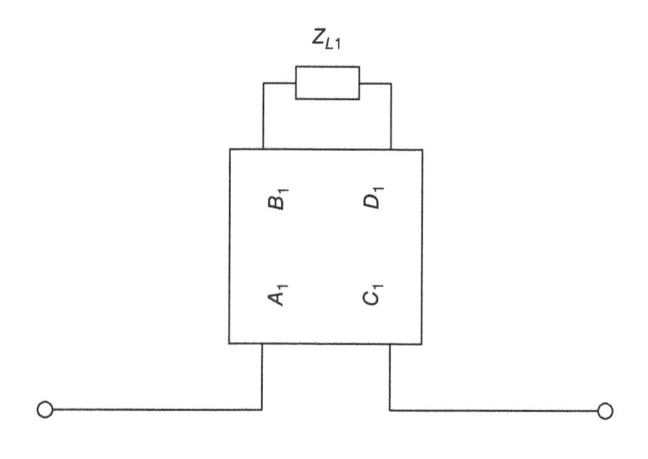

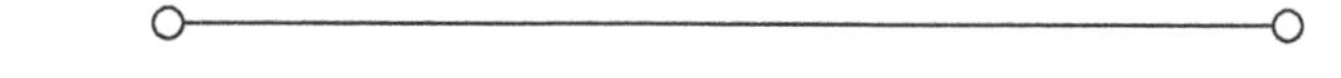

Figure 1.24 Two-port system made of impedance loaded two-port connected in series.

1.7.9 ABCD *Matrix of Impedance Loaded Two-Port Connected in Series*

In this case, a two-port loaded on impedance Z_{L1} is connected in series to form the two-port network as shown in Figure 1.24.

The *ABCD* matrix is that of an impedance in series and where the impedance is equal to the input impedance of the two-port loaded on Z_{L1}. It is given by

$$\begin{pmatrix} A & B \\ C & D \end{pmatrix} = \begin{pmatrix} 1 & \dfrac{A_1 Z_{L1} + B_1}{C_1 Z_{L1} + D_1} \\ 0 & 1 \end{pmatrix} \quad (1.52)$$

1.7.10 Conversion Between Scattering and ABCD *Matrices*

It is often needed to convert from scattering parameters to *ABCD* parameters and vice versa. Referring to Figure 1.3, the *ABCD* matrix of the two-port is given in terms of the scattering parameters by

$$\begin{pmatrix} A & B \\ C & D \end{pmatrix} = \begin{pmatrix} \dfrac{(1+S_{11})(1-S_{22})+S_{21}S_{12}}{2S_{21}} & Z_0\dfrac{(1+S_{11})(1+S_{22})-S_{21}S_{12}}{2S_{21}} \\ Y_0\dfrac{(1-S_{11})(1-S_{22})-S_{21}S_{12}}{2S_{21}} & \dfrac{(1-S_{11})(1+S_{22})+S_{21}S_{12}}{2S_{21}} \end{pmatrix} \quad (1.53)$$

In return, the scattering matrix of the two-port can be expressed in terms of the *ABCD* parameters using the conversion formulas:

$$\begin{pmatrix} S_{11} & S_{12} \\ S_{21} & S_{22} \end{pmatrix} = \begin{pmatrix} \dfrac{A+BY_0-CZ_0-D}{A+BY_0+CZ_0+D} & \dfrac{2(AD-BC)}{A+BY_0+CZ_0+D} \\ \dfrac{2}{A+BY_0+CZ_0+D} & \dfrac{-A+BY_0-CZ_0+D}{A+BY_0+CZ_0+D} \end{pmatrix} \tag{1.54}$$

1.8 Distributed Network Elements

For planar technologies, it is often more practical to obtain circuit layouts given in terms of distributed elements rather than given in terms of lumped elements. This section provides a background review of key aspects needed to better understand the subsequent chapters in this book.

1.8.1 Uniform Transmission Line

A uniform transmission line is a fundamental element when defining distributed networks. There are many ways to symbolize it, and in this book, we will generally represent it as shown in Figure 1.25.

The transmission line is defined by the characteristic impedance of the line Z_0, the propagation constant of the transmission line γ, and the physical length of the transmission line l.

The *ABCD* matrix of a uniform transmission line with physical length l is given by

$$\begin{pmatrix} A & B \\ C & D \end{pmatrix} = \begin{pmatrix} \cosh(\gamma l) & Z_0 \sinh(\gamma l) \\ \dfrac{\sinh(\gamma l)}{Z_0} & \cosh(\gamma l) \end{pmatrix} \tag{1.55}$$

Note that the propagation constant can be expressed in terms of the attenuation constant α and the phase constant β such that

$$\gamma = \alpha + j\beta \tag{1.56}$$

The electrical length is defined as the product of the phase constant and the physical length:

$$\theta = \beta l \tag{1.57}$$

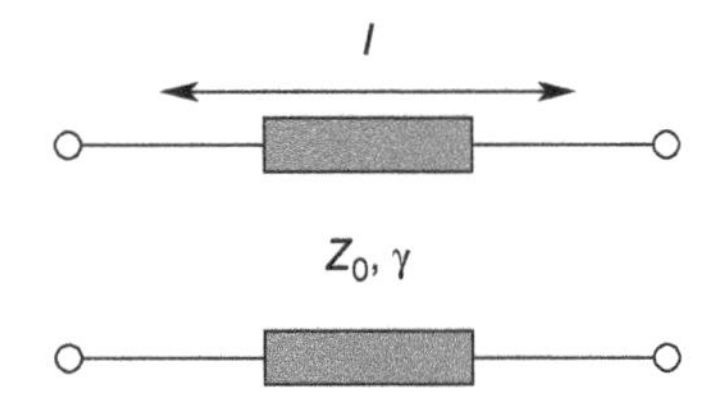

Figure 1.25 Uniform transmission line.

Since the phase rotates by 2π for one wavelength λ, the phase constant can be expressed as

$$\beta = \frac{2\pi}{\lambda} \tag{1.58}$$

The phase constant can also be expressed in terms of the operating radian frequency ω and the phase velocity v such that

$$\beta = \frac{\omega}{v} \tag{1.59}$$

For transmission lines that do not contain ferromagnetic material, for a TEM mode

$$v = \frac{c}{\sqrt{\varepsilon_r}} \tag{1.60}$$

where c is the speed of light and ε_r is the relative permittivity of the medium.

In the case of an ideal lossless transmission line, the attenuation constant α is equal to 0 and the *ABCD* matrix of the transmission line can be expressed in terms of the electrical length θ such that

$$\begin{pmatrix} A & B \\ C & D \end{pmatrix} = \begin{pmatrix} \cos\theta & jZ_0 \sin\theta \\ jY_0 \sin\theta & \cos\theta \end{pmatrix}$$

where $Y_0 = 1/Z_0$.

1.8.2 Unit Element

A unit element (UE) is a two-port made of a single section of a uniform lossless transmission line with a fixed length l and characteristic impedance Z_0 as shown in Figure 1.26.

The *ABCD* matrix of this two-port is given by

$$\begin{pmatrix} A & B \\ C & D \end{pmatrix} = \begin{pmatrix} \cos\theta & jZ_0 \sin\theta \\ jY_0 \sin\theta & \cos\theta \end{pmatrix} \tag{1.61}$$

where $\theta = \beta l$ is the electrical length of the transmission line.

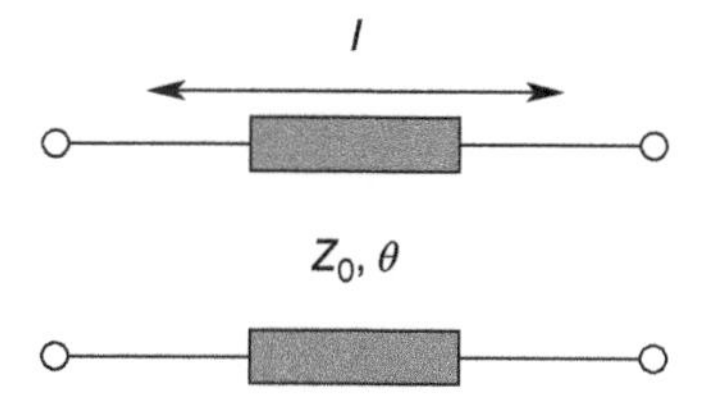

Figure 1.26 The unit element.

1.8.3 Input Impedance and Input Admittance

First, a UE is connected to load impedance Z_L as shown in Figure 1.27, and we are interested in the loaded input impedance of the UE.

The input impedance is given in terms of the load impedance Z_L and *ABCD* parameters of the two-port by

$$Z_{IN} = \left.\frac{V_1}{I_1}\right|_{Z_L} = \frac{AZ_L + B}{CZ_L + D}$$

with the *ABCD* parameters of the UE given by

$$\begin{pmatrix} \cos\theta & jZ_0\sin\theta \\ jY_0\sin\theta & \cos\theta \end{pmatrix}$$

so that

$$Z_{IN} = \frac{\cos\theta Z_L + jZ_0\sin\theta}{jY_0\sin\theta Z_L + \cos\theta} = \frac{\cos\theta(Z_L + jZ_0\tan\theta)}{\cos\theta(1 + jY_0\tan\theta Z_L)} = \frac{(Z_L + jZ_0\tan\theta)}{Y_0(Z_0 + jZ_L\tan\theta)} = Z_0\frac{Z_L + jZ_0\tan\theta}{Z_0 + jZ_L\tan\theta}$$

and the input impedance of a UE loaded on impedance Z_L is given by

$$Z_{IN} = Z_0\frac{Z_L + jZ_0\tan\theta}{Z_0 + jZ_L\tan\theta} \tag{1.62}$$

Note that when $Z_L = Z_0$, then $Z_{IN} = Z_0$ regardless of the electrical length θ.

Second, a UE is connected to load admittance Y_L as shown in Figure 1.28, and we are interested in the loaded input admittance of the UE.

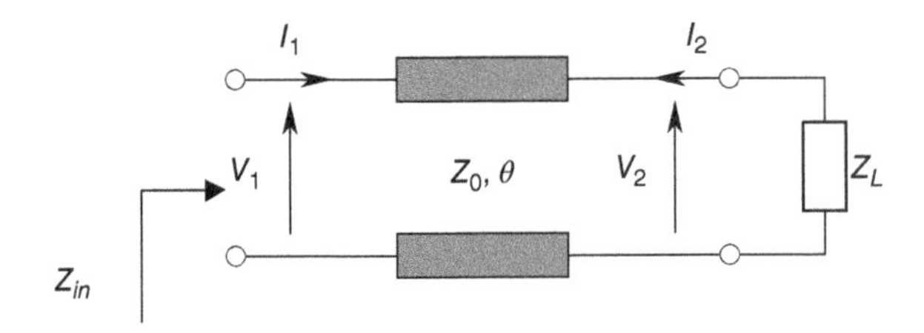

Figure 1.27 Input impedance of a UE loaded on impedance Z_L.

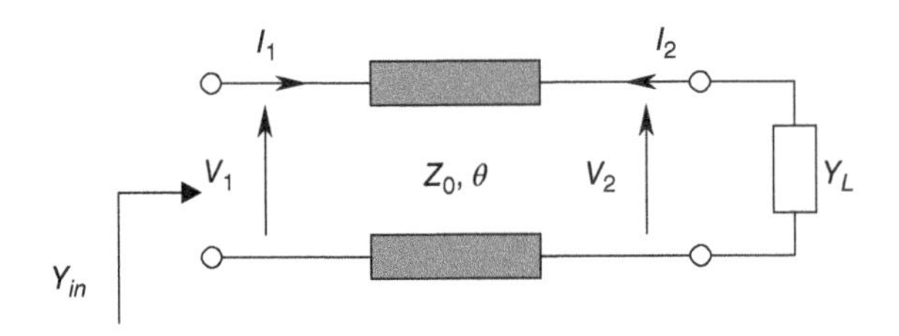

Figure 1.28 Input admittance of a UE loaded on impedance Y_L.

The input admittance is given in terms of the load admittance Y_L and *ABCD* parameters of the two-port by

$$Y_{IN} = \left.\frac{I_1}{V_1}\right|_{Y_L} = \frac{C + DY_L}{A + BY_L}$$

This gives

$$Y_{IN} = \frac{C + DY_L}{A + BY_L} = \frac{jY_0 \sin\theta + \cos\theta Y_L}{\cos\theta + jZ_0 \sin\theta Y_L} = \frac{\cos\theta(jY_0 \tan\theta + Y_L)}{\cos\theta(1 + jZ_0 \tan\theta Y_L)} = \frac{(jY_0 \tan\theta + Y_L)}{Z_0(Y_0 + j\tan\theta Y_L)} = Y_0 \frac{jY_0 \tan\theta + Y_L}{Y_0 + j\tan\theta Y_L}$$

and the input admittance of a UE loaded on admittance Y_L is given by

$$Y_{IN} = Y_0 \frac{Y_L + jY_0 \tan\theta}{Y_0 + j\tan\theta Y_L} \tag{1.63}$$

1.8.4 Short-Circuited Stub Placed in Series

In this case, we have a short-circuited stub that is placed in series to form the two-port of Figure 1.29.

In this case, the stub is loaded with impedance $Z_L = 0$ and the input impedance of the short-circuited stub is given by

$$Z_{IN} = Z_0 \frac{0 + jZ_0 \tan\theta}{Z_0 + j0 \tan\theta} = jZ_0 \tan\theta$$

and the *ABCD* matrix of the two-port system made of this short-circuited stub placed as a series impedance is given by

$$\begin{pmatrix} A & B \\ C & D \end{pmatrix} = \begin{pmatrix} 1 & Z_{IN} \\ 0 & 1 \end{pmatrix} = \begin{pmatrix} 1 & jZ_0 \tan\theta \\ 0 & 1 \end{pmatrix} \tag{1.64}$$

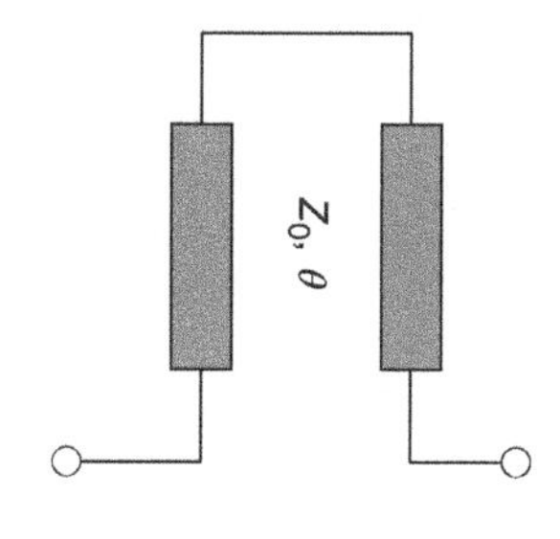

Figure 1.29 Short-circuited stub placed in series.

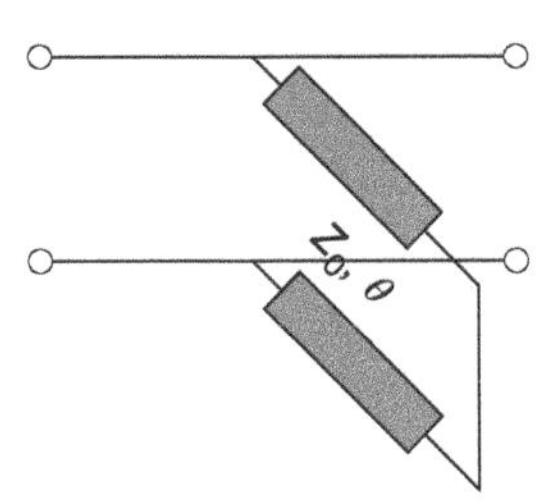

Figure 1.30 Short-circuited stub placed in parallel.

1.8.5 Short-Circuited Stub Placed in Parallel

In this case, we have a short-circuited stub that is placed in parallel to form the two-port of Figure 1.30.

In this case, the stub is loaded by admittance $Y_L = \infty$ and the input admittance of the short-circuited stub is given by

$$Y_{IN} = Y_0 \frac{Y_L + jY_0 \tan\theta}{Y_0 + j\tan\theta Y_L}\bigg|_{Y_L \to \infty} = Y_0 \frac{Y_L}{j\tan\theta Y_L} = \frac{Y_0}{j\tan\theta}$$

and the *ABCD* matrix of the two-port system made of this short-circuited stub placed as a admittance in parallel is given by

$$\begin{pmatrix} A & B \\ C & D \end{pmatrix} = \begin{pmatrix} 1 & 0 \\ Y_{IN} & 1 \end{pmatrix} = \begin{pmatrix} 1 & 0 \\ \frac{Y_0}{j\tan\theta} & 1 \end{pmatrix} \tag{1.65}$$

1.8.6 Open-Circuited Stub Placed in Series

In this case, we have an open-circuited stub that is placed in series to form the two-port of Figure 1.31.

In this case, the stub is loaded by impedance $Z_L = \infty$ and the input impedance of the open-circuited stub is given by

$$Z_{IN} = Z_0 \frac{Z_L + jZ_0 \tan\theta}{Z_0 + jZ_L \tan\theta}\bigg|_{Z_L = \infty} = Z_0 \frac{Z_L}{jZ_L \tan\theta} = \frac{Z_0}{j\tan\theta}$$

and the *ABCD* matrix of the two-port system made of this open-circuited stub placed as a series impedance is given by

$$\begin{pmatrix} A & B \\ C & D \end{pmatrix} = \begin{pmatrix} 1 & Z_{IN} \\ 0 & 1 \end{pmatrix} = \begin{pmatrix} 1 & \frac{Z_0}{j\tan\theta} \\ 0 & 1 \end{pmatrix} \tag{1.66}$$

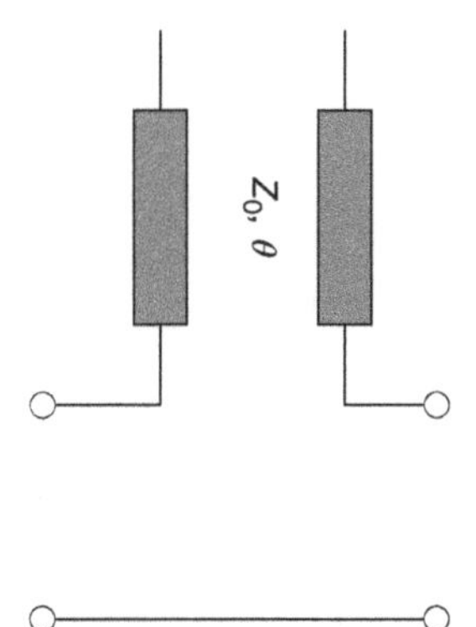

Figure 1.31 Open-circuited stub placed in series.

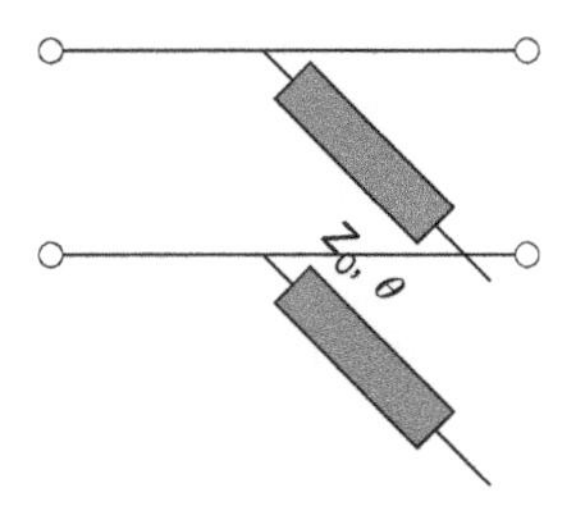

Figure 1.32 Open-circuited stub placed in parallel.

1.8.7 Open-Circuited Stub Placed in Parallel

In this case we have an open-circuited stub that is placed in parallel to form the two-port of Figure 1.32.

In this case the stub is loaded with admittance $Y_L = 0$ and the input admittance of the open-circuited stub is given by

$$Y_{IN} = Y_0 \frac{0 + jY_0 \tan\theta}{Y_0 + j\tan\theta 0} = jY_0 \tan\theta$$

and the *ABCD* matrix of the two-port system made of this open-circuited stub placed as an admittance in parallel is given by

$$\begin{pmatrix} A & B \\ C & D \end{pmatrix} = \begin{pmatrix} 1 & 0 \\ Y_{IN} & 1 \end{pmatrix} = \begin{pmatrix} 1 & 0 \\ jY_0 \tan\theta & 1 \end{pmatrix} \tag{1.67}$$

1.8.8 Richard's Transformation

It is not always easy to see the frequency dependency in the *ABCD* matrix of a UE when expressed in terms of electrical length θ. When the length of the transmission line is $\lambda/4$, there are special properties that can be exploited for distributed network design and synthesis.

The Richard's variable t is defined as the hyperbolic tangent of the delay τ_0 and the Laplace variable s as shown below [7]:

$$t = \tanh(\tau_0 s) \tag{1.68}$$

In (1.68), τ_0 is the delay of the UE and is proportional to the physical length l of the UE and the phase velocity v such that

$$\tau_0 = \frac{l}{v} \tag{1.69}$$

The UEs are assumed to be lossless so that $s \to j\omega$ and

$$t = \tanh(j\tau_0\omega) = j\tan(\tau_0\omega) = j\tan\left(\frac{\omega}{v}l\right) = j\tan(\beta l) = j\tan\theta = j\Omega \tag{1.70}$$

For example, in the case of quarter-wavelength UEs, we have

$$l = \frac{\lambda_0}{4} \tag{1.71}$$

where λ_0 is the wavelength given by the ratio of the velocity in the medium divided by a normalization frequency f_0 such that

$$\lambda_0 = \frac{v}{f_0} \tag{1.72}$$

so that

$$\tau_0 = \frac{\frac{1}{4}\lambda_0}{v} = \frac{\frac{1}{4}\frac{v}{f_0}}{v} = \frac{1}{4}\frac{1}{f_0} = \frac{1}{4}\frac{2\pi}{\omega_0}$$

$$\tau_0 = \frac{\pi}{2}\frac{1}{\omega_0}$$

where ω_0 is the normalization radian frequency.

So that in the case of quarter-wavelength UEs, we have

$$\Omega = \tan(\tau_0\omega) = \tan\left(\frac{\pi}{2}\frac{\omega}{\omega_0}\right) \tag{1.73}$$

1.8.8.1 *ABCD* Matrix of a UE in terms of Richard's Variable

The *ABCD* matrix of a general UE is given by

$$\begin{pmatrix} A & B \\ C & D \end{pmatrix} = \begin{pmatrix} \cos\theta & jZ_0\sin\theta \\ jY_0\sin\theta & \cos\theta \end{pmatrix}$$

and can be rewritten as

$$\begin{pmatrix} A & B \\ C & D \end{pmatrix} = \cos\theta \begin{pmatrix} 1 & jZ_0 \tan\theta \\ jY_0 \tan\theta & 1 \end{pmatrix}$$

and since $t = j\tan\theta$, this matrix can be expressed in terms of the Richard's variable t to give the *ABCD* matrix of a UE to be

$$\begin{pmatrix} A & B \\ C & D \end{pmatrix} = \frac{1}{\sqrt{1-t^2}} \begin{pmatrix} 1 & tZ_0 \\ tY_0 & 1 \end{pmatrix} \tag{1.74}$$

1.8.8.2 *ABCD* Matrices in Terms of Richard's Variable

The input impedance Z_{IN} of a quarter-wavelength UE loaded on impedance Z_L (see Fig. 1.27) can be expressed in terms of the Richard's variable by

$$Z_{IN} = Z_0 \frac{Z_L + jZ_0 \tan\theta}{Z_0 + jZ_L \tan\theta} = Z_0 \frac{Z_L + tZ_0}{Z_0 + tZ_L} \tag{1.75}$$

The input admittance Y_{IN} of a quarter-wavelength UE loaded on impedance Y_L (see Fig. 1.28) can be expressed in terms of the Richard's variable by

$$Y_{IN} = Y_0 \frac{Y_L + jY_0 \tan\theta}{Y_0 + j\tan\theta Y_L} = Y_0 \frac{Y_L + tY_0}{Y_0 + tY_L} \tag{1.76}$$

The *ABCD* matrix of a two-port system made of a short-circuited stub placed as series impedance (see Fig. 1.29) can be expressed in terms of the Richard's variable by

$$\begin{pmatrix} A & B \\ C & D \end{pmatrix} = \begin{pmatrix} 1 & Z_{IN} \\ 0 & 1 \end{pmatrix} = \begin{pmatrix} 1 & jZ_0 \tan\theta \\ 0 & 1 \end{pmatrix} = \begin{pmatrix} 1 & tZ_0 \\ 0 & 1 \end{pmatrix} \tag{1.77}$$

The *ABCD* matrix of a two-port system made of a short-circuited stub placed as parallel admittance (see Fig. 1.30) can be expressed in terms of the Richard's variable by

$$\begin{pmatrix} A & B \\ C & D \end{pmatrix} = \begin{pmatrix} 1 & 0 \\ Y_{IN} & 1 \end{pmatrix} = \begin{pmatrix} 1 & 0 \\ \dfrac{Y_0}{j\tan\theta} & 1 \end{pmatrix} = \begin{pmatrix} 1 & 0 \\ \dfrac{Y_0}{t} & 1 \end{pmatrix} \tag{1.78}$$

The *ABCD* matrix of a two-port system made of an open-circuited stub placed as series impedance (see Fig. 1.31) can be expressed in terms of the Richard's variable by

$$\begin{pmatrix} A & B \\ C & D \end{pmatrix} = \begin{pmatrix} 1 & Z_{IN} \\ 0 & 1 \end{pmatrix} = \begin{pmatrix} 1 & \dfrac{Z_0}{j\tan\theta} \\ 0 & 1 \end{pmatrix} = \begin{pmatrix} 1 & \dfrac{Z_0}{t} \\ 0 & 1 \end{pmatrix} \tag{1.79}$$

The *ABCD* matrix of a two-port system made of an open-circuited stub placed as parallel admittance (see Fig. 1.32) can be expressed in terms of the Richard's variable by

$$\begin{pmatrix} A & B \\ C & D \end{pmatrix} = \begin{pmatrix} 1 & 0 \\ Y_{IN} & 1 \end{pmatrix} = \begin{pmatrix} 1 & 0 \\ jY_0 \tan\theta & 1 \end{pmatrix} = \begin{pmatrix} 1 & 0 \\ tY_0 & 1 \end{pmatrix} \tag{1.80}$$

1.8.9 *Kuroda Identities*

There are situations when a synthesis technique can lead to the direct cascade of stubs. This situation is often impossible to replicate in an actual circuit. There are also times where one might prefer one type of stub to the other. In these situations, one can transform a given set of stub configurations into another set of stub configurations. In this subsection, we recall the four classic Kuroda identities that can help solve these problems.

1.8.9.1 First Low-Pass Kuroda Identity

The first low-pass Kuroda identity consists of a UE with characteristic impedance Z_{01} placed left of a series short-circuited stub with characteristic impedance Z_{02}. This is equivalent to a UE with characteristic impedance Z_{04} placed right of a parallel open-circuited stub with characteristic impedance Z_{03} as shown in Figure 1.33 [7].

The two systems are equivalent if

$$\begin{cases} Z_{04} = Z_{01} + Z_{02} \\ Z_{03} = \dfrac{Z_{01}}{Z_{02}}(Z_{01} + Z_{02}) \end{cases} \tag{1.81}$$

The proof is obtained by equating the *ABCD* matrix of the system on the left

$$\begin{pmatrix} \cos\theta & jZ_{01}\sin\theta \\ jY_{01}\sin\theta & \cos\theta \end{pmatrix}\begin{pmatrix} 1 & jZ_{02}\tan\theta \\ 0 & 1 \end{pmatrix} = \begin{pmatrix} \cos\theta & jZ_{01}\sin\theta + jZ_{02}\tan\theta\cos\theta \\ jY_{01}\sin\theta & \cos\theta + jY_{01}\sin\theta jZ_{02}\tan\theta \end{pmatrix}$$

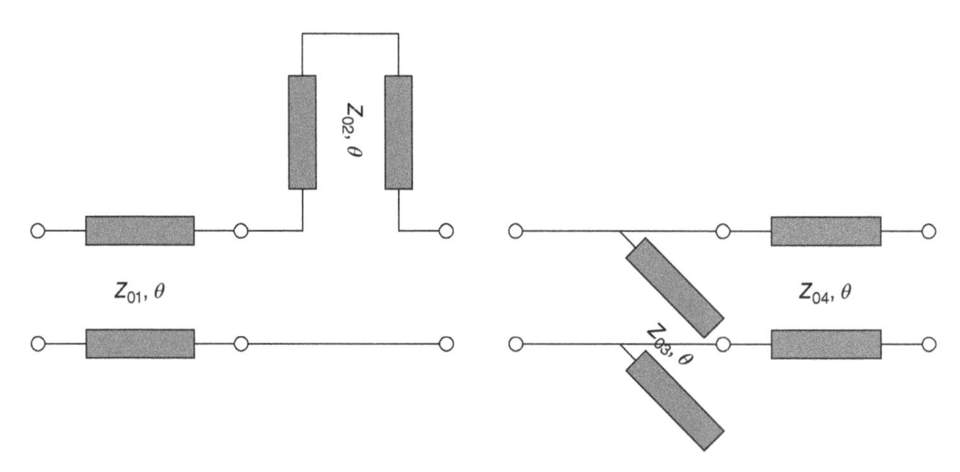

Figure 1.33 First low-pass Kuroda identity.

to the *ABCD* matrix of the system on the right:

$$\begin{pmatrix} 1 & 0 \\ jY_{03}\tan\theta & 1 \end{pmatrix}\begin{pmatrix} \cos\theta & jZ_{04}\sin\theta \\ jY_{04}\sin\theta & \cos\theta \end{pmatrix} = \begin{pmatrix} \cos\theta & jZ_{04}\sin\theta \\ jY_{04}\sin\theta + jY_{03}\tan\theta\cos\theta & \cos\theta + jY_{03}\tan\theta jZ_{04}\sin\theta \end{pmatrix}$$

Equating *B* parameters gives

$$jZ_{01}\sin\theta + jZ_{02}\tan\theta\cos\theta = jZ_{04}\sin\theta$$

and dividing both sides by $\sin\theta$ gives

$$Z_{01} + Z_{02} = Z_{04}$$

Equating *C* parameters gives

$$jY_{01}\sin\theta = jY_{04}\sin\theta + jY_{03}\tan\theta\cos\theta$$

and dividing both sides by $\sin\theta$ gives

$$Y_{01} = Y_{04} + Y_{03}$$

which can be expressed in terms of impedances by

$$Z_{01} = \frac{Z_{03}Z_{04}}{Z_{03} + Z_{04}}$$

which can also be expressed as

$$Z_{03} = \frac{Z_{01}Z_{04}}{Z_{04} - Z_{01}}$$

and replacing Z_{04} by $Z_{01} + Z_{02}$ gives

$$Z_{03} = \frac{Z_{01}(Z_{01} + Z_{02})}{Z_{01} + Z_{02} - Z_{01}} = \frac{Z_{01}}{Z_{02}}(Z_{01} + Z_{02})$$

1.8.9.2 Second Low-Pass Kuroda Identity

The second low-pass Kuroda identity consists of a UE with characteristic impedance Z_{02} placed right of a series short-circuited stub with characteristic impedance Z_{01}. This is equivalent to a UE with characteristic impedance Z_{03} placed left of a parallel open-circuited stub with characteristic impedance Z_{04} as shown in Figure 1.34.

The two systems are equivalent if

$$\begin{cases} Z_{03} = Z_{01} + Z_{02} \\ Z_{04} = \dfrac{Z_{02}}{Z_{01}}(Z_{01} + Z_{02}) \end{cases} \tag{1.82}$$

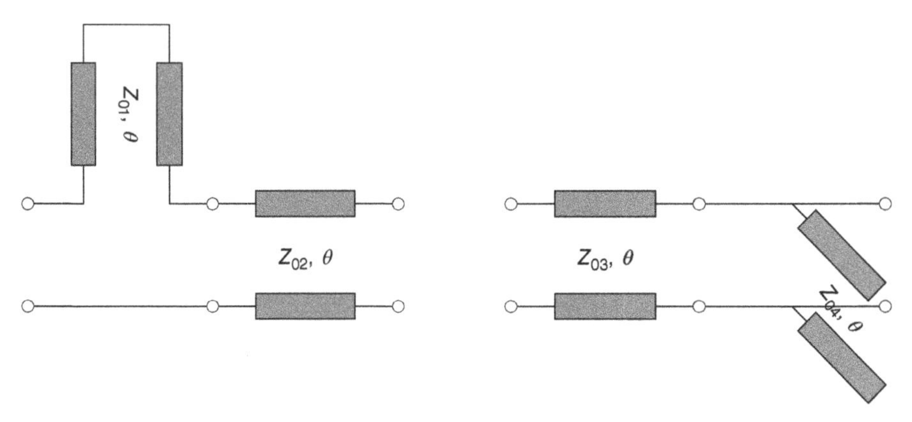

Figure 1.34 Second low-pass Kuroda identity.

The proof is obtained by equating the *ABCD* matrix of the system on the left

$$\begin{pmatrix} 1 & jZ_{01}\tan\theta \\ 0 & 1 \end{pmatrix}\begin{pmatrix} \cos\theta & jZ_{02}\sin\theta \\ jY_{02}\sin\theta & \cos\theta \end{pmatrix} = \begin{pmatrix} \cos\theta + jY_{02}\sin\theta jZ_{01}\tan\theta & jZ_{02}\sin\theta + jZ_{01}\tan\theta\cos\theta \\ jY_{02}\sin\theta & \cos\theta \end{pmatrix}$$

to the *ABCD* matrix of the system on the right:

$$\begin{pmatrix} \cos\theta & jZ_{03}\sin\theta \\ jY_{03}\sin\theta & \cos\theta \end{pmatrix}\begin{pmatrix} 1 & 0 \\ jY_{04}\tan\theta & 1 \end{pmatrix} = \begin{pmatrix} \cos\theta + jY_{04}\tan\theta jZ_{03}\sin\theta & jZ_{03}\sin\theta \\ jY_{03}\sin\theta + jY_{04}\tan\theta\cos\theta & \cos\theta \end{pmatrix}$$

Equating *B* parameters gives

$$jZ_{02}\sin\theta + jZ_{01}\tan\theta\cos\theta = jZ_{03}\sin\theta$$

and dividing both sides by $\sin\theta$ gives

$$Z_{02} + Z_{01} = Z_{03}$$

Equating *C* parameters gives

$$jY_{02}\sin\theta = jY_{03}\sin\theta + jY_{04}\tan\theta\cos\theta$$

and dividing both sides by $\sin\theta$ gives

$$Y_{02} = Y_{03} + Y_{04}$$

which can be expressed in terms of impedances by

$$Z_{02} = \frac{Z_{03}Z_{04}}{Z_{03} + Z_{04}}$$

which can also be expressed as

$$Z_{04} = \frac{Z_{02}Z_{03}}{Z_{03} - Z_{02}}$$

and replacing Z_{03} by $Z_{01} + Z_{02}$ gives

$$Z_{04} = \frac{Z_{02}(Z_{01} + Z_{02})}{Z_{01} + Z_{02} - Z_{02}} = \frac{Z_{02}}{Z_{01}}(Z_{01} + Z_{02})$$

1.8.9.3 First High-Pass Kuroda Identity

The first high-pass Kuroda identity consists of a UE with characteristic impedance Z_{02} placed right of a parallel short-circuited stub with characteristic impedance Z_{01}. This is equivalent to a UE with characteristic impedance Z_{03} placed left of a parallel short-circuited stub with characteristic impedance Z_{04} followed by an ideal transformer n as shown in Figure 1.35.

The two systems are equivalent if

$$\begin{cases} n = 1 + \dfrac{Z_{02}}{Z_{01}} \\ Z_{03} = \dfrac{Z_{01}Z_{02}}{Z_{01} + Z_{02}} \\ Z_{04} = \dfrac{Z_{01}{}^{2}}{Z_{01} + Z_{02}} \end{cases} \tag{1.83}$$

The proof is obtained by equating the *ABCD* matrix of the system on the left

$$\begin{pmatrix} 1 & 0 \\ \dfrac{Y_{01}}{j\tan\theta} & 1 \end{pmatrix} \begin{pmatrix} \cos\theta & jZ_{02}\sin\theta \\ jY_{02}\sin\theta & \cos\theta \end{pmatrix} = \begin{pmatrix} \cos\theta & jZ_{02}\sin\theta \\ jY_{02}\sin\theta - jY_{01}\dfrac{\cos^2\theta}{\sin\theta} & \cos\theta + Z_{02}Y_{01}\cos\theta \end{pmatrix}$$

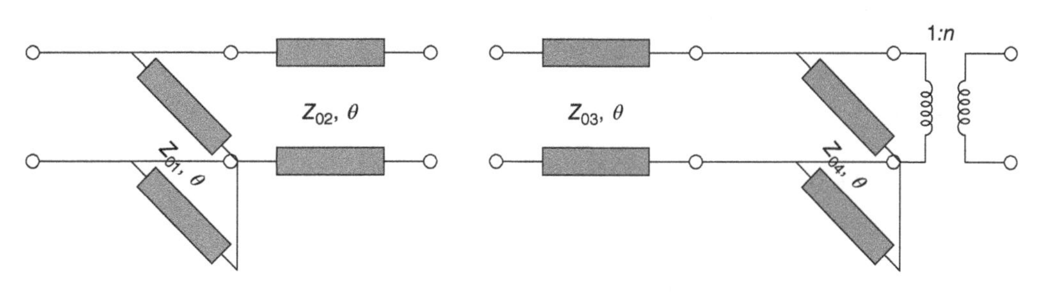

Figure 1.35 First high-pass Kuroda identity.

to the *ABCD* matrix of the system on the right:

$$\begin{pmatrix} \cos\theta & jZ_{03}\sin\theta \\ jY_{03}\sin\theta & \cos\theta \end{pmatrix} \begin{pmatrix} 1 & 0 \\ \dfrac{Y_{04}}{j\tan\theta} & 1 \end{pmatrix} \begin{pmatrix} \dfrac{1}{n} & 0 \\ 0 & n \end{pmatrix} = \begin{pmatrix} \dfrac{\cos\theta + Z_{03}Y_{04}\cos\theta}{n} & jnZ_{03}\sin\theta \\ \dfrac{jY_{03}\sin\theta - jY_{04}\dfrac{\cos^2\theta}{\sin\theta}}{n} & n\cos\theta \end{pmatrix}$$

Equating *D* parameters gives

$$\cos\theta + Z_{02}Y_{01}\cos\theta = n\cos\theta$$

and dividing both sides by $\cos\theta$ gives

$$1 + Z_{02}Y_{01} = n \quad \text{or} \quad n = 1 + \frac{Z_{02}}{Z_{01}}$$

Equating *B* parameters gives

$$jZ_{02}\sin\theta = jnZ_{03}\sin\theta$$

and dividing both sides by $\sin\theta$ gives

$$jZ_{02} = jnZ_{03}$$

then replacing n gives

$$Z_{03} = \frac{Z_{02}}{n} = \frac{Z_{02}}{1 + \dfrac{Z_{02}}{Z_{01}}} = \frac{Z_{01}Z_{02}}{Z_{01} + Z_{02}}$$

Equating *A* parameters gives

$$\cos\theta = \frac{\cos\theta + Z_{03}Y_{04}\cos\theta}{n}$$

and dividing both sides by $\cos\theta$ gives

$$1 = \frac{1 + Z_{03}Y_{04}}{n}$$

from which we find that

$$1 + \frac{Z_{03}}{Z_{04}} = n = 1 + \frac{Z_{02}}{Z_{01}} \quad \text{or} \quad Z_{04} = \frac{Z_{01}}{Z_{02}}Z_{03}$$

then replacing Z_{03} gives

$$Z_{04}=\frac{Z_{01}}{Z_{02}}\frac{Z_{01}Z_{02}}{Z_{01}+Z_{02}}=\frac{Z_{01}{}^2}{Z_{01}+Z_{02}}$$

1.8.9.4 Second High-Pass Kuroda Identity

The second high-pass Kuroda identity consists of a UE with characteristic impedance Z_{02} placed right of a series open-circuited stub with characteristic impedance Z_{01}. This is equivalent to a UE with characteristic impedance Z_{03} placed left of a series open-circuited stub with characteristic impedance Z_{04} followed by an ideal transformer n as shown in Figure 1.36.

The two systems are equivalent if

$$\begin{cases}\frac{1}{n}=1+\frac{Z_{01}}{Z_{02}}\\ Z_{03}=Z_{01}+Z_{02}\\ Z_{04}=\frac{Z_{01}}{Z_{02}}(Z_{01}+Z_{02})\end{cases} \tag{1.84}$$

The proof is obtained by equating the *ABCD* matrix of the system on the left

$$\begin{pmatrix}1 & \frac{Z_{01}}{j\tan\theta}\\ 0 & 1\end{pmatrix}\begin{pmatrix}\cos\theta & jZ_{02}\sin\theta\\ jY_{02}\sin\theta & \cos\theta\end{pmatrix}=\begin{pmatrix}\cos\theta+Z_{01}Y_{02}\cos\theta & jZ_{02}\sin\theta-jZ_{01}\frac{\cos^2\theta}{\sin\theta}\\ jY_{02}\sin\theta & \cos\theta\end{pmatrix}$$

to the *ABCD* matrix of the system on the right:

$$\begin{pmatrix}\cos\theta & jZ_{03}\sin\theta\\ jY_{03}\sin\theta & \cos\theta\end{pmatrix}\begin{pmatrix}1 & \frac{Z_{04}}{j\tan\theta}\\ 0 & 1\end{pmatrix}\begin{pmatrix}\frac{1}{n} & 0\\ 0 & n\end{pmatrix}=\begin{pmatrix}\frac{\cos\theta}{n} & n\left(jZ_{03}\sin\theta-jZ_{04}\frac{\cos^2\theta}{\sin\theta}\right)\\ \frac{jY_{03}\sin\theta}{n} & n(\cos\theta+Z_{04}Y_{03}\cos\theta)\end{pmatrix}$$

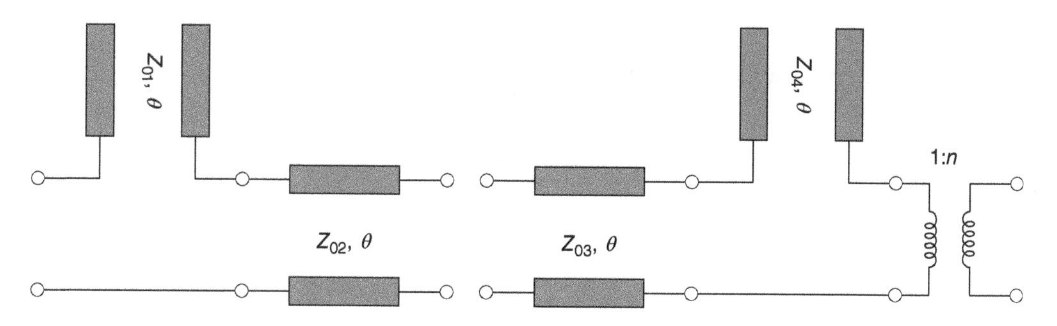

Figure 1.36 Second high-pass Kuroda identity.

Equating A parameters gives

$$\cos\theta + Z_{01} Y_{02} \cos\theta = \frac{\cos\theta}{n}$$

and dividing both sides by $\cos\theta$ gives

$$1 + Z_{01} Y_{02} = \frac{1}{n} \quad \text{or} \quad \frac{1}{n} = 1 + \frac{Z_{01}}{Z_{02}}$$

Equating C parameters gives

$$jY_{02} \sin\theta = \frac{jY_{03} \sin\theta}{n}$$

and dividing both sides by $\sin\theta$ gives

$$jY_{02} = \frac{jY_{03}}{n} \quad \text{or} \quad Z_{03} = \frac{1}{n} Z_{02}$$

and replacing n gives

$$Z_{03} = \left(1 + \frac{Z_{01}}{Z_{02}}\right) Z_{02} = Z_{01} + Z_{02}$$

Equating D parameters gives

$$\cos\theta = n(\cos\theta + Z_{04} Y_{03} \cos\theta)$$

and dividing both sides by $\cos\theta$ gives

$$1 = n(1 + Z_{04} Y_{03})$$

from which we find that

$$1 + \frac{Z_{04}}{Z_{03}} = \frac{1}{n} = 1 + \frac{Z_{01}}{Z_{02}} \quad \text{or} \quad Z_{04} = \frac{Z_{01}}{Z_{02}} Z_{03}$$

and replacing Z_{03} gives

$$Z_{04} = \frac{Z_{01}}{Z_{02}} Z_{03} = \frac{Z_{01}}{Z_{02}} (Z_{01} + Z_{02})$$

References

[1] I. J. Bahl, Fundamental of RF and Microwave Transistors Amplifiers, Wiley-Interscience, Hoboken, NJ, 2009.
[2] S. J. Mason, "Feedback theory—some properties of signal flow graphs," Proc. IRE, vol. 41, pp. 1144–1156, 1953.
[3] K. Chang, Microwave Solid-State Circuits and Applications, Wiley-Interscience, New York, 1994.
[4] T. T. Ha, Solid-State Microwave Amplifier Design, Wiley-Interscience, New York, 1981.
[5] H. T. Friis, "Noise figures of radio receivers," Proc. IRE, vol. 32, pp. 419–422, 1944.
[6] P. Jarry, J. Beneat, Advanced Design Techniques and Realizations of Microwave and RF Filters, Wiley-IEEE Press, Hoboken, NJ, 2008.
[7] H. Baher, Synthesis of Electrical Networks, Wiley-Interscience, Chichester, NY, 1984.

2

Introduction to the Real Frequency Technique

Multistage Lumped Amplifier Design

2.1 Introduction

The real frequency technique (RFT) was introduced by Carlin [1] and is an optimum solution to broadband amplifier equalization problems [2, 3]. A simplified RFT was extended to multistage microwave amplifier design [4]. The technique is a computer-aided design procedure that uses

Microwave Amplifier and Active Circuit Design Using the Real Frequency Technique, First Edition.
Pierre Jarry and Jacques N. Beneat.

measured scattering parameters of transistor devices and a Levenberg–Marquardt optimization algorithm for solving least-squares problems [5, 6]. B.S. Yarman summarizes these RFT results in Ref. [7].

The RFT is probably the best circuit synthesis method available for the design of broadband amplifiers. When applied to the design of multistage microwave amplifiers, the procedure contrary to purely analytical design techniques accepts measured scattering parameters for characterizing the active devices. It does not require an a priori choice of equalizer topology nor an analytical expression for the transfer function of the system.

In this chapter, a complete description of the RFT is provided for the case of multistage microwave amplifier design for the simultaneous optimization of transducer gain and voltage standing wave ratio (VSWR). In this chapter, the equalizers are represented as realizable transfer functions in the Laplace variable ($s = \sigma + j\omega$) with transmission zeros only at $\omega = 0$ and $\omega = \infty$.

In subsequent chapters, the technique is extended to the optimization of multistage transducer gain, VSWR, noise figure, group delay, and transimpedance gain. The equalizer is extended to lossless distributed and lossy distributed networks. The technique is also extended to the case of multistage active filters with finite transmission zeros.

The RFT has the advantage of generality, being applicable to all matching problems, and universality, as it does not require a predefined equalizer topology. The simple formalism of the RFT allows optimizing many performance parameters of single- or multistage microwave amplifiers.

2.2 Multistage Lumped Amplifier Representation

The multistage lumped amplifier is represented in Figure 2.1 [8]. It assumes that there are k transistors. An equalizer is associated with each transistor. There is one final equalizer placed after the kth transistor. The source and load impedances are assumed arbitrary, complex, and frequency dependent. In the text descriptions, a stage will refer to the combination of a transistor and its associated equalizer. The transistor can represent a transistor device alone or a transistor device with added but known circuitry as long as scattering parameter data of a corresponding two-port are available.

The equalizer of the kth stage is characterized by the scattering matrix

$$S_{E_k} = \begin{pmatrix} e_{11,k}(s) & e_{12,k}(s) \\ e_{21,k}(s) & e_{22,k}(s) \end{pmatrix} \tag{2.1}$$

The transistor of the kth stage is characterized by the scattering matrix

$$S_{T_k} = \begin{pmatrix} S_{11,k}(s) & S_{12,k}(s) \\ S_{21,k}(s) & S_{22,k}(s) \end{pmatrix} \tag{2.2}$$

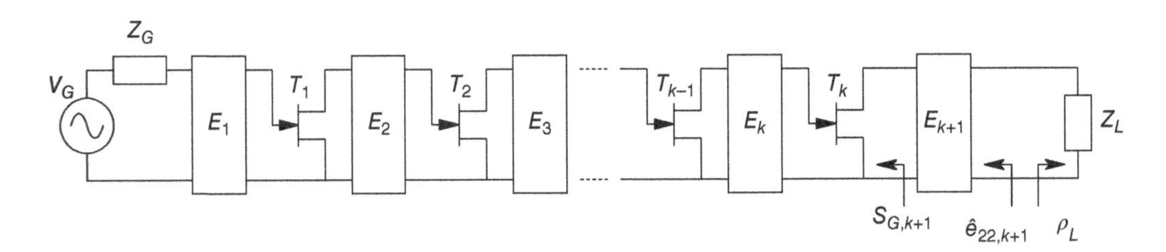

Figure 2.1 Multistage lumped amplifier representation.

The problem consists of defining the scattering parameter functions $e_{ij,k}(s)$ of an equalizer k that optimize some amplifier characteristic(s) when transistor scattering parameters are given at several frequency points in a band of interest. The scattering parameters of the transistor are numerical and are either computed from models or tabulated from manufacturer's datasheets or data files. The equalizers are optimized independently one after the other.

In this chapter, the technique optimizes simultaneously the multistage transducer gain and input and output VSWR over a set of m frequency points. The equalizers are assumed lossless and reciprocal two-ports that can be synthesized as lumped circuits. In other chapters, additional characteristics will be optimized such as noise figure and group delay, and the equalizers will be represented by lossless or lossy distributed networks.

In this chapter, the equalizer is assumed to be a lossless and reciprocal lumped two-port so that its scattering matrix can be expressed as [9]

$$S_E = \begin{pmatrix} e_{11}(s) = \dfrac{h(s)}{g(s)} & e_{12}(s) = \dfrac{f(s)}{g(s)} \\ e_{21}(s) = \dfrac{f(s)}{g(s)} & e_{22}(s) = \dfrac{-\varepsilon h(-s)}{g(s)} \end{pmatrix} \tag{2.3}$$

where $f(s)$ is an even or odd function. If $f(s)$ is even, then $\varepsilon = 1$; and if $f(s)$ is odd, then $\varepsilon = -1$. In addition, the lossless condition $|e_{11}|^2 + |e_{12}|^2 = 1$ leads to

$$g(s)g(-s) = h(s)h(-s) + f(s)f(-s) \tag{2.4}$$

$h(s)$ and $g(s)$ are polynomials of order n with real-valued coefficients:

$$h(s) = h_0 + h_1 s + \cdots + h_{n-1}s^{n-1} + h_n s^n$$
$$g(s) = g_0 + g_1 s + \cdots + g_{n-1}s^{n-1} + g_n s^n$$

and in this chapter, $f(s)$ is taken as

$$f(s) = \pm s^k \quad \text{and} \quad \varepsilon = (-1)^k \tag{2.5}$$

where k represents the number of transmission zeros at $\omega = 0$ with $\alpha \leq n$, leading to

$$g(s)g(-s) = h(s)h(-s) + (-1)^k s^{2k} \tag{2.6}$$

The problem consists of defining unknown polynomial coefficients $h_0, h_1, h_2, \ldots, h_n$ of an equalizer that optimize the transducer gain and/or VSWR. Once $h(s)$ is known, $g(s)g(-s)$ can be computed using

$$G(s^2) = g(s)g(-s) = G_0 + G_1 s^2 + \cdots + G_n s^{2n}$$

where the polynomial coefficients of $G(s^2)$ are given by

$$
\begin{aligned}
G_0 &= h_0^2 \\
G_1 &= -h_1^2 + 2h_2h_0 \\
&\vdots \\
G_i &= (-1)^i h_i^2 + 2h_{2i}h_0 + 2\sum_{j=2}^{i}(-1)^{j-1}h_{j-1}h_{2j-i+1} \\
&\vdots \\
G_k &= G_i|_{i=k} + (-1)^k \\
&\vdots \\
G_n &= (-1)^n h_n^2
\end{aligned}
$$

The left-half roots of $g(s)g(-s)$ are retained to form a Hurwitz polynomial $g(s)$. At this point, the equalizer scattering parameters $e_{11}(s)$, $e_{12}(s)$, $e_{21}(s)$, and $e_{22}(s)$ can be computed. During optimization, they will be needed for computing the error function to be optimized, and after optimization, for example, $e_{11}(s)$ can be used to realize the equalizer using the Darlington theorem [10].

2.3 Overview of the RFT

The main concept of the RFT is shown in Figure 2.2. The design of equalizer E_i is a double-matching problem on complex loads $Z_{\mathrm{left},i}$ and $Z_{\mathrm{right},i}$. Referring to Figure 2.1, $Z_{\mathrm{left},i}$ represents the source load Z_G with equalizer–transistor stages 1 through $i-1$ and $Z_{\mathrm{right},i}$ represents transistor T_i with equalizer–transistor stages $i+1$ through k and equalizer E_{k+1} and load Z_L.

The problem consists of designing one equalizer at a time to improve one or several frequency characteristics. For example, in the case of Figure 2.2, from (1.24) the transducer gain is given by

$$
G_T = \frac{\left(1-|S_{G,i}|^2\right)\left(1-|S_{L,i}|^2\right)}{|1-e_{11,i}S_{G,i}|^2|1-S_{L,i}\hat{e}_{22,i}|^2}|e_{21,i}|^2 \tag{2.7}
$$

with

$$
S_{G,i} = \frac{Z_{\mathrm{left,i}} - Z_0}{Z_{\mathrm{left,i}} + Z_0} \tag{2.8}
$$

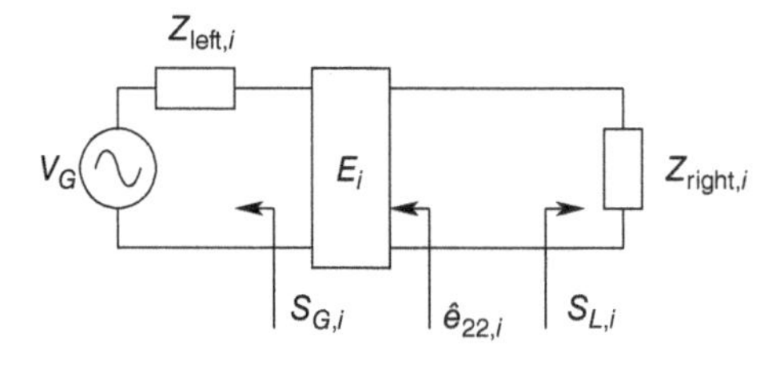

Figure 2.2 Equalizer design as the double matching problem.

$$S_{L,i}=\frac{Z_{\text{right},i}-Z_0}{Z_{\text{right},i}+Z_0} \tag{2.9}$$

$$\hat{e}_{22,i}=e_{22,i}+\frac{e_{21,i}^2 S_{G,i}}{1-e_{11,i}S_{G,i}} \tag{2.10}$$

These quantities are assumed frequency dependent. For example, the $e'_{ij}s$ are polynomials described in Section 2.1, and the complex impedances $Z_{\text{left},i}$ and $Z_{\text{right},i}$ include the frequency dependence of the other equalizers as well as of the transistors.

For a multistage amplifier, where there is more than one equalizer, there are three main approaches. The first approach is to start on the left with E_1T_1 and then move to E_2T_2, and so forth. The second approach is to start on the right with E_{k+1}, then E_kT_k, and then move to $E_{k-1}T_{k-1}$, and so forth. The third would be to start somewhere in the middle. In this chapter, the first approach will be used. In later chapters, the second approach will be more suitable.

The design of an equalizer involves the use of an optimization process based on the Levenberg–Marquardt–More optimization technique and optimizing a single equalizer at a given time makes for very efficient optimization.

2.4 Multistage Transducer Gain

The overall transducer gain T_{GT} of the multistage amplifier when terminated on arbitrary terminations Z_G and Z_L as shown in Figure 2.1 is given by [11, 12]

$$T_{GT}=T_k\frac{|e_{21,k+1}|^2\left(1-|\rho_L|^2\right)}{|1-e_{11,k+1}S_{G,k+1}|^2|1-\hat{e}_{22,k+1}\rho_L|^2} \tag{2.11}$$

with

$$\hat{e}_{22,k+1}=e_{22,k+1}+\frac{e_{21,k+1}^2 S_{G,k+1}}{1-e_{11,k+1}S_{G,k+1}} \tag{2.12}$$

$$S_{G,k+1}=S_{22,k}+\frac{S_{12,k}S_{21,k}\hat{e}_{22,k}}{1-S_{11,k}\hat{e}_{22,k}} \tag{2.13}$$

$$\rho_L=\frac{Z_L-Z_0}{Z_L+Z_0} \tag{2.14}$$

where

$S_{G,k+1}$ is the output reflection coefficient of transistor k when loaded on input
$\hat{e}_{22,k+1}$ is the output reflection of equalizer $k+1$ when loaded on input
ρ_L is the reflection coefficient of the load impedance Z_L
T_k is the transducer gain of the first k equalizer–transistor stages terminated on load Z_0 as shown in Figure 2.3. Z_0 is the reference used for the scattering parameters. For example $Z_0 = 50\Omega$ is often used to provide the measured scattering parameters of a transistor

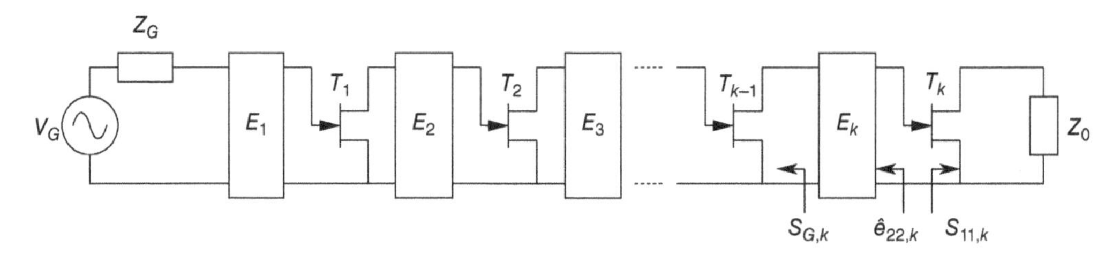

Figure 2.3 Transducer gain for k equalizer–transistor stages terminated on Z_0.

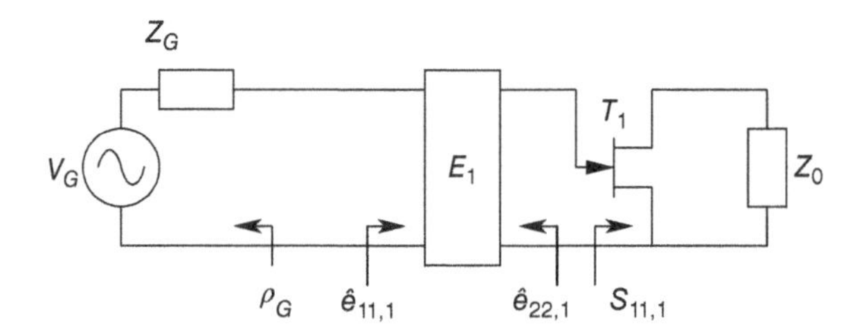

Figure 2.4 Transducer gain for the first stage terminated on Z_0.

The transducer gain T_k of the first k equalizer–transistor stages is computed recursively by

$$T_k = T_{k-1}\frac{|e_{21,k}|^2|S_{21,k}|^2}{|1-e_{11,k}S_{G,k}|^2|1-\hat{e}_{22,k}S_{11,k}|^2} \quad \text{for } k>1 \tag{2.15}$$

with

$$\hat{e}_{22,k} = e_{22,k} + \frac{e_{21,k}^2 S_{G,k}}{1-e_{11,k}S_{G,k}} \tag{2.16}$$

$$S_{G,k} = S_{22,k-1} + \frac{S_{12,k-1}S_{21,k-1}\hat{e}_{22,k-1}}{1-S_{11,k-1}\hat{e}_{22,k-1}} \tag{2.17}$$

where

T_{k-1} is the transducer gain of the first $k-1$ equalizer–transistor stages
$e_{ij,k}$ is the scattering parameters of equalizer k
$S_{ij,k}$ is the scattering parameters of transistor k
$\hat{e}_{22,k}$ is the output reflection coefficient of equalizer k when loaded on input
$S_{G,k}$ is the output reflection coefficient of transistor $k-1$ when loaded on input

The recursion starts with the first stage ($k=1$) as shown in Figure 2.4.

The transducer gain of the first stage is given by

$$T_1 = \left(1-|\rho_G|^2\right)\frac{|e_{21,1}|^2|S_{21,1}|^2}{|1-e_{11,1}S_{G,1}|^2|1-\hat{e}_{22,1}S_{11,1}|^2} \tag{2.18}$$

with

$$\hat{e}_{22,1} = e_{22,1} + \frac{e_{21,1}^2 S_{G,1}}{1 - e_{11,1} S_{G,1}} \tag{2.19}$$

$$S_{G,1} = \rho_G \tag{2.20}$$

$$\rho_G = \frac{Z_G - Z_0}{Z_G + Z_0} \tag{2.21}$$

where

$S_{G,1}$ would be the output reflection coefficient of a transistor 0, but since there is no transistor 0, it is equal to the reflection coefficient of the source ρ_G
$\hat{e}_{22,1}$ is the output reflection of equalizer 1 when loaded on input
ρ_G is the reflection coefficient of the source impedance Z_G

The transducer gain equations are demonstrated in Appendix A.

2.5 Multistage VSWR

The input VSWR corresponding to the k equalizer–transistor stages terminated on Z_0 as shown in Figure 2.3 is computed [11, 12] using

$$\mathrm{VSWR}_{IN} = \frac{1 + |\hat{e}_{11,1}|}{1 - |\hat{e}_{11,1}|} \tag{2.22}$$

As transistor–equalizer stages are added to form the situation of Figure 2.3, the input VSWR is calculated recursively using

$$\begin{aligned}
S_{L,k} &= S_{11,k} \\
\hat{e}_{11,k} &= e_{11,k} + \frac{e_{21,k}^2 S_{L,k}}{1 - e_{22,k} S_{L,k}} \\
S_{L,k-1} &= S_{11,k-1} + \frac{S_{12,k-1} S_{21,k-1} \hat{e}_{11,k}}{1 - S_{22,k-1} \hat{e}_{11,k}} \\
\hat{e}_{11,k-1} &= e_{11,k-1} + \frac{e_{21,k-1}^2 S_{L,k-1}}{1 - e_{22,k-1} S_{L,k-1}} \\
&\vdots \\
\hat{e}_{11,1} &= e_{11,1} + \frac{e_{21,1}^2 S_{L,1}}{1 - e_{22,1} S_{L,1}}
\end{aligned} \tag{2.23}$$

where

$\hat{e}_{11,k}$ is the input reflection coefficient of equalizer k when loaded on output
$S_{L,k}$ is the input reflection coefficient of transistor k when loaded on output

After the last $(k+1)$th equalizer has been added for compensating arbitrary load impedance Z_L as shown in Figure 2.1, the input VSWR is computed using the recursive formula starting with

$$\hat{e}_{11,k+1} = e_{11,k+1} + \frac{e_{21,k+1}^2 \rho_L}{1 - e_{22,k+1}\rho_L}$$
$$S_{L,k} = S_{11,k} + \frac{S_{12,k} S_{21,k} \hat{e}_{11,k+1}}{1 - S_{22,k}\hat{e}_{11,k+1}} \tag{2.24}$$
$$\rho_L = \frac{Z_L - Z_0}{Z_L + Z_0}$$

The output VSWR including the $(k+1)$th equalizer and terminated on Z_L as shown in Figure 2.1 is given by

$$\text{VSWR}_{OUT} = \frac{1 + |\hat{e}_{22,k+1}|}{1 - |\hat{e}_{22,k+1}|} \tag{2.25}$$

with

$$\hat{e}_{22,k+1} = e_{22,k+1} + \frac{e_{21,k+1}^2 S_{G,k+1}}{1 - e_{11,k+1} S_{G,k+1}}$$
$$S_{G,k+1} = S_{22,k} + \frac{S_{12,k} S_{21,k} \hat{e}_{22,k}}{1 - S_{11,k}\hat{e}_{22,k}} \tag{2.26}$$

where

$\hat{e}_{22,k+1}$ is the output reflection coefficient of equalizer $k+1$ when loaded on input
$S_{G,k+1}$ is the output reflection coefficient of transistor k when loaded on input

One uses (2.16) and (2.17) for recursively computing the output VSWR.

2.6 Optimization Process

The goal of the optimization process is to find the polynomial coefficients that will minimize a single-error function that represents the error between one or several desired target functions and the actual response of the structure at a number of frequency points in a frequency band of interest. Clearly, there exist many optimization routines but one that was found to work well in the case of microwave amplifier design is to use a single-valued error function based on a sum of squares and an optimization algorithm based on a Levenberg–Marquardt–More technique for solving least-squares problems.

2.6.1 *Single-Valued Error and Target Functions*

The error function returns a single value and is based on the sum of individual errors at m frequency points ω_i. The expression of the error in the case of one target function is given by

$$E_r = \sum_{i=1}^{m} \left(\frac{f_{\text{actual}}(\omega_i, \boldsymbol{X})}{f_{\text{target}}(\omega_i)} - 1 \right)^2 \tag{2.27}$$

where

$f_{\text{target}}(\omega_i)$ represents the desired characteristic over m frequency points ω_i
$f_{\text{actual}}(\omega_i, \boldsymbol{X})$ represents the actual response of the system at the m frequency points

Since the actual system response depends on the unknown parameters to be optimized, $f_{\text{actual}}(\omega_i, \boldsymbol{X})$ includes a $\boldsymbol{X}$ vector notation. For example, a system response of interest is the overall transducer gain. Since there are $(n+1)$ polynomial coefficients for one polynomial $h(s)$ and there are $k+1$ equalizers, the $\boldsymbol{X}$ vector would represent $(k+1)\times(n+1)$ polynomial coefficients. However, the power of the RFT is that it does not require simultaneous optimization of all $(k+1)\times(n+1)$ parameters. The optimization process is progressive and only optimizes one equalizer at a time. Therefore, the $\boldsymbol{X}$ vector contains at most $(n+1)$ parameters corresponding to the coefficients of a single $h(s)$ polynomial.

In this chapter, the error function will be made of three desired target functions using a weighted sum of squares:

$$E = \sum_{i=1}^{m} \left[W_1 \left(\frac{G_{T,k}(\omega_i, \boldsymbol{H})}{G_{T0,k}} - 1 \right)^2 + W_2 \left(\frac{\text{VSWR}_{IN,k}(\omega_i, \boldsymbol{H})}{\text{VSWR}_{IN0,k}} - 1 \right)^2 + W_3 \left(\frac{\text{VSWR}_{OUT,k}(\omega_i, \boldsymbol{H})}{\text{VSWR}_{OUT0,k}} - 1 \right)^2 \right] \tag{2.28}$$

where the vector $\boldsymbol{X}$ is replaced by column vector $\boldsymbol{H}$ representing the unknown polynomial coefficients $h_0, h_1, h_2, \ldots, h_n$ of equalizer k being optimized, and W_1, W_2, and W_3 are different factors.

$G_{T,k}(\omega_i, \boldsymbol{H})$ is the transducer power gain for k equalizer–transistor stages given by the equations in Section 2.4 and used to optimize the kth equalizer. The transducer power gain depends on frequency through the scattering parameters $e_{ij}(j\omega)$ of the equalizer and the scattering parameters $S_{ij}(j\omega)$ of the transistor. The target transducer power gain $G_{T0,k}$ for optimizing the kth equalizer is taken as a constant in the frequency band being optimized. The expressions for the target transducer gain are provided at the end of this section and applied in Section 2.8.

$\text{VSWR}_{IN,k}(\omega_i, \boldsymbol{H})$ is the input VSWR at stage k given by the equations in Section 2.5. The target $\text{VSWR}_{IN0,k}$ is a constant over the frequency band being optimized. The value is often between 1 and 2. Note that the input VSWR is generally optimized using the first equalizer.

$\text{VSWR}_{OUT,k}(\omega_i, \boldsymbol{H})$ is the output VSWR at stage k given by the equations in Section 2.5. The target $\text{VSWR}_{OUT0,k}$ is a constant over the frequency band being optimized. The value is often between 1 and 2. Note that the output VSWR is generally optimized using the last equalizer.

In this chapter, the target function for the transducer gain across the frequency band of interest is taken as a constant value. The constant value is based on the theoretical maximum gain one could expect and is explained in Appendix A. The results are summarized below.

The target transducer gain for the first stage is given by

$$T_{0,1} = \min \left\{ \frac{|S_{21,1}|^2}{\left| 1 - |S_{11,1}|^2 \right|} \right\} \tag{2.29}$$

The target transducer gain for the kth stage is given by

$$T_{0,k} = \min\left\{ T_{MAX,1} \times T_{MAX,2} \times \cdots \times T_{MAX,k-1} \frac{|S_{21,k}|^2}{\left|1-|S_{11,k}|^2\right|} \right\} \tag{2.30}$$

where

$$T_{MAX,i} = \frac{|S_{21,i}|^2}{\left|1-|S_{11,i}|^2\right|\left|1-|S_{22,i}|^2\right|} \frac{1}{\left|1-\dfrac{S_{12,i}S_{21,i}S_{11,i}^*S_{22,i}^*}{\left(1-|S_{11,i}|^2\right)\left(1-|S_{22,i}|^2\right)}\right|^2} \tag{2.31}$$

The target transducer gain for the $(k+1)$th equalizer is given by

$$T_{0,k+1} = \min\{ T_{MAX,1} \times T_{MAX,2} \times \cdots \times T_{MAX,k-1} \times T_{MAX,k} \} \tag{2.32}$$

An example of how the weights and target functions are used is given in Section 2.8.

2.6.2 *Levenberg–Marquardt–More Optimization*

This algorithm is used to define n unknown parameters that minimize a single-valued error function $E(\boldsymbol{X})$ made of a sum of squares:

$$E(\boldsymbol{X}) = \sum_{i=1}^{m} [e_i(\boldsymbol{X})]^2 \quad \text{with} \quad e_i(\boldsymbol{X}) = \frac{f_{\text{actual}}(\omega_i, \boldsymbol{X})}{f_{\text{target}}(\omega_i)} - 1 \quad \text{and} \quad \boldsymbol{X} = \begin{bmatrix} x_1 \\ x_2 \\ \vdots \\ x_n \end{bmatrix} \tag{2.33}$$

where $\boldsymbol{X}$ is the parameter vector made of the n parameters to optimize.

By defining $e(\boldsymbol{X})$ as the error vector made of the m single errors between actual and target functions at the m frequencies

$$e^T(\boldsymbol{X}) = [e_1(\omega_1, \boldsymbol{X}), e_2(\omega_2, \boldsymbol{X}), e_3(\omega_3, \boldsymbol{X}), \ldots, e_m(\omega_m, \boldsymbol{X})] \tag{2.34}$$

The single-valued error function $E(\boldsymbol{X})$ can be written as the norm squared of the error vector $e(\boldsymbol{X})$:

$$E(\boldsymbol{X}) = e^T(\boldsymbol{X}) e(\boldsymbol{X}) = \|e(\boldsymbol{X})\|^2$$

In order to minimize the single-valued error function $E(\boldsymbol{X})$, the gradient of $E(\boldsymbol{X})$ is a useful quantity and is defined as

$$\nabla E(\boldsymbol{X}) = \nabla\left(\|e(\boldsymbol{X})\|^2\right) = 2\boldsymbol{J}^T e(\boldsymbol{X})$$

where $\boldsymbol{J}$ is the Jacobian matrix defined by

$$J = \begin{bmatrix} \frac{\partial e_1(X)}{\partial x_1} & \frac{\partial e_1(X)}{\partial x_2} & \cdots & \frac{\partial e_1(X)}{\partial x_n} \\ \frac{\partial e_2(X)}{\partial x_1} & \frac{\partial e_2(X)}{\partial x_2} & \cdots & \frac{\partial e_2(X)}{\partial x_n} \\ \frac{\partial e_3(X)}{\partial x_1} & \frac{\partial e_3(X)}{\partial x_2} & \cdots & \frac{\partial e_3(X)}{\partial x_n} \\ \frac{\partial e_4(X)}{\partial x_1} & \frac{\partial e_4(X)}{\partial x_2} & \cdots & \frac{\partial e_4(X)}{\partial x_n} \\ \vdots & \vdots & \vdots & \vdots \\ \frac{\partial e_m(X)}{\partial x_1} & \frac{\partial e_m(X)}{\partial x_2} & \cdots & \frac{\partial e_m(X)}{\partial x_n} \end{bmatrix} \tag{2.35}$$

For a slight displacement vector ΔX in the parameter vector X, the error vector at $X + \Delta X$ can be approximated using a first-order Taylor series such that

$$e(X + \Delta X) \approx e(X) + J\Delta X$$

The gradient of the single-valued error function at $X + \Delta X$ can then be approximated using

$$\nabla E(X + \Delta X) = \nabla \|e(X + \Delta X)\|^2 = 2J^T e(X + \Delta X) \approx 2J^T [e(X) + J\Delta X]$$

To define the n unknown parameters, the single-valued error function $E(X)$ is minimized using an iterative method based on making the gradient $\nabla E(X + \Delta X)$ equal to 0. A key feature of the technique is to define the displacement vector ΔX that makes the gradient equal to 0. This is done using the Taylor approximation results provided previously:

$$\nabla E(X + \Delta X) \approx 2J^T [e(X) + J\Delta X] = J^T e(X) + J^T J \Delta X = 0$$

from which the displacement vector that makes the gradient equal to 0 is given by

$$\Delta X = -\left[J^T J\right]^{-1} J^T e(X) \tag{2.36}$$

Levenberg and Marquardt [13, 14] proposed the addition of a parameter λ in the expression of the displacement vector to improve convergence issues of the iterative process. In this book, the practical implementation of the technique proposed by More [15] is used. In this case, the expression for the displacement vector is given by

$$\Delta X = -\left[J^T J + \lambda D^T D\right]^{-1} J^T e(X) \tag{2.37}$$

where D is a $n \times n$ diagonal matrix and the parameter λ is computed to satisfy the Taylor series approximation assumption so that

$$\|D\Delta X\| \leq \Delta \tag{2.38}$$

where Δ is a constant greater than 0. In other words, $D\,\Delta X$ is an n column vector and its norm should be kept small at every iteration.

The details for computing ΔX at some iteration j in the optimization program are given in Appendix B. It is shown that λ is typically computed so that $\left\| D_j\left[J_j^T J_j + \lambda_j D_j^T D_j\right]^{-1} J_j^T e(X_j)\right\| = \Delta_j$ at every iteration. Finding the single parameter λ_j that solves the aforementioned equation is done using a modified Newton zero crossing method to avoid some special cases of nonconvergence. The value of Δ_j is also computed at each iteration using some rule. The diagonal elements of the $n \times n$ D matrix are typically computed using the derivative of the error vector with respect to the unknown parameters such that

$$d_i^j = \left\| \frac{\partial e(X_j)}{\partial x_i} \right\|$$

The X vector will be taken as column vector H made of the unknown coefficients $h_0, h_1, h_2, \ldots, h_n$ of equalizer k being optimized.

2.7 Design Procedures

The design of a multistage amplifier starts with a first equalizer–transistor stage terminated on Z_0 as shown in Figure 2.5. Typically, the polynomial order of an equalizer might depend on how much it contributes to the target functions. For instance, the first equalizer might have a high order to satisfy both transducer gain and input VSWR requirements. The order of the equalizer can be increased if the optimization converges to an insufficient level of performance.

The error function uses the transducer gain equation provided in Section 2.4. The target function for the transducer gain is taken as a constant representing the theoretical maximum transducer gain and is given in Section 2.6 and demonstrated in Appendix A. The error function also uses the input VSWR equation provided in Section 2.5, and the target function for the VSWR is typically a constant less than 2.

The Jacobian matrix used in the Levenberg optimization algorithm relies on the derivative of the error function. For the transducer gain, an analytical expression can be developed but it is directly computed into the program. On the other side, the error function for the input VSWR is rather tedious and one can always revert to numerical derivatives such that

$$\frac{\partial \mathrm{VSWR}_{IN,k}(\omega_j, H)}{\partial h_i} = \frac{\mathrm{VSWR}_{IN,k}(\omega_j, H + \Delta h_i) - \mathrm{VSWR}_{IN,k}(\omega_j, H)}{\Delta h_i}$$

It was found that a step of $\Delta h_i = 10^{-5}$ was satisfactory when starting with $h(t)$ polynomial coefficients initialized at $h_i = \pm 1$ when using impedance and frequency normalization. For more

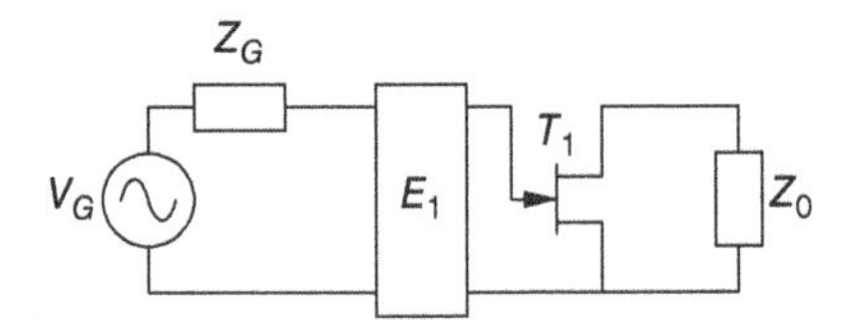

Figure 2.5 The design procedure starts from the source with a first equalizer–transistor stage.

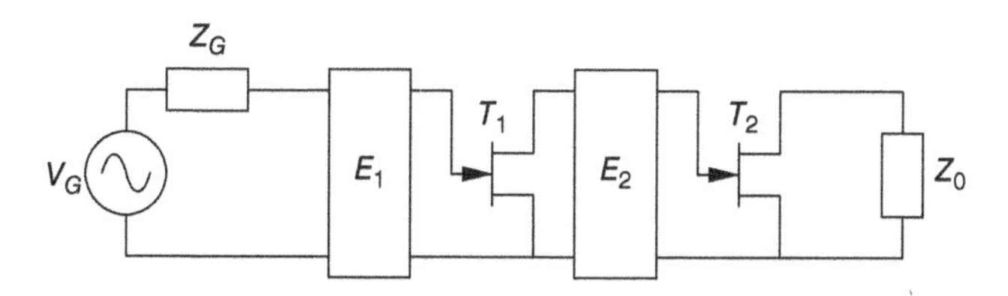

Figure 2.6 Once the first equalizer is defined, the second equalizer–transistor stage is added.

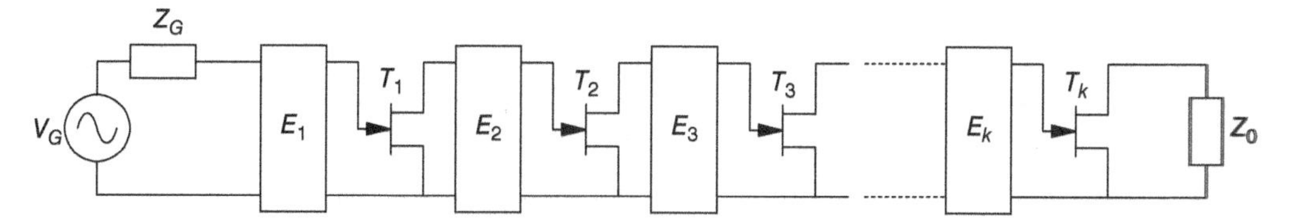

Figure 2.7 The procedure is repeated until the kth equalizer–transistor stage is added.

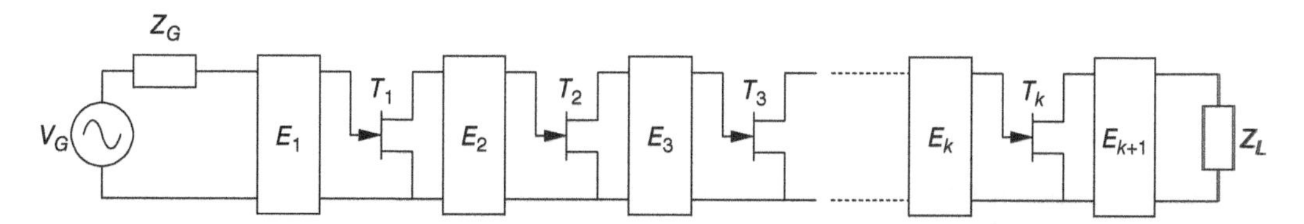

Figure 2.8 The final step is to add an additional equalizer to compensate for an arbitrary load Z_L.

information, see Section 2.8. One should note that numerical derivatives are not time prohibitive since only one equalizer is optimized at a given time for very few unknown parameters.

For the first few stages, the input VSWR has more weight than the transducer gain, and for the last equalizer, the output VSWR has more weight.

Once the first equalizer has been optimized, the second equalizer–transistor stage is placed between the first transistor and the load Z_0 as shown in Figure 2.6. At this time, the first equalizer is no longer an unknown quantity, so that when optimizing the second equalizer only $e_{11}(s)$, $e_{22}(s)$, and $e_{21}(s)$ of the second equalizer are variable when computing transducer gain and VSWR using the recursive formulas of Sections 2.4 and 2.5.

The procedure is then repeated until the kth equalizer has been optimized when terminated on Z_0 as shown in Figure 2.7.

Once the kth equalizer has been optimized, a last equalizer is added to compensate for the arbitrary load Z_L as shown in Figure 2.8.

For this last equalizer, the transducer gain formula includes the effect of the arbitrary load Z_L as given in Section 2.4 and one uses the output VSWR formula of Section 2.5 rather than the input VSWR formula.

2.8 Four-Stage Amplifier Design Example

In this example, the four-stage amplifier of Figure 2.9 is to be optimized for gain and VSWR. The source and load terminations are $Z_G = Z_L = Z_0 = 50\Omega$.

The transistors are assumed to be identical, and Table 2.1 provides a few of the measured scattering parameters of the transistor with respect to $Z_0 = 50\Omega$.

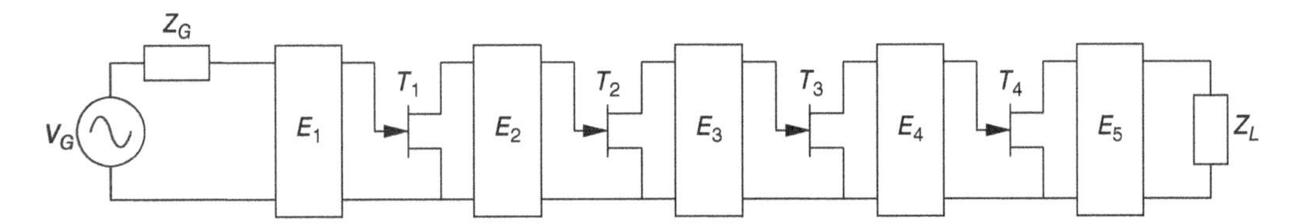

Figure 2.9 Four-stage amplifier structure used in the example.

Table 2.1 Measured scattering parameters of the transistor

Frequency (GHz)	S_{11}		S_{12}		S_{21}		S_{22}	
	Magnitude	Phase	Magnitude	Phase	Magnitude	Phase	Magnitude	Phase
18.0	0.721	−143.0	0.0752	7.9	1.16	57.6	0.559	−73.9
19.0	0.719	−144.7	0.0752	6.8	1.10	54.4	0.568	−76.2
20.0	0.717	−146.2	0.0752	5.8	1.05	51.4	0.578	−78.4
21.0	0.716	−147.6	0.0752	4.9	0.996	48.4	0.588	−80.6

For this design, the equalizers are taken as second-order transfer functions ($n = 2$) to limit the complexity of the equalizer circuitry. In addition, the equalizers are taken as having all-pole characteristics to simplify the synthesis:

$$e_{21}(s) = \frac{f(s)}{g(s)} = \frac{1}{g(s)}$$

Since $f(s) = \pm s^k$, this gives $k = 0$ and $\varepsilon = 1$. The $g(s)$ polynomial is computed using

$$g(s)g(-s) = h(s)h(-s) + 1$$

The all-pole characteristic places some constraints on the polynomial coefficients to satisfy the aforementioned equation. For this example, we need to use $h_0 = 0$ and $g_0 = 1$. Using a frequency scaling to 21 GHz that corresponds to the maximum frequency in the band of interest, a suitable initial guess for $h(s)$ is $h_1 = h_2 \cdots = h_n = 1$. Then, $g(s)$ is computed selecting the left-haft plane roots of the aforementioned equation. Knowledge of $h(s)$, $g(s)$ is all that is needed to define $e_{11,1}(s)$, $e_{12,1}(s)$, $e_{21,1}(s)$, and $e_{22,1}(s)$.

The optimization process starts with the first equalizer–transistor stage as shown in Figure 2.4. In this case, the actual transducer gain is given by

$$T_1 = \left(1 - |S_G|^2\right) \frac{|e_{21,1}|^2 |S_{21,1}|^2}{|1 - e_{11,1} S_{G,1}|^2 |1 - \hat{e}_{22,1} S_{11,1}|^2}$$

with

$$\hat{e}_{22,1} = e_{22,1} + \frac{e_{21,1}^2 S_{G,1}}{1 - e_{11,1} S_{G,1}}$$

$$S_{G,1} = S_G$$

$$S_G = \frac{Z_G - Z_0}{Z_G + Z_0}$$

since $Z_G = Z_0 = 50\Omega$, $S_G = 0 = S_{G,1}$, and $\hat{e}_{22,1} = e_{22,1}$ and the transducer gain of the first stage reduces to

$$T_1 = \frac{|e_{21,1}|^2 |S_{21,1}|^2}{|1 - e_{22,1} S_{11,1}|^2}$$

The target transducer gain is based on the maximum theoretical transducer gain across the band of interest as explained in Appendix A. The target gain is taken as a constant representing the minimum value of this theoretical maximum gain seen in the band of interest as shown below:

$$T_{0,1} = \min\left\{\frac{|S_{21,1}|^2}{1 - |S_{11,1}|^2}\right\}$$

The input VSWR is given by

$$\text{VSWR}_{IN} = \frac{1 + |\hat{e}_{11,1}|}{1 - |\hat{e}_{11,1}|}$$

with

$$S_{L,1} = S_{11,1}$$

$$\hat{e}_{11,1} = e_{11,1} + \frac{e_{21,1}^2 S_{L,1}}{1 - e_{22,1} S_{L,1}}$$

The target input VSWR is taken as a constant less than 2.

For the first stage, the input VSWR is the main concern and the impact of the output VSWR is minimal so the weighting functions are chosen as $W_1 = 1$, $W_2 = 5$, and $W_3 = 0$.

After running the Levenberg–Marquardt–More optimization program, the coefficients stabilize around the solution $h_1 = -0.445$, $h_2 = -0.668$ and the first equalizer has for $h(s)$ polynomial

$$h(s) = 0 - 0.445s - 0.668s^2$$

and the $g(s)$ polynomial is computed using

$$g(s)g(-s) = h(s)h(-s) + 1 = 1 - 0.198s^2 + 0.446s^4$$

Using `roots.m` in MATLAB, the roots are found as

```
-0.9271 + 0.7985i
-0.9271 - 0.7985i
0.9271 + 0.7985i
0.9271 - 0.7985i
```

from which the left-half plane roots are selected for a strictly Hurwitz polynomial so that

$$g(s)=K(s-(-0.9271+j0.7985))(s-(-0.9271-j0.7985))=K\left(s^2+1.8541+1.4970\right)$$

The constraint $g_0=1$ must be respected so that

$$g(s)=\frac{1}{1.4970}\left(s^2+1.8542+1.4970\right)=0.668s^2+1.239s+1$$

and

$$e_{11,1}(s)=\frac{h(s)}{g(s)}=\frac{0-0.445s-0.668s^2}{1+1.239s+0.668s^2}$$

and the complete scattering matrix of the first equalizer is given by

$$S_{E_1}=\begin{pmatrix} \frac{h(s)}{g(s)}=\frac{0-0.445s-0.668s^2}{1+1.239s+0.668s^2} & \frac{f(s)}{g(s)}=\frac{1}{1+1.239s+0.668s^2} \\ \frac{f(s)}{g(s)}=\frac{1}{1+1.239s+0.668s^2} & \frac{-\varepsilon h(-s)}{g(s)}=\frac{-(0+0.445s-0.668s^2)}{1+1.239s+0.668s^2} \end{pmatrix}$$

The transducer gain when adding the first equalizer is shown in Figure 2.10. It is seen that the gain has improved over the intrinsic gain of the transistor.

At this point, the second equalizer–transistor is added to the circuit. The actual transducer gain is given in this case by

$$T_2=T_1\frac{|e_{21,2}|^2|S_{21,2}|^2}{|1-e_{11,2}S_{G,2}|^2|1-\hat{e}_{22,2}S_{11,2}|^2}$$

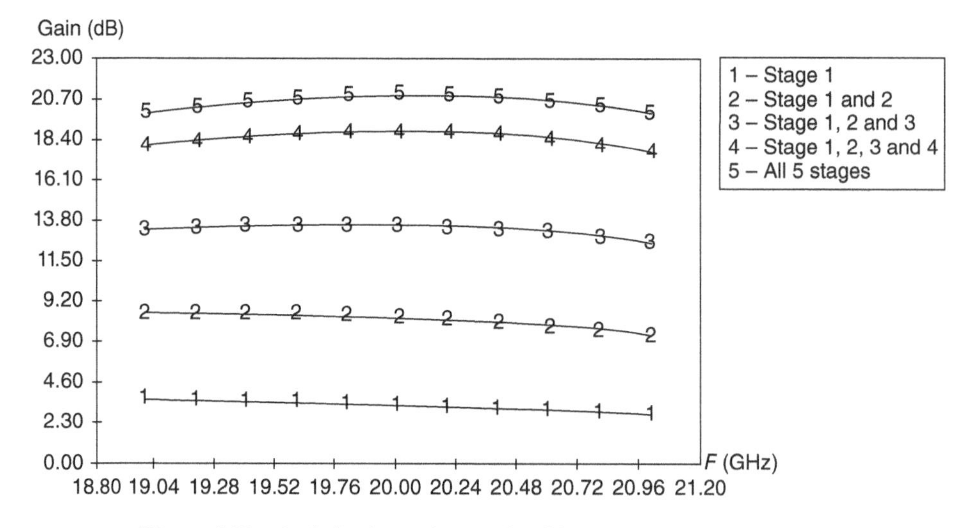

Figure 2.10 Optimized transducer gain of intermediary and final stages.

with

$$\hat{e}_{22,2} = e_{22,2} + \frac{e_{21,2}^2 S_{G,2}}{1 - e_{11,2} S_{G,2}}$$

$$S_{G,2} = S_{22,1} + \frac{S_{12,1} S_{21,1} \hat{e}_{22,1}}{1 - S_{11,1} \hat{e}_{22,1}}$$

and where T_1 no longer varies and is computed at each of the m frequency points using the $h(s)$ and $g(s)$ polynomials computed during the optimization of the first stage.

The target transducer gain for the second stage is given by

$$T_{0,2} = \min\left\{ T_{MAX,1} \frac{|S_{21,2}|^2}{\left|1 - |S_{11,2}|^2\right|} \right\}$$

with

$$T_{MAX,1} = \frac{|S_{21,1}|^2}{\left|1 - |S_{11,1}|^2\right| \left|1 - |S_{22,1}|^2\right|} \frac{1}{\left|1 - \dfrac{S_{12,1} S_{21,1} S_{11,1}^* S_{22,1}^*}{\left(1 - |S_{11,1}|^2\right)\left(1 - |S_{22,1}|^2\right)}\right|^2}$$

The input VSWR is given by

$$\text{VSWR}_{IN} = \frac{1 + |\hat{e}_{11,1}|}{1 - |\hat{e}_{11,1}|}$$

with

$$S_{L,2} = S_{11,2}$$

$$\hat{e}_{11,2} = e_{11,2} + \frac{e_{21,2}^2 S_{L,2}}{1 - e_{22,2} S_{L,2}}$$

$$S_{L,1} = S_{11,1} + \frac{S_{12,1} S_{21,1} \hat{e}_{11,2}}{1 - S_{22,1} \hat{e}_{11,2}}$$

$$\hat{e}_{11,1} = e_{11,1} + \frac{e_{21,1}^2 S_{L,1}}{1 - e_{22,1} S_{L,1}}$$

The target input VSWR is again taken as a constant less than 2.

For the second stage, the input VSWR is again a concern and the impact of the output VSWR is minimal, so the weighting functions are chosen as $W_1 = 1$, $W_2 = 5$, and $W_3 = 0$.

After running the Levenberg–Marquardt–More optimization program, the second equalizer has for $h(s)$ polynomial

$$h(s) = 0 - 0.287s - 0.484s^2$$

from which $g(s)$ is constructed and the transfer function of the input reflection coefficient of the second equalizer is

$$e_{11,2}(s)=\frac{0-0.287s-0.484s^2}{1+1.025s+0.484s^2}$$

The improvement in the overall transducer gain is shown in Figure 2.10.

The process is repeated for the third and fourth stages. For these stages, the main concern is the gain so the weighting functions are chosen as $W_1=1$, $W_2=0$, and $W_3=0$. After optimization, the input reflection coefficients for the third and fourth equalizers are found to be

$$e_{11,3}(s)=\frac{0+0.090s-0.251s^2}{1+0.715s+0.251s^2}$$

$$e_{11,4}(s)=\frac{0+0.136s-0.248s^2}{1+0.717s+0.248s^2}$$

Here again, the transducer gain improves with each additional equalizer–transistor stage as seen in Figure 2.10.

For the last equalizer, the procedure is slightly different since only the equalizer is added with no pairing transistor and the termination is no longer Z_0 but some arbitrary load Z_L. In this case, the actual transducer gain is given by

$$T_{GT}=T_4\frac{|e_{21,5}|^2\left(1-|\rho_L|^2\right)}{|1-e_{11,5}S_{G,5}|^2|1-\hat{e}_{22,5}\rho_L|^2}$$

where

$$\hat{e}_{22,5}=e_{22,5}+\frac{e_{21,5}^2S_{G,5}}{1-e_{11,5}S_{G,5}}$$

$$S_{G,5}=S_{22,4}+\frac{S_{12,4}S_{21,4}\hat{e}_{22,4}}{1-S_{11,4}\hat{e}_{22,4}}$$

$$\rho_L=\frac{Z_L-Z_0}{Z_L+Z_0}$$

and the target transducer gain is given by

$$T_{0,k}=\min\{T_{MAX,1}\times T_{MAX,2}\times T_{MAX,3}\times T_{MAX,4}\}$$

where

$$T_{MAX,i}=\frac{|S_{21,i}|^2}{\left|1-|S_{11,i}|^2\right|\left|1-|S_{22,i}|^2\right|}\frac{1}{\left|1-\dfrac{S_{12,i}S_{21,i}S_{11,i}^*S_{22,i}^*}{\left(1-|S_{11,i}|^2\right)\left(1-|S_{22,i}|^2\right)}\right|^2}\quad \text{for } i=1,2,3,4$$

The output VSWR is given by

$$\text{VSWR}_{OUT} = \frac{1 + |\hat{e}_{22,5}|}{1 - |\hat{e}_{22,5}|}$$

with

$$\hat{e}_{22,5} = e_{22,5} + \frac{e_{21,5}^2 S_{G,5}}{1 - e_{11,5} S_{G,5}}$$

$$S_{G,5} = S_{22,4} + \frac{S_{12,4} S_{21,4} \hat{e}_{22,4}}{1 - S_{11,4} \hat{e}_{22,4}}$$

Note that the output VSWR is computed recursively using quantities also computed for the transducer gain.

The target output VSWR is taken as a constant less than 2.

The main purpose of the last equalizer is to adapt the load so that the weighting functions are chosen as $W_1 = 0$, $W_2 = 0$, and $W_3 = 1$. After optimization, the input reflection coefficient for the last equalizer is found to be

$$e_{11,5}(s) = \frac{0 + 0.271s + 0.480s^2}{1 + 1.016s + 0.480s^2}$$

Figure 2.10 shows that the last equalizer has improved the overall transducer gain even if no additional transistor was added.

The input and output VSWR for the entire system with the four transistors and the five equalizers is shown in Figure 2.11.

Overall, the four-stage amplifier example has an overall transducer gain of 20.5 ± 0.6 dB, and an input VSWR that is less than 2.25 and an output VSWR that is less than 2.1. This was accomplished using second-order equalizers.

The equalizers can then be realized as lumped ladders using traditional polynomial divisions of the input impedance. The case of the first equalizer is shown below.

Starting with the transfer function of the input reflection coefficient

$$e_{11,1}(s) = \frac{0 - 0.445s - 0.668s^2}{1 + 1.239s + 0.668s^2}$$

the input impedance is calculated such that

$$Z_{IN,1}(s) = Z_0 \frac{1 + e_{11,1}(s)}{1 - e_{11,1}(s)} = Z_0 \frac{1 + 0.794s}{1 + 1.684s + 1.336s^2}$$

The input reflection transfer function coefficients provided in this example were already normalized to 1 Ω. It is seen that the input impedance starts on a shunt capacitor so one uses the admittance instead of the impedance to start the synthesis:

$$Y_1(s) = \frac{1.336s^2 + 1.684s + 1}{0.794s + 1}$$

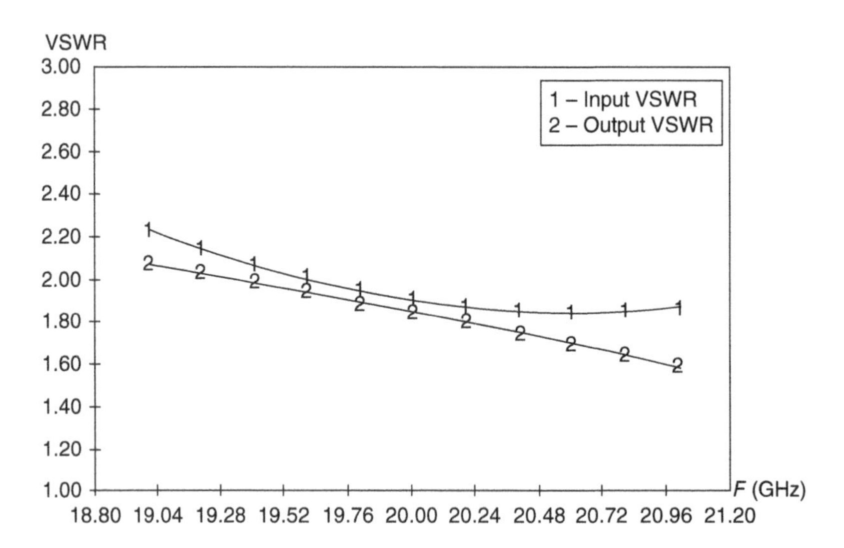

Figure 2.11 Optimized input and output VSWR.

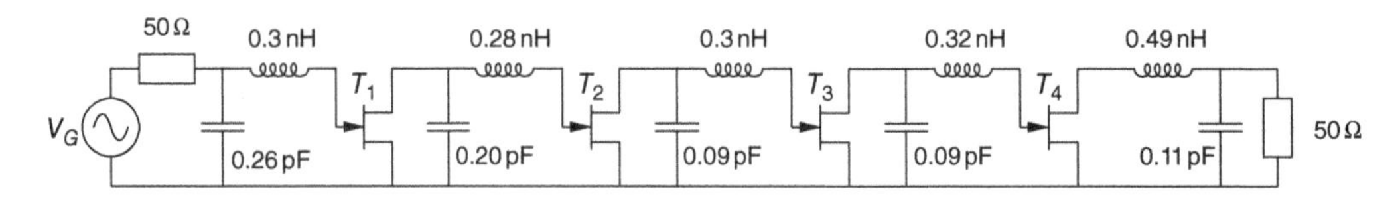

Figure 2.12 Lumped realization of the four-transistor amplifier example.

A succession of polynomial divisions is conducted as shown below.

This represents a normalized shunt capacitor of 1.684 followed by a normalized series inductor of 0.794 on a normalized 1 Ω load. The element's values are then denormalized for 50 Ω terminations:

$$C_0 = \frac{1.684}{Z_0} = \frac{1.684}{50} = 0.0336$$

$$L_0 = Z_0 \times 0.794 = 50 \times 0.794 = 39.7$$

The equalizer transfer functions in this example were also normalized to 21 GHz so that the actual capacitor and inductor values are

$$L = \frac{L_0}{\omega_0} = \frac{39.7}{2\pi \times 21 \times 10^9} = 3.00 \times 10^{-10}\,\mathrm{H}$$

$$C = \frac{C_0}{\omega_0} = \frac{0.0336}{2\pi \times 21 \times 10^9} = 2.54 \times 10^{-13}\,\mathrm{F}$$

The procedure is repeated for the other equalizers. The complete multistage lumped amplifier circuit is given in Figure 2.12.

2.9 Transistor Feedback Block for Broadband Amplifiers

At low frequencies, microwave GaAs MESFETs are unstable and are virtually impossible to match over large bandwidths. This is mainly due to a high input reflection coefficient and the high gain of the device. It is thus necessary to use either a lossy matching or RL feedback technique to reduce terminal impedances and stabilize the transistor.

There are several topologies that can be used for adapting or compensating the drop in transistor gain at high frequencies and for a compromise between gain, VSWR, and noise.

2.9.1 Resistive Adaptation

Figure 2.13 shows parallel resistors at the input and output of the transistor.

The resistors will attenuate the gain at low frequencies than at high frequencies. They improve the reactive impedance matching to the transistor by dissipating a large part of the available power.

The simultaneous use of two resistors for an amplifier stage is rare. The transistor output VSWR is usually low due to the typical values of the conductance at the output of the transistor and using a resistor for that purpose will not be necessary. Another problem is that the two resistors will degrade the noise performance of the transistor, especially the resistor at the input of the transistor.

As a result, an input resistor can be used for very large frequency band applications but only for the transistors at the intermediary to final stages of the amplifier, and not in the first transistor stage that is so critical for the overall noise performance of the amplifier. In the absence of an input resistor, the output resistor can be used with the first transistor stage to improve stability at low frequencies.

2.9.2 Resistive Feedback

Figure 2.14 shows how a resistor can be placed in a parallel configuration between the drain and gate.

The resistor like in the case of resistive adaptation reduces the gain at low frequencies but has a less severe gain reduction at high frequencies. It is also used to improve matching between transistor and source, and the stability of the transistor. It does, however, contribute to noise at the input of the transistor stage and that in turn degrades the noise performance of the amplifier. Another problem is that it is difficult to realize in hybrid technology.

2.9.3 Reactive Feedback

Figure 2.15 shows how an inductor can be placed in a series configuration in the source.

This topology can be used to improve the matching between output reflection coefficient of an equalizer placed on the left of the transistor stage and the reflection coefficient $\rho_{G\min}$ that

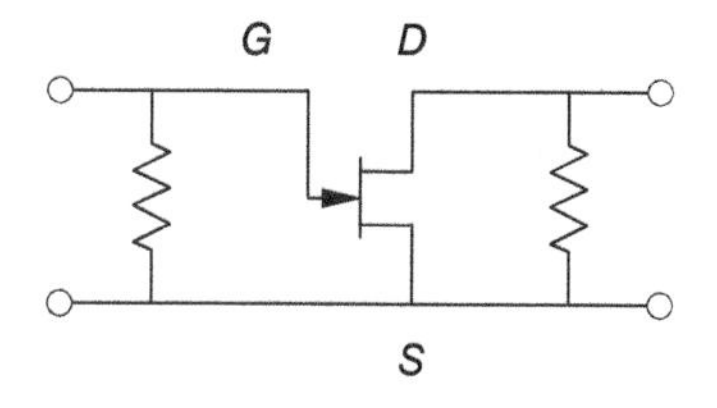

Figure 2.13 Resistive adaptation at the input and output of the transistor.

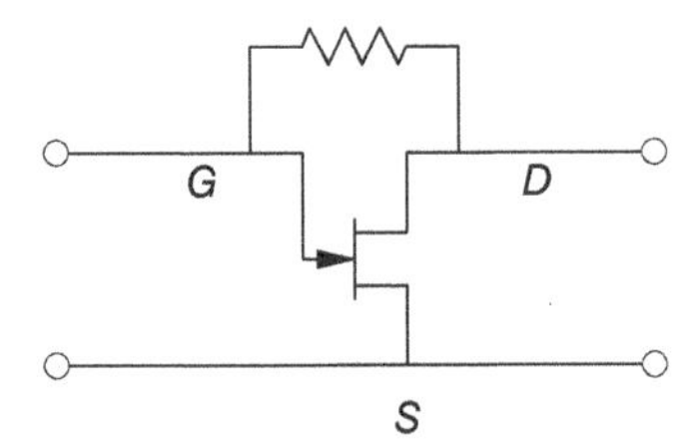

Figure 2.14 Transistor with resistive feedback.

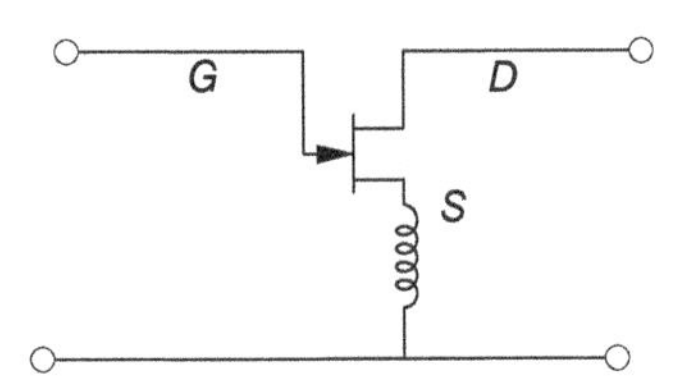

Figure 2.15 Transistor with reactive feedback.

minimizes the noise factor of the transistor stage. It also reduces slightly the minimum noise factor $F_{\min}$ and improves the stability of the transistor but it also reduces the maximum gain of the transistor stage. The reduction in maximum gain limits the operating frequency band of the amplifier.

2.9.4 Transistor Feedback Block

For example, in order to limit the drop in maximum transistor gain at high frequencies, an inductor is added in series to the resistor in the resistive feedback topology. Two other inductors, one in series with the gate and the other in series with the drain, can be used to increase the gain at high frequencies. This modified restive feedback transistor stage is given in Figure 2.16.

The inductors also slightly reduce the minimum noise factor $F_{\min}$.

Finally, taking into account all possible variations of a transistor stage based on the previous topologies, a general transistor feedback block (TFB) has been defined and is shown in Figure 2.17.

The TFB of Figure 2.17 is very general and is rarely used but was built into the program for future possible needs.

A more realistic TFB used for extending the range for broadband applications includes a parallel feedback loop, a drain series inductance, and a gate shunt resistance as shown in Figure 2.18 [16, 17].

This block is designed using the following approach:

- R_F is chosen such that it provides a perfect match (i.e., $R_F = g_m Z_0^2$).
- L_F is chosen to reduce feedback at higher frequencies since in general $R_F \geq g_m Z_0^2$.
- L_D is introduced to compensate for the transistor C_{DS}.
- L_G is computed for input matching and high-frequency gain.
- C is a decoupling capacitor introduced in the feedback loop.
- R_G is a damping resistor that prevents low-frequency oscillations.

The scattering parameters of the TFB are computed prior to RFT optimization using the measured parameters of the transistor and relying on *ABCD* matrix computations of two-ports

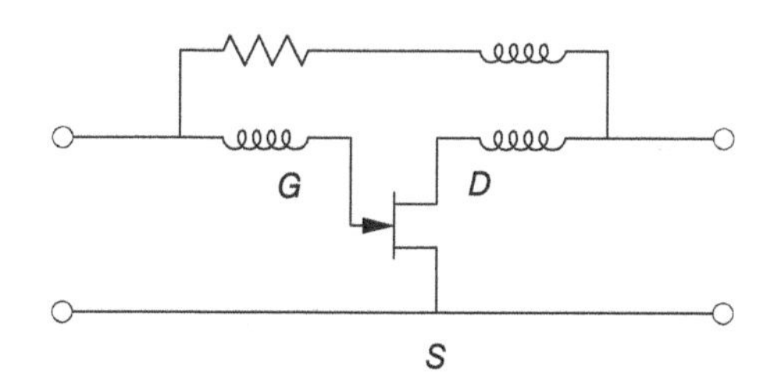

Figure 2.16 Modified resistive feedback.

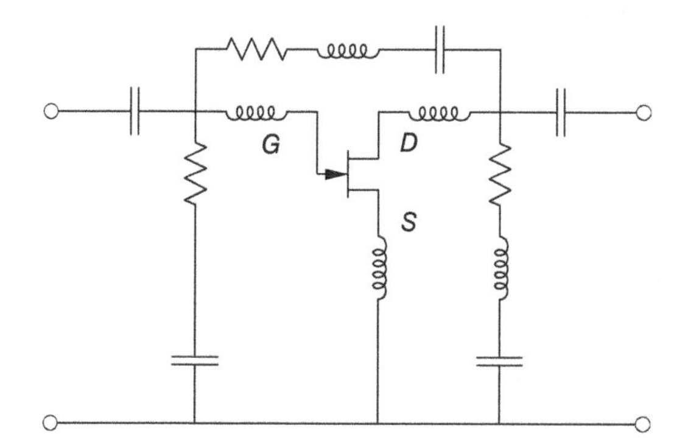

Figure 2.17 General transistor feedback block.

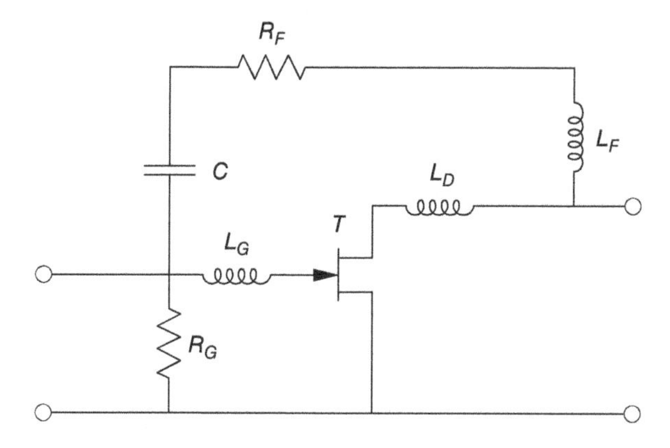

Figure 2.18 Transistor feedback block.

representing the transistor and the added circuitry. An example of such computation where the bias circuit effects are added to the S parameters of the transistor is given in Section 3.7.

The TFB circuit itself is optimized independently of the RFT for maximum gain flatness, VSWR around unity and low noise.

2.10 Realizations

2.10.1 Three-Stage Hybrid Amplifier

The RFT was applied to the design of a 4 MHz to 6 GHz three-stage broadband amplifier. The amplifier was intended for a wideband fiber optic receiver [16]. The transistor used was the

RTC-CFX31X GaAs FETs provided by the CNET (France). The scattering parameters of this transistor are given in Table 2.2. The multistage amplifier structure is shown in Figure 2.14. As mentioned in Section 2.9, a TFB was used to extend the bandwidth of the transistors. The scattering parameters of the transistor given in Table 2.2 were combined with the TFB circuitry and circuitry from the bias circuit. These were computed numerically prior to running the optimization of the RFT and are given in Table 2.3. The equalizers were optimized for gain and VSWR. The synthesized element values did not present any realization problems as seen in Figure 2.19.

The multistage amplifier was realized in hybrid form fabricated on an alumina substrate. The alumina substrate was 0.635 mm thick and 1 inch wide. Lumped capacitors were made with layered ceramics, and inductors were realized by gold wires. A picture of the amplifier is given in Figure 2.20.

The measured and theoretical gain and VSWR are given in Figure 2.21. The small-signal gain is about 16 dB with a good flatness (±1 dB) across the 4 MHz to 6 GHz frequency range. Input and output VSWRs are less than 3:1. The 1 dB gain compression was measured as 9 dB m at 2 and 6 GHz. The noise figure was less than 9 dB except for a local peak at 4 GHz.

Table 2.2 Scattering parameters of the CFX31X GaAs FET

Frequency (GHz)	S_{11}		S_{12}		S_{21}		S_{22}	
	Magnitude	Phase	Magnitude	Phase	Magnitude	Phase	Magnitude	Phase
0.05	0.999	−1.4	0.001	89.2	3.87	179.0	0.660	−0.5
0.90	0.989	−25.1	0.021	76.1	3.77	162.0	0.650	−9.2
1.75	0.956	−48.4	0.040	62.6	3.50	145.0	0.631	−18.9
2.60	0.892	−69.0	0.056	50.0	3.68	127.0	0.600	−30.8
3.45	0.871	−85.1	0.064	31.5	2.71	115.0	0.579	−38.5
4.30	0.871	−99.8	0.069	34.6	2.39	103.0	0.569	−44.8
5.15	0.826	−111.0	0.072	30.1	2.12	92.5	0.570	−49.9
6.00	0.810	−120.0	0.073	27.0	1.92	85.0	0.570	−54.0

Table 2.3 Scattering parameters of the transistor feedback block

Frequency (GHz)	S_{11}		S_{12}		S_{21}		S_{22}	
	Magnitude	Phase	Magnitude	Phase	Magnitude	Phase	Magnitude	Phase
0.05	0.376	−11.7	0.151	7.5	2.22	175.0	0.242	−11.7
0.90	0.374	−25.7	0.150	−6.2	2.18	168.0	0.234	1.7
1.75	0.405	−48.5	0.144	−12.5	2.17	155.0	0.250	0.9
2.60	0.440	−68.9	0.133	−17.9	2.12	140.0	0.271	−6.7
3.45	0.500	−85.7	0.121	−23.2	2.06	128.0	0.296	−12.7
4.30	0.562	−102.0	0.106	−26.7	2.00	115.0	0.330	−19.3
5.15	0.605	−115.0	0.092	−27.4	1.91	102.0	0.369	−26.5
6.00	0.641	−127.0	0.079	−25.5	1.83	91.5	0.396	−33.2

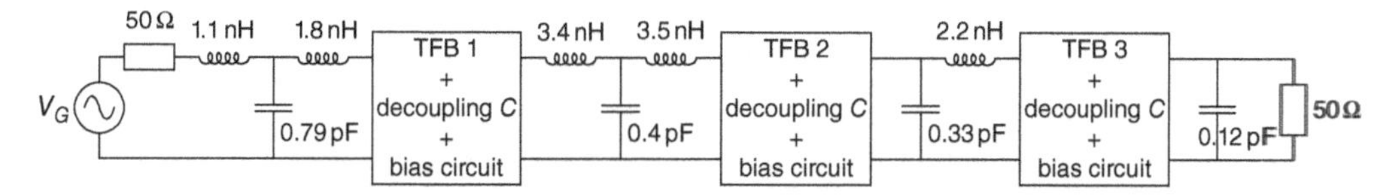

Figure 2.19 Three-stage amplifier structure used in the RFT.

Figure 2.20 Hybrid three-stage 4 MHz to 6 GHz GaAs FET amplifier.

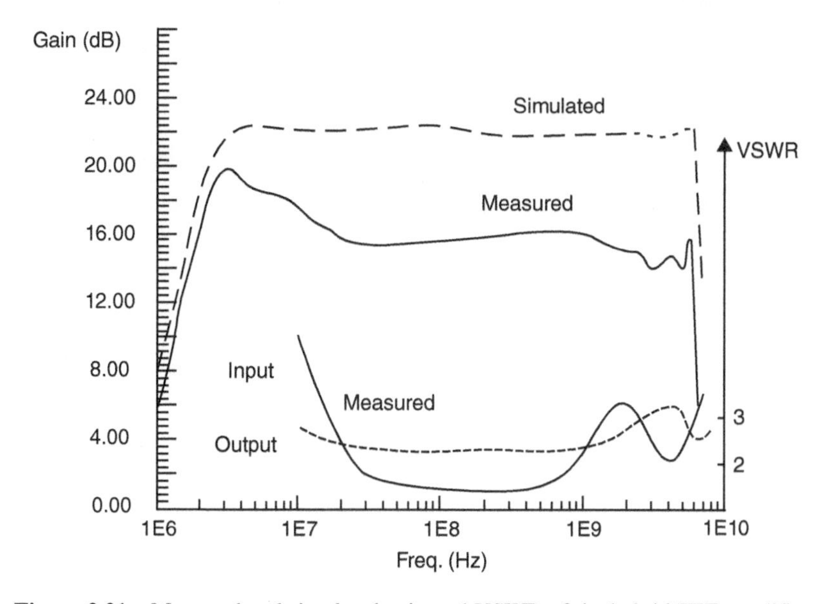

Figure 2.21 Measured and simulated gain and VSWR of the hybrid FET amplifier.

Figure 2.22 shows the amplifier response to a 5 Gbits/s input pulse. The distortion is negligible. This example shows that equalizers obtained using the RFT are realizable. It should be noted that the nonunilateral behavior of the TFB has been taken into account. The experimental results are in good agreement with the theoretical calculations.

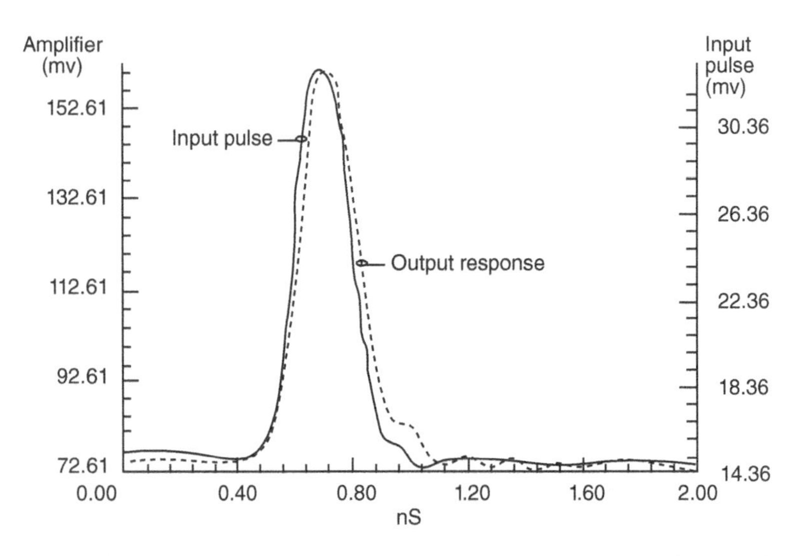

Figure 2.22 Response of the hybrid three-stage FET amplifier to a 5 Gbits/s pulse.

Figure 2.23 Two-stage monolithic amplifier.

2.10.2 Two-Stage Monolithic Amplifier

A monolithic version of the amplifier was realized using the Plessey III-V foundry. The chip size was 2.3 × 1.8 mm [17, 18]. The chip area was optimized by incorporating two independent amplifier stages. The amplifiers are identical, apart from input and output locations, which are necessarily different to enable each amplifier stage to be used independently or to enable the two to be cascaded via an external off-chip capacitor. A photomicrograph of the chip is given in Figure 2.23. The transistor drains are biased via the RF output leads from each stage. The gates are grounded via shunt resistor as in the hybrid case. The two-stage monolithic amplifier is mounted in a miniaturized hybrid alumina substrate and incorporates ferrite bead RF chokes in the bias circuit.

The measured and simulated gain and VSWR are given in Figure 2.24. The gain is 17 ± 1 dB and input and output VSWRs are less than 2:1 over the 5 MHz to 7 GHz frequency range. The noise figure above 200 MHz is less than 5.5 dB. The 1 dB gain compression point is greater than 18 dB m and has been measured at several discrete frequencies.

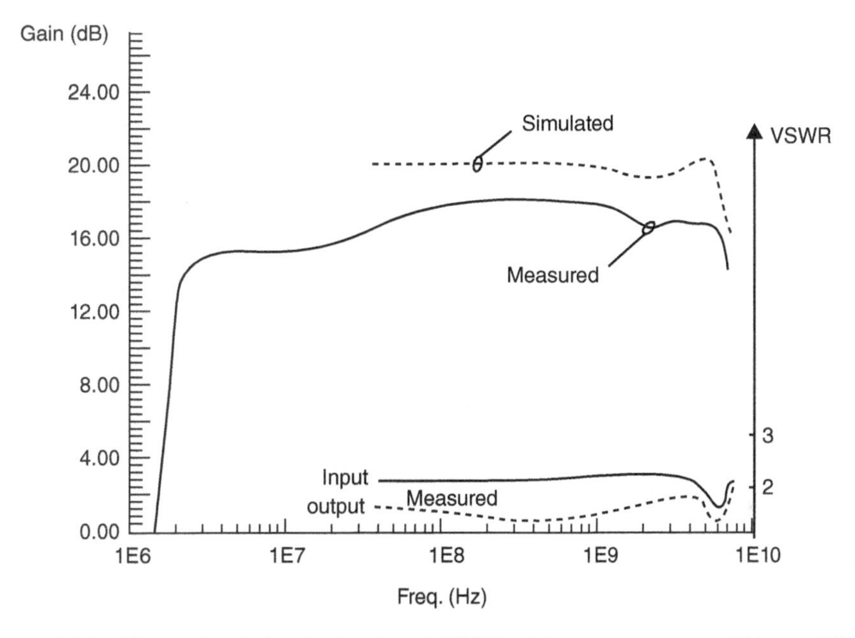

Figure 2.24 Measured and simulated gain and VSWR of the two-stage monolithic amplifier.

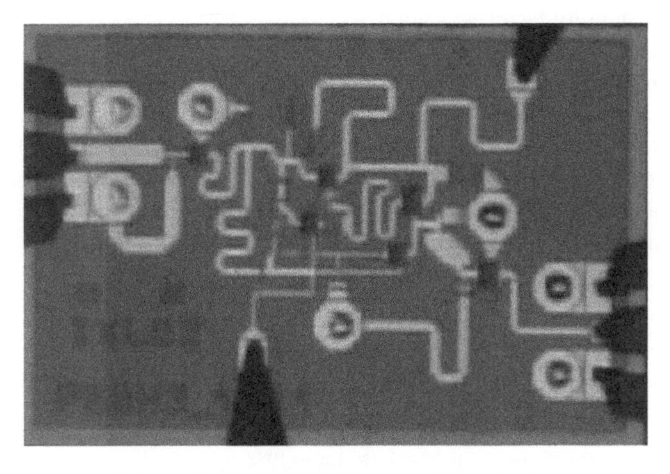

Figure 2.25 HEMT driver amplifier circuit.

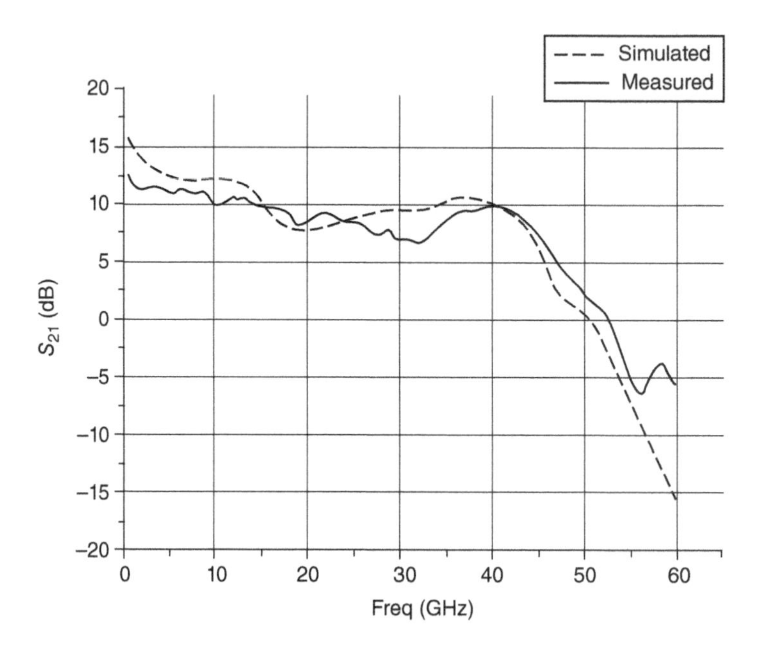

Figure 2.26 Simulated and measured gain of the HEMT driver amplifier circuit.

2.10.3 Single-Stage GaAs Technology Amplifier

The RFT was also successfully applied to the design of a driver amplifier using 0.2 μm GaAs technology by the OMMIC Foundry [19]. This driver amplifier was intended for 40 Gb/s optical communication systems in a 45 MHz to 65 GHz band. Common source stages and current source used depletion high electron-mobility transistors. The surface of the amplifier was $1 \times 1.5\,\text{mm}^2$. A picture of the amplifier is given in Figure 2.25.

The measured and simulated gains of the circuit are given in Figure 2.26. The measured gain is about 9.8 ± 2.5 dB. The return loss is less than −9 dB from DC to 20 GHz and less than −5 dB from 20 to 40 GHz. The stability factor was good.

In conclusion, the RFT is well adapted to the design of microwave amplifiers and in particular to ultra-broadband amplifiers. A three-stage 5 MHz to 6 GHz hybrid amplifier, a two-stage 5 MHz to 7 GHz monolithic amplifier, and a very broadband 0–40 GHz MMIC have been realized using this synthesis theory. The results obtained confirm the value of the synthesis approach used and a computer program called FREEL was created. The amplifiers are used for optical detection at very high data rates [19]. The amplifiers were made for PLESSY, OMMIC (PHILIPS), and CNET factories with commercial transistors.

References

[1] H. J. Carlin, "A new approach to gain-bandwidth problems," IEEE Trans. Circ. Syst., vol. 24, no. 4, pp. 170–175, April 1977.

[2] H. J. Carlin, J. J. Komiak, "A new method of broad-band equalization applied to microwave amplifiers," IEEE Trans. Microw. Theory Tech., vol. 27, no. 2, pp. 93–99, December 1982.

[3] H. J. Carlin, P. Amstutz, "On optimum broad-band matching," IEEE Trans. Circ. Syst., vol. 28, no. 5, pp. 401–405, February 1979.

[4] B. S. Yarman, H. J. Carlin, "A simplified real frequency technique applied to broadband multistage microwave amplifiers," IEEE Trans. Microw. Theory Tech., vol. 30, no. 12, pp. 2216–2222, December 1982.
[5] B. S. Yarman, "A simplified real technique for broadband matching a complex generator to a complex load," RCA Rev., vol. 43, pp. 529–541, 1982.
[6] H. J. Carlin, B. S. Yarman, "The double matching problem: Analytic and real frequency solutions," IEEE Trans. Circ. Syst., pp. 15–28, January 1983.
[7] B.S. Yarman, Design of Ultra Wideband Power Transfer Networks, John Wiley & Sons, Inc., Chichester, UK, 2010.
[8] A. Perennec, Synthesis and Realizations of Microwave Amplifiers by the Real Frequency Method, Ph.D. dissertation, Brest University, July 1988.
[9] V. Belevitch, "Elementary application of the scattering formalism to network design," IRE Trans. Circ. Theory, vol. CT 3, no. 2, pp. 97–104, June 1956.
[10] S. Darlington, "Theory of reactance 4-poles which produce prescribed insertion loss characteristics," J. Math. Phys., vol. 18, pp. 257–353, 1939.
[11] A. Perennec, P. Jarry, "Real frequency broadband microwave matching using a modified Marquardt optimization," IEEE Melecom'87, Rome, Italy, pp. 319–322, March 1987.
[12] P. Jarry, A. Perennec, "Optimization of gain and VSWR in multistage microwave amplifiers using the real frequency method," European Conference on Circuit Theory and Design, ECCTD'87, Paris, pp. 203–208, September 1987.
[13] K. Levenberg, "A method for the solution of certain non-linear problems in least squares," Q. Appl. Math. 2, pp. 164–168, 1944.
[14] D. Marquardt, "An algorithm for least-squares estimation of nonlinear parameters," J. Soc. Indus. Appl. Math., Vol. 11, no 2, pp. 431–441, June 1963.
[15] J. J. More, "The Levenberg-Marquardt algorithm: implementation and theory," Lecture Notes in Mathematics 630, Springer Verlag, New York, 1978.
[16] A. Perennec, R. Soares, P. Jarry, "Application of circuit synthesis to the design of a three-stage, 4MHz to 6GHz-amplifier for a wide band fiber optics receiver," 18th European Microwave Conference, EMC'87, Stockholm, September 1988.
[17] A. Perennec, R. Soares, P. Jarry, P. Legaud, M. Goloubkoff, "Computer-aided-design of hybrid and monolithic broad-band amplifiers for optoelectronic receivers," IEEE Trans. Microw. Theory Tech., vol. MTT 37, pp. 1475–1478, September 1989.
[18] P. Jarry "Synthesis and CAD of microwave filters and amplifiers," International Microwave Symposium, SBMO, Sao-Paulo, pp. 657–666, July 24–27, 1989.
[19] E. Kerherve, C. P. Moreira, P. Jarry, L. Courcelle, "40-Gb/s wide-band MMIC pHEMT modulator driver amplifier design with the real frequency technique," IEEE Trans. Microw. Theory Tech., vol. 53, no. 6, pp. 2145–2152, June 2005.

3

Multistage Distributed Amplifier Design

3.1 Introduction

This chapter presents how the real frequency technique (RFT) is used in the design of multistage distributed amplifiers [1–4]. The equalizers are made of transmission lines. A new variable and formalism better suited for the distributed medium are introduced. The modified RFT is able

Microwave Amplifier and Active Circuit Design Using the Real Frequency Technique, First Edition.
Pierre Jarry and Jacques N. Beneat.

to optimize and synthesize the transmission lines. A method is also provided for optimizing the multistage amplifier noise figure. A technique is given to solve the problem of too high or too low characteristic impedance requirements for the microstrip realizations. The technique was added to the program called FREEL.

The chapter concludes with realizations of a 1.15–1.5 GHz, a 2–8 GHz, and a 5.925–6.425 GHz distributed amplifiers.

3.2 Multistage Distributed Amplifier Representation

The multistage distributed amplifier is represented as shown in Figure 3.1 [5, 6]. It assumes that there are k transistors. An equalizer is associated with each transistor. There is one final equalizer placed after the kth transistor. The source and load impedances are assumed arbitrary, complex, and frequency dependent. The transistor can represent a transistor alone or a transistor with added but known circuitry as long as the S parameters can be provided.

The equalizer of the kth stage is characterized by the scattering matrix

$$S_{E_k} = \begin{pmatrix} e_{11,k}(t) & e_{12,k}(t) \\ e_{21,k}(t) & e_{22,k}(t) \end{pmatrix} \tag{3.1}$$

The transistor circuit of the kth stage is characterized by the scattering matrix

$$S_{T_k} = \begin{pmatrix} S_{11,k}(t) & S_{12,k}(t) \\ S_{21,k}(t) & S_{22,k}(t) \end{pmatrix} \tag{3.2}$$

where t is the Richard's variable [7] and is described in Chapter 1.

In this chapter, the equalizer is assumed to be a lossless reciprocal distributed two-port with a scattering matrix such that

$$S_E = \begin{pmatrix} e_{11}(t) = \dfrac{h(t)}{g(t)} & e_{12}(t) = \dfrac{f(t)}{g(t)} \\ e_{21}(t) = \dfrac{f(t)}{g(t)} & e_{22}(t) = \dfrac{-h(-t)}{g(t)} \end{pmatrix} \tag{3.3}$$

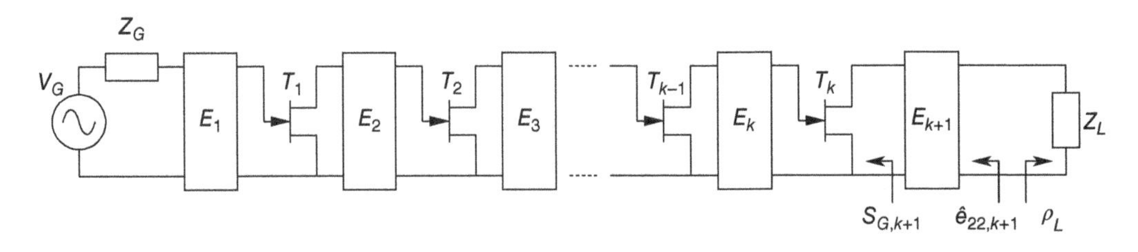

Figure 3.1 Multistage distributed amplifier representation.

where $h(t)$ and $g(t)$ are polynomials of order n with real-valued coefficients:

$$h(s) = h_0 + h_1 t + \cdots + h_{n-1} t^{n-1} + h_n t^n$$
$$g(s) = g_0 + g_1 t + \cdots + g_{n-1} t^{n-1} + g_n t^n$$

and where $f(t)$ is a function of the form

$$f(t) = t^r \left(1 - t^2\right)^{\frac{n-q}{2}} \tag{3.4}$$

where q represents the number of possible low-pass stubs and r represents the number of possible high-pass stubs.

The lossless condition $|e_{11}|^2 + |e_{12}|^2 = 1$ leads to

$$g(t)g(-t) = h(t)h(-t) + f(t)f(-t) \tag{3.5}$$

so that

$$g(t)g(-t) = h(t)h(-t) + t^{2r}\left(1 - t^2\right)^{n-q} \tag{3.6}$$

For a given equalizer, the problem consists of defining the unknown polynomial coefficients h_0, $h_1, h_2, \ldots, h_n$ that optimize the response of the system. Once $h(t)$ is known, $g(t)g(-t)$ can be computed using

$$G(t^2) = g(t)g(-t) = G_0 + G_1 t^2 + \cdots + G_n t^{2n}$$

where

$$\begin{cases} G_0 = h_0^2 + 1 \\ G_1 = -h_1^2 + 2h_2 h_0 - C_{n-q}^1 \\ \ldots \\ G_i = (-1)^i h_i^2 + 2h_{2i} h_0 + 2\sum_{j=2}^{i} (-1)^{j-1} h_{j-1} h_{2i-j+1} + (-1)^i C_{n-q}^i \\ \ldots \\ G_{n-q} = (-1)^{n-q} h_{n-q}^2 + 2h_{2(n-q)} h_0 + 2\sum_{j=2}^{n-q} (-1)^{j-1} h_{j-1} h_{2(n-q)-j+1} + (-1)^{n-q} \\ \ldots \\ G_n = (-1)^n h_n^2 \end{cases} \tag{3.7}$$

with

$$C_n^p = \frac{n!}{p!(n-p)!}$$

The left-half roots of $g(t)g(-t)$ are retained to form a Hurwitz polynomial $g(t)$. At this point, the equalizer scattering parameters $e_{11}(t)$, $e_{12}(t)$, $e_{21}(t)$, and $e_{22}(t)$ can be computed. During optimization, they will be needed for computing the error function to be optimized, and after optimization, for example, $e_{11}(t)$ can be used to realize the equalizer using the synthesis procedures given in Section 3.8.

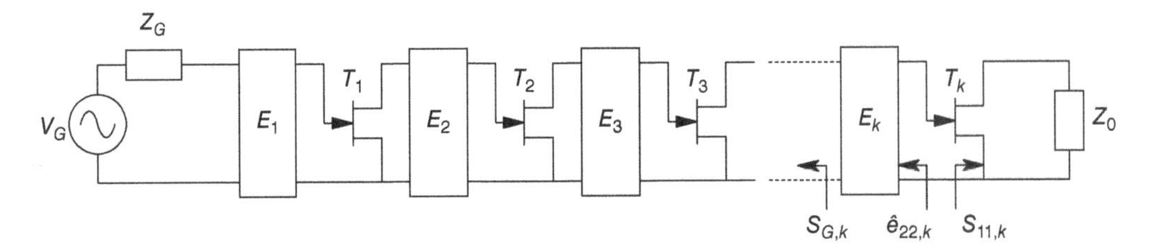

Figure 3.2 Transducer gain for k equalizer–transistor stages terminated on Z_0.

3.3 Multistage Transducer Gain

The overall transducer gain T_{GT} of the multistage amplifier when terminated on arbitrary terminations Z_G and Z_L as shown in Figure 3.1 is given by [5, 6]

$$T_{GT}=T_k\frac{|e_{21,k+1}|^2\left(1-|\rho_L|^2\right)}{|1-e_{11,k+1}S_{G,k+1}|^2|1-\hat{e}_{22,k+1}\rho_L|^2} \tag{3.8}$$

where

$$\hat{e}_{22,k+1}=e_{22,k+1}+\frac{e_{21,k+1}^2S_{G,k+1}}{1-e_{11,k+1}S_{G,k+1}} \tag{3.9}$$

$$S_{G,k+1}=S_{22,k}+\frac{S_{12,k}S_{21,k}\hat{e}_{22,k}}{1-S_{11,k}\hat{e}_{22,k}} \tag{3.10}$$

$$\rho_L=\frac{Z_L-Z_0}{Z_L+Z_0} \tag{3.11}$$

where

$S_{G,k+1}$ is the output reflection coefficient of transistor k when loaded on input
$\hat{e}_{22,k+1}$ is the output reflection of equalizer $k+1$ when loaded on input
ρ_L is the reflection coefficient of the load impedance Z_L
T_k is the transducer gain of the first k equalizer–transistor stages terminated on Z_0 as shown in Figure 3.2. Z_0 is the reference used for the scattering parameters. For example, $Z_0=50\ \Omega$ is often used to provide the measured scattering parameters of a transistor.

The transducer gain T_k of the first k equalizer–transistor stages is computed recursively by

$$T_k=T_{k-1}\frac{|e_{21,k}|^2|S_{21,k}|^2}{|1-e_{11,k}S_{G,k}|^2|1-\hat{e}_{22,k}S_{11,k}|^2}\quad \text{for } k>1 \tag{3.12}$$

with

$$\hat{e}_{22,k}=e_{22,k}+\frac{e_{21,k}^2S_{G,k}}{1-e_{11,k}S_{G,k}} \tag{3.13}$$

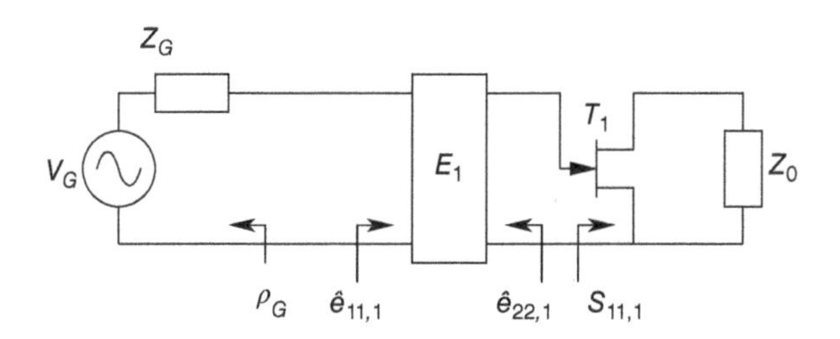

Figure 3.3 Transducer gain for the first stage terminated on Z_0.

$$S_{G,k} = S_{22,k-1} + \frac{S_{12,k-1}S_{21,k-1}\hat{e}_{22,k-1}}{1 - S_{11,k-1}\hat{e}_{22,k-1}} \tag{3.14}$$

and where

T_{k-1} is the transducer gain of the first $k-1$ equalizer–transistor stages
$e_{ij,k}$ are the scattering parameters of equalizer k
$S_{ij,k}$ are the scattering parameters of transistor k
$\hat{e}_{22,k}$ is the output reflection coefficient of equalizer k when loaded on input
$S_{G,k}$ is the output reflection coefficient of transistor $k-1$ when loaded on input

The recursion starts with the first stage ($k = 1$) as shown in Figure 3.3.
The transducer gain of the first stage is given by

$$T_1 = \left(1 - |\rho_G|^2\right) \frac{|e_{21,1}|^2 |S_{21,1}|^2}{|1 - e_{11,1}S_{G,1}|^2 |1 - \hat{e}_{22,1}S_{11,1}|^2} \tag{3.15}$$

with

$$\hat{e}_{22,1} = e_{22,1} + \frac{e_{21,1}^2 S_{G,1}}{1 - e_{11,1}S_{G,1}} \tag{3.16}$$

$$S_{G,1} = \rho_G \tag{3.17}$$

$$\rho_G = \frac{Z_G - Z_0}{Z_G + Z_0} \tag{3.18}$$

where

$S_{G,1}$ would be the output reflection coefficient of a transistor 0, but since there is no transistor 0, it is equal to the reflection coefficient of the source ρ_G
$\hat{e}_{22,1}$ is the output reflection of equalizer 1 when loaded on input
ρ_G is the reflection coefficient of the source impedance Z_G

The transducer gain equations are demonstrated in Appendix A.

3.4 Multistage VSWR

The input VSWR corresponding to the k equalizer–transistor stages terminated on Z_0 as shown in Figure 3.3 is computed [5, 6] using

$$\text{VSWR}_{IN} = \frac{1+|\hat{e}_{11,1}|}{1-|\hat{e}_{11,1}|} \tag{3.19}$$

As transistor–equalizer stages are added to form the situation of Figure 3.2, the input VSWR is calculated recursively using

$$\begin{aligned}
S_{L,k} &= S_{11,k} \\
\hat{e}_{11,k} &= e_{11,k} + \frac{e_{21,k}^2 S_{L,k}}{1-e_{22,k}S_{L,k}} \\
S_{L,k-1} &= S_{11,k-1} + \frac{S_{12,k-1}S_{21,k-1}\hat{e}_{11,k}}{1-S_{22,k-1}\hat{e}_{11,k}} \\
\hat{e}_{11,k-1} &= e_{11,k-1} + \frac{e_{21,k-1}^2 S_{L,k-1}}{1-e_{22,k-1}S_{L,k-1}} \\
&\vdots \\
\hat{e}_{11,1} &= e_{11,1} + \frac{e_{21,1}^2 S_{L,1}}{1-e_{22,1}S_{L,1}}
\end{aligned} \tag{3.20}$$

where

$\hat{e}_{11,k}$ is the input reflection coefficient of equalizer k when loaded on output
$S_{L,k}$ is the input reflection coefficient of transistor k when loaded on output

After the last $(k+1)$th equalizer has been added for compensating arbitrary load impedance Z_L as shown in Figure 3.1, the input VSWR is computed using the recursive formula starting with

$$\begin{aligned}
\hat{e}_{11,k+1} &= e_{11,k+1} + \frac{e_{21,k+1}^2 \rho_L}{1-e_{22,k+1}\rho_L} \\
S_{L,k} &= S_{11,k} + \frac{S_{12,k}S_{21,k}\hat{e}_{11,k+1}}{1-S_{22,k}\hat{e}_{11,k+1}} \\
\rho_L &= \frac{Z_L - Z_0}{Z_L + Z_0}
\end{aligned} \tag{3.21}$$

The output VSWR of the overall network including the $(k+1)$th equalizer terminated on Z_L as shown in Figure 3.1 is given by

$$\text{VSWR}_{OUT} = \frac{1+|\hat{e}_{22,k+1}|}{1-|\hat{e}_{22,k+1}|} \tag{3.22}$$

with

$$\begin{aligned}
\hat{e}_{22,k+1} &= e_{22,k+1} + \frac{e_{21,k+1}^2 S_{G,k+1}}{1-e_{11,k+1}S_{G,k+1}} \\
S_{G,k+1} &= S_{22,k} + \frac{S_{12,k}S_{21,k}\hat{e}_{22,k}}{1-S_{11,k}\hat{e}_{22,k}}
\end{aligned} \tag{3.23}$$

where

$\hat{e}_{22,k+1}$ is the output reflection coefficient of equalizer $k+1$ when loaded on input
$S_{G,k+1}$ is the output reflection coefficient of transistor k when loaded on input

and one uses (3.13) and (3.14) for recursively computing the output VSWR.

3.5 Multistage Noise Figure

Figure 3.4 shows the case of the first equalizer–transistor stage. The transistor T_1 is the active device with noise parameters $F_{\min,1}$, $R_{n,1}$, and $\rho_{G\min,1}$.

The transistor T_1 is looking at a source reflection coefficient $\hat{e}_{22,1}$ so that according to (1.37) in Chapter 1 the noise figure of this first stage is given by

$$F_1 = F_{\min,1} + 4R_{n,1}Y_0 \frac{\left|\hat{e}_{22,1} - \rho_{G\min,1}\right|^2}{\left(1 - \left|\hat{e}_{22,1}\right|^2\right)\left|1 + \rho_{G\min,1}\right|^2} \tag{3.24}$$

where

$F_{\min,1}$ is the minimum noise figure of transistor 1
$R_{n,1}$ is the noise resistance of transistor 1
$\rho_{\min,1}$ is the optimum noise reflection coefficient of transistor 1
Y_0 is the admittance used to define the noise parameters of transistor 1 (e.g., $Y_0 = 1/50\,\Omega$)
$\hat{e}_{22,1}$ is the output reflection coefficient of equalizer 1 when loaded on input

In the case of a cascade of k equalizer–transistor stages (see Fig. 3.2), according to (1.40), (3.24), (1.41), and (1.27), the noise of the cascaded system is given by [5, 6]

$$F_k = F_{k-1} + \frac{1}{T_{A,k-1}}\left[F_{\min,k} + 4R_{n,k}Y_0 \frac{\left|\hat{e}_{22,k} - \rho_{G\min,k}\right|^2}{\left(1 - \left|\hat{e}_{22,k}\right|\right)^2\left|1 + \rho_{G\min,k}\right|^2} - 1\right] \tag{3.25}$$

where $T_{A,k-1}$ is the available power gain given by

$$T_{A,k-1} = \frac{T_{k-1}}{1 - \left|S_{G,k}\right|^2} \tag{3.26}$$

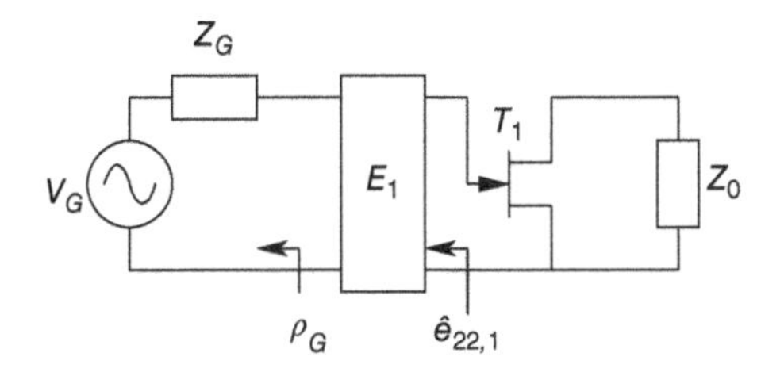

Figure 3.4 Noise figure of the first stage.

where

$F_{\min,k}$ is the minimum noise figure of transistor k
$R_{n,k}$ is the noise resistance of transistor k
$\rho_{\min,k}$ is the optimum noise reflection coefficient of transistor k
Y_0 is the admittance used to define the noise parameters of transistor k (e.g., $Y_0 = 1/50\,\Omega$)
$\hat{e}_{22,k}$ is the output reflection coefficient of equalizer k given by (3.13)
$S_{G,k}$ is the output reflection coefficient of transistor $k-1$ given by (3.14)
T_{k-1} is the transducer power gain after $k-1$ equalizer–transistor stages given by (3.12)

3.6 Optimization Process

The optimization algorithm is used to define polynomial coefficients $h_0, h_1, \ldots, h_n$ of $h(t)$ for an equalizer that minimize a least mean square error function that includes desired target functions with different weight factors.

For the distributed amplifiers of this chapter, the error function consists of the transducer gain, the VSWR, and the noise figure.

$$E = \sum_{i=1}^{m}\left[W_1\left(\frac{G_{T,k}(\Omega_i,\boldsymbol{H})}{G_{T0,k}}-1\right)^2 + W_2\left(\frac{\mathrm{VSWR}_k(\Omega_i,\boldsymbol{H})}{\mathrm{VSWR}_{0,k}}-1\right)^2 + W_3\left(\frac{F_k(\Omega_i,\boldsymbol{H})}{F_{0,k}}-1\right)^2\right] \quad (3.27)$$

where the functions are given in terms of Richard's frequency $\Omega_i = \tan\left(\frac{\pi}{2}\frac{\omega_i}{\omega_0}\right)$ and where ω_i are the actual frequencies and ω_0 is the normalization frequency used in the Richard's transform.

$G_{T,k}(\Omega_i, \boldsymbol{H})$ is the transducer power gain for k equalizer–transistor stages given by the equations in Section 3.3 and used to optimize the kth equalizer. The transducer power gain depends on frequency through the scattering parameters $e_{ij}(j\Omega)$ of the equalizer and the scattering parameters $S_{ij}(j\Omega)$ of the transistor. The target transducer power gain $G_{T0,k}$ for optimizing the kth equalizer is taken as a constant in the frequency band being optimized. The expressions for the target transducer gain are provided in Section 3.9.

$\mathrm{VSWR}_{IN,k}(\Omega_i, \boldsymbol{H})$ is the input VSWR at stage k given by the equations in Section 3.4. The target $\mathrm{VSWR}_{IN0,k}$ is a constant over the frequency band being optimized. The value is often between 1 and 2. Note that the input VSWR is generally optimized using the first equalizer.

$\mathrm{VSWR}_{OUT,k}(\Omega_i, \boldsymbol{H})$ is the output VSWR at stage k given by the equations in Section 3.4. The target $\mathrm{VSWR}_{OUT0,k}$ is a constant over the frequency band being optimized. The value is often between 1 and 2. Note that the output VSWR is generally optimized using the last equalizer.

$F_k(\Omega_i, \boldsymbol{H})$ is the noise figure for k equalizer–transistor stages given by the equations in Section 3.5 and used to optimize the kth equalizer. The noise figure $F_{0,k}$ for optimizing the kth equalizer is taken as a constant in the frequency band being optimized. The expressions for the target noise figure are provided in Section 3.9. Note that due to the importance of the source impedance on the noise figure the noise figure is generally optimized using the first equalizer.

The vector $\boldsymbol{H}$ includes the polynomial coefficients $h_0, h_1, \ldots, h_n$ and is used as the $\boldsymbol{X}$ vector in the Levenberg–More optimization algorithms of Chapter 2 and Appendix B.

The main computation in the optimization routine is to define the step $\Delta\boldsymbol{H}$ to add to the elements of vector $\boldsymbol{H}$ at the next iteration given by

$$\Delta H = -\left[J^T J + \lambda D^T D\right]^{-1} J^T e(H) \tag{3.28}$$

where $e(\boldsymbol{H})$ is the error vector, J is the Jacobian matrix of the error with elements $\partial e_j / \partial h_i$, D is a diagonal matrix, and λ is the Levenberg–Marquardt–More parameter described in Appendix B.

An important element is the Jacobian matrix. It is based on derivatives of the error functions. For the transducer power gain, analytical expressions for the derivatives that can be determined are rather tedious and lengthy to express and are written into a program. In the case of the input VSWR, the analytical expressions become very cumbersome and it is preferred to use a numerical derivative such that

$$\frac{\partial \text{VSWR}_{IN}(\Omega_j, \boldsymbol{H})}{\partial h_i} = \frac{\text{VSWR}_{IN}(\Omega_j, \boldsymbol{H} + \Delta h_i) - \text{VSWR}_{IN}(\Omega_j, \boldsymbol{H})}{\Delta h_i} \tag{3.29}$$

where a step of $\Delta h_i = 10^{-5}$ was found satisfactory when starting with $h(t)$ polynomial coefficients initialized at $h_i = \pm 1$.

3.7 Transistor Bias Circuit Considerations

This section provides a possible transistor bias circuit and shows how it can be used during the optimization process to provide more accurate results. Figure 3.5 shows the bias circuit. It is made of a high impedance $\lambda/4$ transmission line with a low impedance $\lambda/4$ open-circuited stub in parallel. This combination brings an open circuit on the gate of the transistor and reduces the effects of the RC polarization on the response of the amplifier [7].

Including the effects of the bias circuit on the S parameters representing the transistor will improve the accuracy of the equalizers optimized using the RFT.

The effects of the bias circuit on the S parameters are included by using the *ABCD* matrices of the circuitry and of the transistor.

To illustrate the technique, we provide the case of the bias circuit of Figure 3.5, but the technique can be used for any circuitry suitably described by *ABCD* matrices.

The bias circuit is first modeled as a two-port characterized by an $A'B'C'D'$ matrix is cascaded with the transistor as shown in Figure 3.6.

This particular bias circuit ($A'B'C'D'$) can be thought of a two-port (*ABCD*) loaded on admittance Y_{L1} as shown in Figure 3.7.

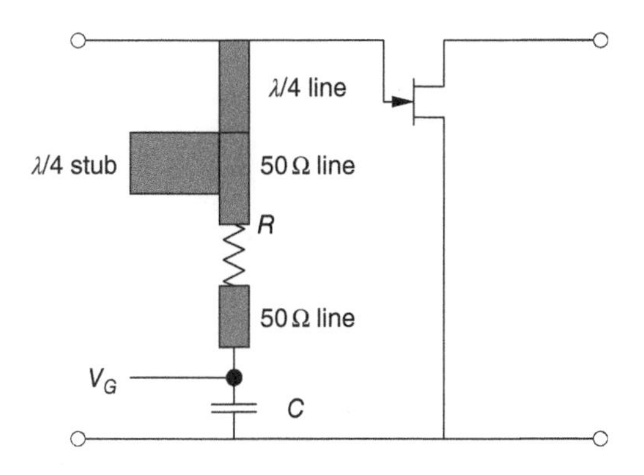

Figure 3.5 Bias circuit.

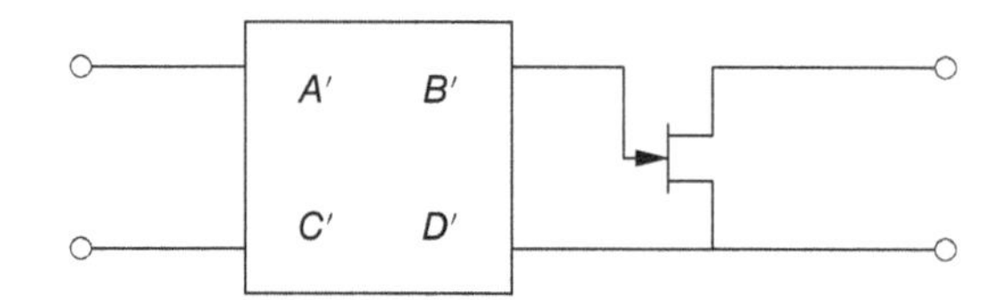

Figure 3.6 Bias circuit modeled as a two-port.

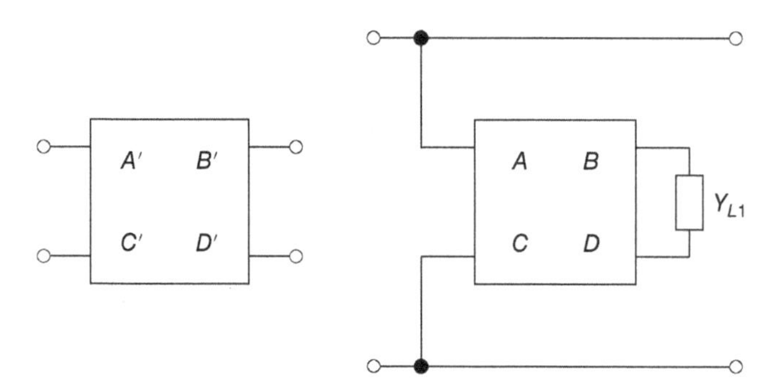

Figure 3.7 Loaded two-port to model the bias circuit.

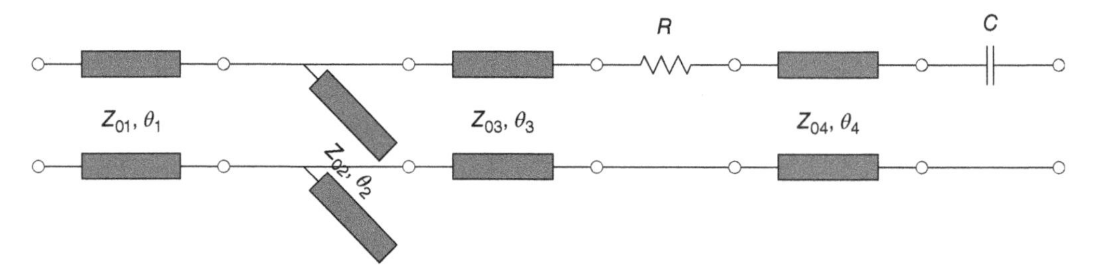

Figure 3.8 Cascade of the elements in the bias circuit.

The loaded two-port ($ABCD$) represents the cascade of the $\lambda/4$ line, the $\lambda/4$ stub, the 50 Ω line, the resistor, the second 50 Ω line, and the capacitor as shown in Figure 3.8, and the load admittance Y_{L1} is assumed to be equal to infinity.

According to (1.51) in Chapter 1, the two-port ($A'B'C'D'$) in Figure 3.7 has an $ABCD$ matrix such that

$$\begin{pmatrix} A' & B' \\ C' & D' \end{pmatrix} = \begin{pmatrix} 1 & 0 \\ \dfrac{C+DY_{L1}}{A+BY_{L1}} & 1 \end{pmatrix} = \begin{pmatrix} 1 & 0 \\ \dfrac{D}{B} & 1 \end{pmatrix}$$

and where

$$\begin{pmatrix} A & B \\ C & D \end{pmatrix} = \begin{pmatrix} A_1 & B_1 \\ C_1 & D_1 \end{pmatrix}\begin{pmatrix} A_2 & B_2 \\ C_2 & D_2 \end{pmatrix}\begin{pmatrix} A_3 & B_3 \\ C_3 & D_3 \end{pmatrix}\begin{pmatrix} A_4 & B_4 \\ C_4 & D_4 \end{pmatrix}\begin{pmatrix} A_5 & B_5 \\ C_5 & D_5 \end{pmatrix}\begin{pmatrix} A_6 & B_6 \\ C_6 & D_6 \end{pmatrix}$$

with

$$\begin{pmatrix} A_1 & B_1 \\ C_1 & D_1 \end{pmatrix} = \begin{pmatrix} \cos\theta_1 & jZ_{01}\sin\theta_1 \\ jY_{01}\sin\theta_1 & \cos\theta_1 \end{pmatrix}$$

Z_{01} is a high impedance and $\theta_1 = 90°$

$$\begin{pmatrix} A_2 & B_2 \\ C_2 & D_2 \end{pmatrix} = \begin{pmatrix} 1 & 0 \\ jY_{02}\tan\theta_2 & 1 \end{pmatrix}$$

Z_{02} is a low impedance and $\theta_2 = 90°$

$$\begin{pmatrix} A_3 & B_3 \\ C_3 & D_3 \end{pmatrix} = \begin{pmatrix} \cos\theta_3 & jZ_{03}\sin\theta_3 \\ jY_{03}\sin\theta_3 & \cos\theta_3 \end{pmatrix}$$

$Z_{03} = 50\,\Omega$ and θ_3 is free

$$\begin{pmatrix} A_4 & B_4 \\ C_4 & D_4 \end{pmatrix} = \begin{pmatrix} 1 & R \\ 0 & 1 \end{pmatrix}$$

R comes from bias conditions

$$\begin{pmatrix} A_5 & B_5 \\ C_5 & D_5 \end{pmatrix} = \begin{pmatrix} \cos\theta_4 & jZ_{04}\sin\theta_4 \\ jY_{04}\sin\theta_4 & \cos\theta_4 \end{pmatrix}$$

$Z_{04} = 50\,\Omega$ and θ_4 is free

$$\begin{pmatrix} A_6 & B_6 \\ C_6 & D_6 \end{pmatrix} = \begin{pmatrix} 1 & \dfrac{1}{jC\omega} \\ 0 & 1 \end{pmatrix}$$

C comes from bias conditions

The procedure is then to convert the S parameters of the transistor to $ABCD$ parameters using (1.53) and to multiply the $ABCD$ matrix of the bias circuit with that of the transistor, and finally, to convert back the $ABCD$ parameters to S parameters using (1.54). These will be the $S_{ij,k}$ parameters used in the RFT to represent the transistor k and its bias circuit. Note that one needs to compute the $S_{ij,k}$ parameters for each frequency used in the optimization frequency band. The method is, therefore, numerical rather than analytical.

The same $ABCD$ matrix approach is used to characterize the scattering parameters of the feedback transistor block using broadband applications described in Chapter 2.

The same $ABCD$ matrix approach is also used to define the noise parameters of the transistor with bias circuit [5]. It is based on a noise correlation matrix for various cascade, parallel, and series combinations of $ABCD$ matrices so that the noise parameters of the transistor with the bias circuit can be computed. Since the treatment is rather lengthy, it was left for Appendix C.

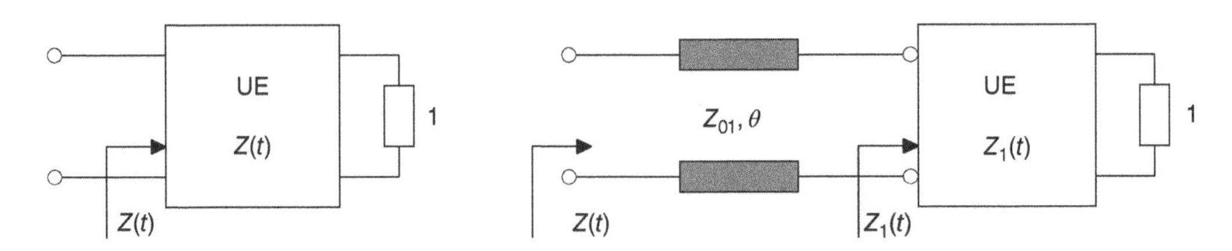

Figure 3.9 Extraction of one UE.

3.8 Distributed Equalizer Synthesis

The first step consists of computing the input impedance from the input reflection coefficient of the equalizer.

$$Z(t) = \frac{1+e_{11}(t)}{1-e_{11}(t)} \tag{3.30}$$

Note that during optimization, one usually uses normalized to 1 Ω terminations. Section 3.8.3 shows how to denormalize the synthesized equalizer elements.

3.8.1 Richard's Theorem

Starting from $Z(t)$ the input impedance given in terms of the Richard's variable, it is possible to extract a unit element (UE) that will have for characteristic impedance $Z_{01} = Z(1)$ and the remaining network will have a new input impedance called $Z_1(t)$ as shown in Figure 3.9.

The Richard's theorem provides the equation relating the initial input impedance $Z(t)$, the remaining input impedance $Z_1(t)$, and the characteristic impedance $Z(1)$ of the extracted UE and is given as

$$Z_{01} = Z(1) \tag{3.31}$$

$$Z_1(t) = Z(1)\frac{Z(t)-tZ(1)}{Z(1)-tZ(t)} \tag{3.32}$$

To illustrate the extraction process, we provide an example. It starts with the input impedance of the lossless distributed network, for example,

$$Z(t) = \frac{100t^3+50t^2+300t+30}{9t^3+170t^2+31t+30}$$

First, one computes the value Z_{01} of the characteristic impedance of the first UE using

$$Z_{01} = Z(1) = \frac{100+50+300+30}{9+170+31+30} = \frac{480}{240} = 2$$

Then, the remaining input impedance $Z_1(t)$ is determined:

$$Z_1(t) = 2\frac{Z(t)-2t}{2-tZ(t)} = 2\frac{\frac{100t^3+50t^2+300t+30}{9t^3+170t^2+31t+30}-2t}{2-t\frac{100t^3+50t^2+300t+30}{9t^3+170t^2+31t+30}} = \frac{9t^2+120t+15}{25t^2+8t+15}$$

The process is then repeated to define the characteristic impedance Z_{02} of a second UE:

$$Z_{02} = Z_1(1) = \frac{9+120+15}{25+8+15} = \frac{144}{48} = 3$$

and a remaining input impedance $Z_2(t)$ is calculated:

$$Z_2(t) = 3\frac{Z_1(t)-3t}{3-tZ_1(t)} = 3\frac{\frac{9t^2+120t+15}{25t^2+8t+15}-3t}{3-t\frac{9t^2+120t+15}{25t^2+8t+15}} = \frac{25t+5}{t+5}$$

A third characteristic impedance Z_{03} of a third UE is extracted:

$$Z_{03} = Z_2(1) = \frac{25+5}{1+5} = \frac{30}{6} = 5$$

and a remaining input impedance $Z_3(t)$ is calculated:

$$Z_3(t) = 5\frac{Z_2(t)-5t}{5-tZ_2(t)} = 5\frac{\frac{25t+5}{t+5}-5t}{5-t\frac{25t+5}{t+5}} = 1$$

Since the remaining input impedance is down to a constant 1 Ω, the process is stopped and the circuit layout of the distributed network can be drawn. It is shown in Figure 3.10.

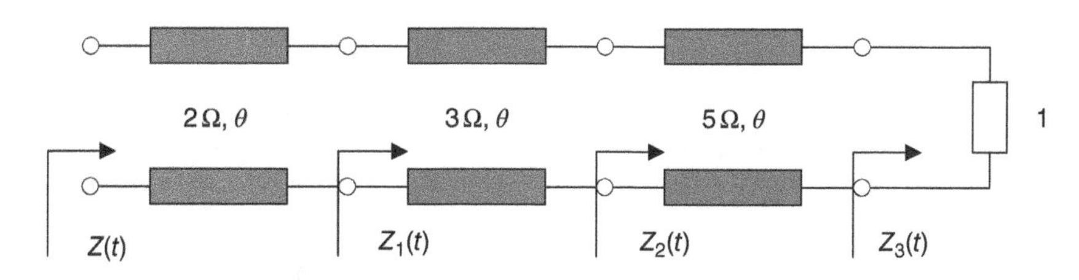

Figure 3.10 Extraction of a cascade of UEs.

A MATLAB program that computes this example is given below:

```
clc;
clear;
close all;
```

```
syms t;

Z = (100*t^3 + 50*t^2 + 300*t + 30)/(9*t^3 + 170*t^2 + 31*t + 30);

t = 1;
Z01 = eval(Z);
syms t;
Z1 = simplify( (Z01*Z - Z01*Z01*t)/(Z01 - t*Z) );

t = 1;
Z02 = eval(Z1);
syms t;
Z2 = simplify( (Z02*Z1 - Z02*Z02*t)/(Z02 - t*Z1) );

t = 1;
Z03 = eval(Z2);
syms t;
Z3 = simplify( (Z03*Z2 - Z03*Z03*t)/(Z03 - t*Z2) );
```

3.8.2 Stub Extraction

The Richard's theorem for the extraction of a cascade of UEs assumes that numerator polynomial and denominator polynomial have the same order:

$$Z(t) = \frac{a_n t^n + a_{n-1} t^{n-1} + \cdots + a_2 t^2 + a_1 t + a_0}{b_n t^n + b_{n-1} t^{n-1} + \cdots + b_2 t^2 + b_1 t + b_0}$$

If they are not of the same order because $a_n = 0$ or $b_n = 0$, then one should extract a stub instead of a cascaded UE.

Case $a_n = 0$ In this case, the denominator polynomial has a higher order than the numerator polynomial and an open-circuited stub placed in parallel must be extracted. The characteristic impedance of the stub will be

$$Z_0 = \frac{a_{n-1}}{b_n} \tag{3.33}$$

and the remaining input impedance is given by

$$\frac{1}{Z_1(t)} = \frac{1}{Z(t)} - \frac{t}{Z_0}$$

$$Z_1(t) = \frac{Z_0 Z(t)}{Z_0 - tZ(t)} \tag{3.34}$$

Case $b_n = 0$ In this case, the numerator polynomial has a higher order than the denominator polynomial and a short-circuited stub placed in series must be extracted. The characteristic impedance of the stub will be

$$Z_0 = \frac{a_n}{b_{n-1}} \tag{3.35}$$

and the remaining input impedance is given by

$$Z_1(t) = Z(t) - tZ_0 \tag{3.36}$$

To illustrate the extraction process, we provide an example. It starts with the input impedance of the lossless distributed network, for example,

$$Z(t) = \frac{18t^2 + 240t + 30}{9t^3 + 170t^2 + 31t + 30}$$

First, we note that $a_n = 0$ (where $n = 3$). Therefore, we should extract an open-circuited stub placed in parallel with characteristic impedance given by

$$Z_{01} = \frac{a_2}{b_3} = \frac{18}{9} = 2$$

and the remaining input impedance $Z_1(t)$ is given by

$$Z_1(t) = \frac{Z_{01}Z(t)}{Z_{01} - tZ(t)} = \frac{2\dfrac{18t^2 + 240t + 30}{9t^3 + 170t^2 + 31t + 30}}{2 - t\dfrac{18t^2 + 240t + 30}{9t^3 + 170t^2 + 31t + 30}} = \frac{9t^2 + 120t + 15}{25t^2 + 8t + 15}$$

Since $a_n \neq 0$ and $b_n \neq 0$, we extract a cascaded UE with characteristic impedance Z_{02} such that

$$Z_{02} = Z_1(1) = \frac{9 + 120 + 15}{25 + 8 + 15} = \frac{144}{48} = 3$$

and the remaining input impedance $Z_2(t)$ is calculated:

$$Z_2(t) = 3\frac{Z_1(t) - 3t}{3 - tZ_1(t)} = 3\frac{\dfrac{9t^2 + 120t + 15}{25t^2 + 8t + 15} - 3t}{3 - t\dfrac{9t^2 + 120t + 15}{25t^2 + 8t + 15}} = \frac{25t + 5}{t + 5}$$

Again $a_n \neq 0$ and $b_n \neq 0$, so we extract a cascaded UE with characteristic impedance Z_{02} such that

$$Z_{03} = Z_2(1) = \frac{25 + 5}{1 + 5} = \frac{30}{6} = 5$$

and the remaining input impedance is calculated:

$$Z_3(t) = 5\frac{Z_2(t) - 5t}{5 - tZ_2(t)} = 5\frac{\dfrac{25t + 5}{t + 5} - 5t}{5 - t\dfrac{25t + 5}{t + 5}} = 1$$

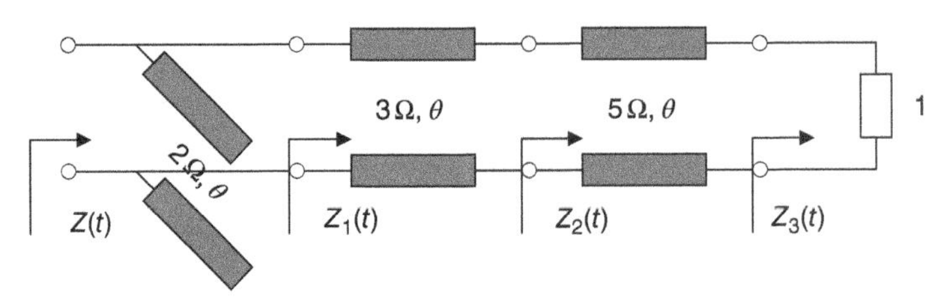

Figure 3.11 Extraction of stub and cascade of UEs.

Since the remaining input impedance is down to a constant 1 Ω, the process is stopped and the circuit layout of the distributed network can be drawn. It is shown in Figure 3.11.

A MATLAB program that computes this example is given below:

```
clc;
clear;
close all;

syms t;

Z = (18*t^2 + 240*t + 30)/(9*t^3 + 170*t^2 + 31*t + 30);

Z01 = 18/9;
syms t;
Z1 = simplify( (Z01*Z)/(Z01 - t*Z) )

t = 1;
Z02 = eval(Z1);
syms t;
Z2 = simplify( (Z02*Z1 - Z02*Z02*t)/(Z02 - t*Z1) );

t = 1;
Z03 = eval(Z2);
syms t;
Z3 = simplify( (Z03*Z2 - Z03*Z03*t)/(Z03 - t*Z2) );
```

Note that in some cases one might extract two consecutive stubs. If that happens, one should use redundant UEs and the Kuroda identities of Section 1.8.9.

3.8.3 *Denormalization*

Throughout the optimization process, the equalizers are normalized to 1 Ω and one should denormalize them after the previous synthesis procedures have been completed. This is simply accomplished by multiplying the normalized characteristic impedances obtained from the extraction procedures of $Z(t)$ by the actual termination R_0 such that

$$Z_{0k} = R_0 Z_{0k} \tag{3.37}$$

For example, if $Z(t)$ was extracted to a final resistor of 1 Ω and one wants to use the extracted UE network into a 50 Ω system, then one uses $R_0 = 50$.

Note that besides impedance normalization, the RFT uses frequency normalization as well so that initial guesses for the polynomial coefficients can be taken as $h_i = \pm 1$. Section 2.8 provides a thorough example of the entire process of optimization and denormalization.

3.8.4 UE Impedances Too Low

When a UE has a characteristic impedance that is too low to be realized practically, it is possible to perform a few circuit manipulations that lead to UEs and stubs with more suitable characteristic impedance values [4, 5]. The first step is to equate a UE with a UE and two admittances in parallel as shown in Figure 3.12.

We equate the two networks through their *ABCD* matrices:

$$\begin{pmatrix} \cos\theta_A & jZ_{0A}\sin\theta_A \\ jY_{0A}\sin\theta_A & \cos\theta_A \end{pmatrix} \text{ and } \begin{pmatrix} \cos\theta_B + jY_L Z_{0B}\sin\theta_B & jZ_{0B}\sin\theta_B \\ 2Y_L\cos\theta_B + jY_{0B}\sin\theta_B + jY_L^2 Z_{0B}\sin\theta_B & \cos\theta_B + jY_L Z_{0B}\sin\theta_B \end{pmatrix}$$

When the physical length of the starting UE is $l_A = \lambda/4$, we find that the two systems have the same *ABCD* matrices if

$$\theta_A = \frac{\pi}{2} \tag{3.38}$$

$$\theta_B = \sin^{-1}\left(\frac{Z_{0A}}{Z_{0B}}\right) \tag{3.39}$$

$$Y_L = jX_L \quad \text{where} \quad X_L = \frac{1}{Z_{0B}}\sqrt{\left(\frac{Z_{0B}}{Z_{0A}}\right)^2 - 1} \tag{3.40}$$

The next step consists of replacing the admittance in parallel by an open-circuited stub in parallel as shown in Figure 3.13.

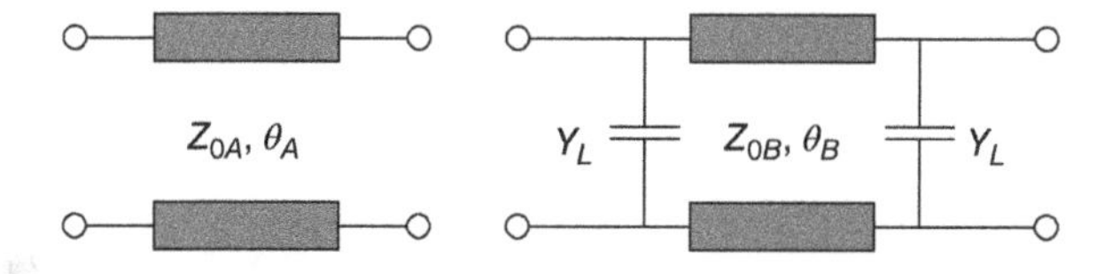

Figure 3.12 Equating a UE with a UE with two admittances in parallel.

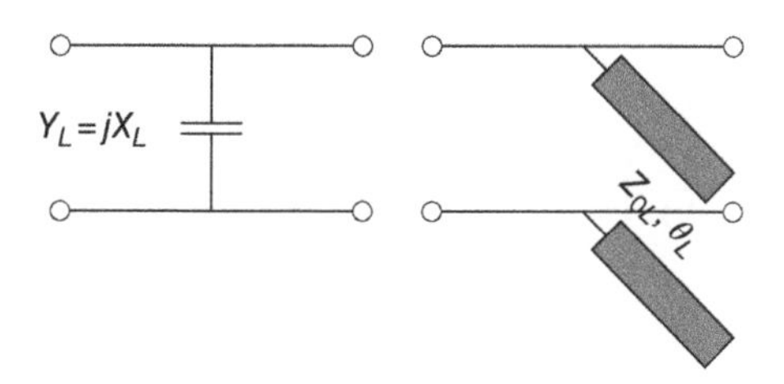

Figure 3.13 Replacing the admittance in parallel with an open-circuited stub in parallel.

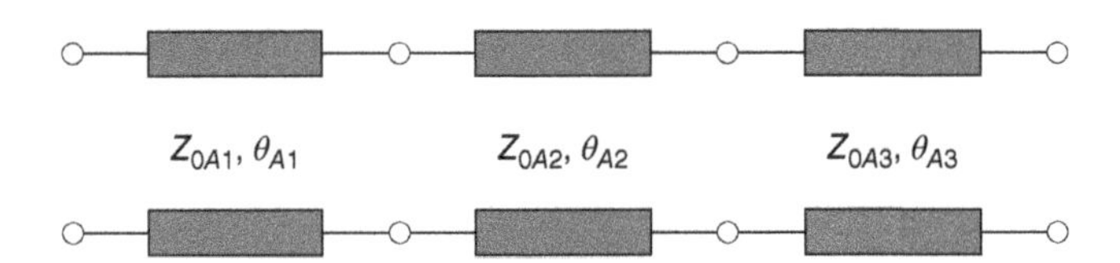

Figure 3.14 UEs with characteristic impedances that are too low.

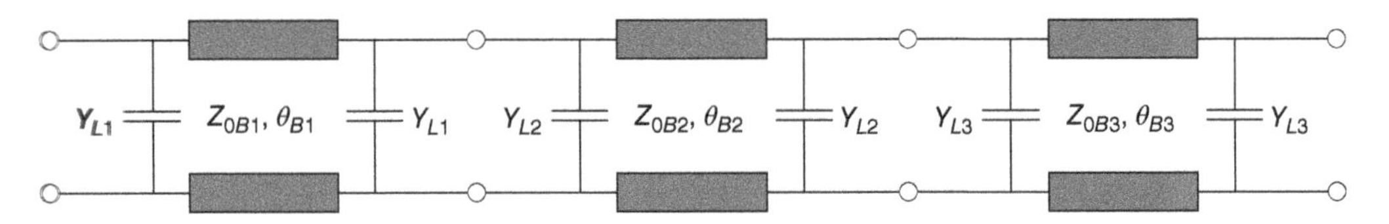

Figure 3.15 Replacing the UEs with a UE with two admittances in parallel.

We again equate their *ABCD* matrices:

$$\begin{pmatrix} 1 & 0 \\ Y_L & 1 \end{pmatrix} \text{ and } \begin{pmatrix} 1 & 0 \\ jY_{0L}\tan\theta_L & 1 \end{pmatrix}$$

The characteristic admittance Y_{0L} and electrical length θ_L of the stub is given by the imaginary part X_L of the admittance in parallel such that

$$Y_{0L}\tan\theta_L = X_L \tag{3.41}$$

To illustrate the method, an example is given in Figure 3.14. It consists of three cascaded UEs with characteristic impedances $Z_{0A1}=26$, $Z_{0A2}=19$, and $Z_{0A3}=18.5$, and electrical lengths $\theta_{A1}=\theta_{A2}=\theta_{A3}=\pi/2$ $(l=\lambda/4)$.

We assume that the smallest achievable characteristic impedance is 40 Ω. This means that we must transform all three UEs. Since they all have a length of $l=\lambda/4$, we can replace each UE by a UE with two admittances in parallel as shown in Figure 3.15.

At this point, we decide to take the characteristic impedance of all UEs to be equal to 40 Ω and to use the electrical length of the UE to satisfy (3.39). This gives

$$Z_{0B1}=Z_{0B2}=Z_{0B3}=40\,\Omega \quad \text{and}$$

$$\theta_{B1}=\sin^{-1}\left(\frac{Z_{0A1}}{Z_{0B1}}\right)=\sin^{-1}\left(\frac{26}{40}\right)=40.6^\circ \quad X_{L1}=\frac{1}{Z_{0B1}}\sqrt{\left(\frac{Z_{0B1}}{Z_{0A1}}\right)^2-1}=\frac{1}{40}\sqrt{\left(\frac{40}{26}\right)^2-1}=0.0292$$

$$\theta_{B2}=\sin^{-1}\left(\frac{Z_{0A2}}{Z_{0B2}}\right)=\sin^{-1}\left(\frac{19}{40}\right)=28.4^\circ \quad X_{L2}=\frac{1}{Z_{0B2}}\sqrt{\left(\frac{Z_{0B1}}{Z_{0A2}}\right)^2-1}=\frac{1}{40}\sqrt{\left(\frac{40}{19}\right)^2-1}=0.0463$$

$$\theta_{B3}=\sin^{-1}\left(\frac{Z_{0A3}}{Z_{0B3}}\right)=\sin^{-1}\left(\frac{18.5}{40}\right)=27.6^\circ \quad X_{L3}=\frac{1}{Z_{0B3}}\sqrt{\left(\frac{Z_{0B3}}{Z_{0A3}}\right)^2-1}=\frac{1}{40}\sqrt{\left(\frac{40}{18.5}\right)^2-1}=0.0479$$

The admittances in parallel are then replaced by open-circuited stubs in parallel as shown in Figure 3.16.

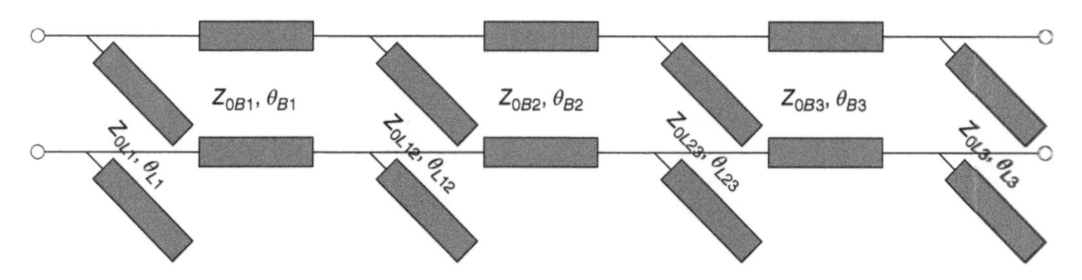

Figure 3.16 Replacing the admittances in parallel by open-circuited stubs in parallel.

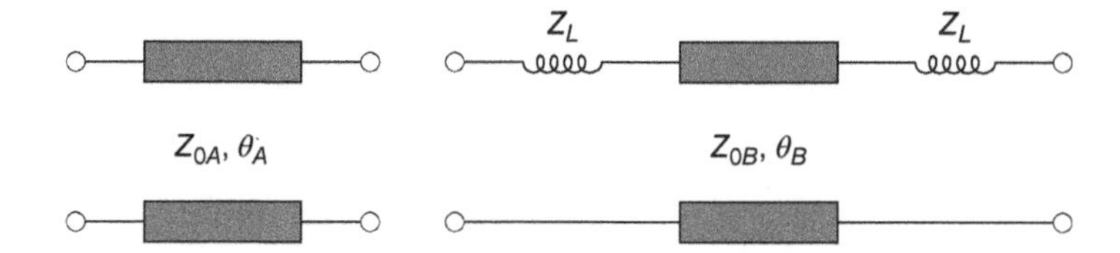

Figure 3.17 Equating the UE with a UE with two impedances in series.

Here again, we decide to take the characteristic impedance of all UEs to be equal to 40 Ω and to use the electrical length of the stub to satisfy (3.41). This gives

$$Z_{0L1} = Z_{0L12} = Z_{0L23} = Z_{0L3} = 40\,\Omega \quad \text{and}$$

$$\begin{aligned}
\theta_{L1} &= \tan^{-1}(Z_{0L1}X_{L1}) = \tan^{-1}(40 \times 0.0292) = 49.4^\circ \\
\theta_{L12} &= \tan^{-1}(Z_{0L12}(X_{L1} + X_{L2})) = \tan^{-1}(40 \times (0.0292 + 0.0463)) = 71.7^\circ \\
\theta_{L23} &= \tan^{-1}(Z_{0L23}(X_{L2} + X_{L3})) = \tan^{-1}(40 \times (0.0463 + 0.0479)) = 75.1^\circ \\
\theta_{L3} &= \tan^{-1}(Z_{0L3}X_{L3}) = \tan^{-1}(40 \times 0.0479) = 62.4^\circ
\end{aligned}$$

3.8.5 UE Impedances Too High

When the characteristic impedance of a UE is too high to be realized practically, it is possible to perform few circuit manipulations that lead to UEs and stubs with more suitable characteristic impedance values [4, 5]. The first step is to equate a UE with a UE and two impedances in series as shown in Figure 3.17.

We equate the two networks through their *ABCD* matrices given below:

$$\begin{pmatrix} \cos\theta_A & jZ_{0A}\sin\theta_A \\ jY_{0A}\sin\theta_A & \cos\theta_A \end{pmatrix} \quad \text{and} \quad \begin{pmatrix} \cos\theta_B + jZ_L Y_{0B}\sin\theta_B & 2Z_L\cos\theta_B + jZ_{0B}\sin\theta_B + jZ_L^2 Y_{0B}\sin\theta_B \\ jY_{0B}\sin\theta_B & \cos\theta_B + jZ_L Y_{0B}\sin\theta_B \end{pmatrix}$$

When the physical length of the starting UE is $l_A = \lambda/4$, we find that the two systems have the same *ABCD* matrices if

$$\theta_A = \frac{\pi}{2} \tag{3.42}$$

$$\theta_B = \sin^{-1}\left(\frac{Z_{0B}}{Z_{0A}}\right) \tag{3.43}$$

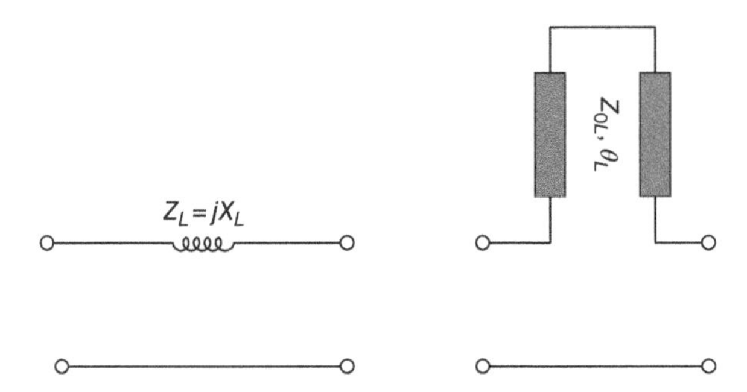

Figure 3.18 Replacing the impedance in series with a short-circuited stub in series.

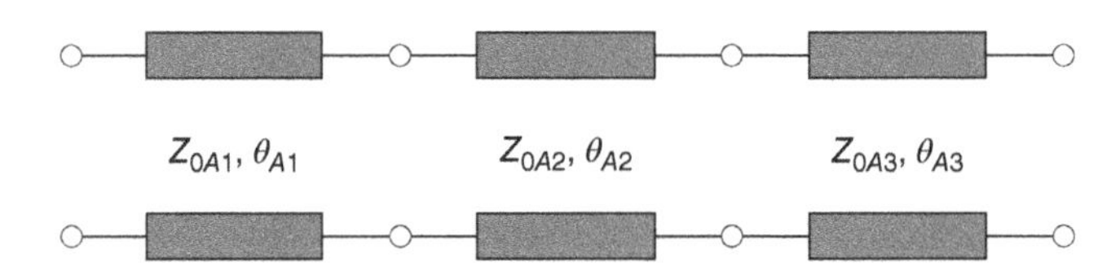

Figure 3.19 UEs with characteristic impedances that are too high.

$$Z_L = jX_L \quad \text{where} \quad X_L = Z_{0B}\sqrt{\left(\frac{Z_{0A}}{Z_{0B}}\right)^2 - 1} \tag{3.44}$$

The next step consists of replacing the impedance in series with a short-circuited stub in series as shown in Figure 3.18.

We again equate their *ABCD* matrices:

$$\begin{pmatrix} 1 & Z_L \\ 0 & 1 \end{pmatrix} \quad \text{and} \quad \begin{pmatrix} 1 & jZ_{0L}\tan\theta_L \\ 0 & 1 \end{pmatrix}$$

The characteristic impedance Z_{0L} and electrical length θ_L of the stub is given by the imaginary part X_L of the impedance in series such that

$$Z_{0L}\tan\theta_L = X_L \tag{3.45}$$

To illustrate the method, an example is given in Figure 3.19. It consists of three cascaded UEs with characteristic impedances $Z_{0A1} = 180$, $Z_{0A2} = 165$, and $Z_{0A3} = 100$, and electrical lengths $\theta_{A1} = \theta_{A2} = \pi/2$ $(l = \lambda/4)$, and $\theta_{A3} = \pi/4$ $(l = \lambda/8)$.

We assume that the highest achievable characteristic impedance is 140 Ω. This means that we must transform the first two UEs. Since these UEs have a length of $l = \lambda/4$, we can replace them by a UE with two impedances in series as shown in Figure 3.20. We leave the third UE untouched.

At this stage, we decide to take the characteristic impedance of the modified UEs to be equal to 140 Ω and to use the electrical length of the UE to satisfy (3.43). This gives

$Z_{0B1} = Z_{0B2} = 140\ \Omega$, and we keep $Z_{0A3} = 100\ \Omega$ since it is already below 140 Ω.

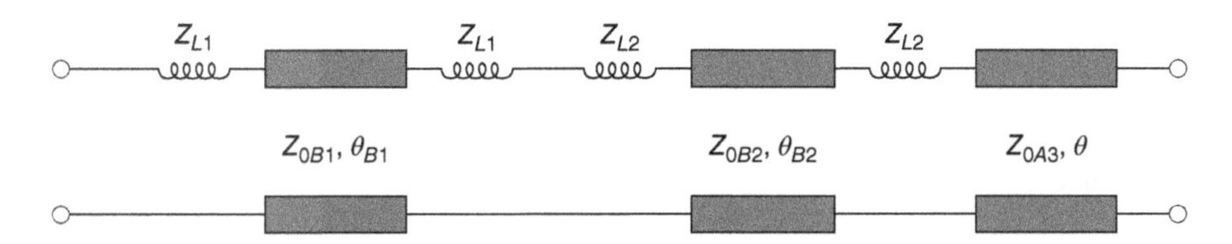

Figure 3.20 Replacing UEs with a UE with two impedances in series.

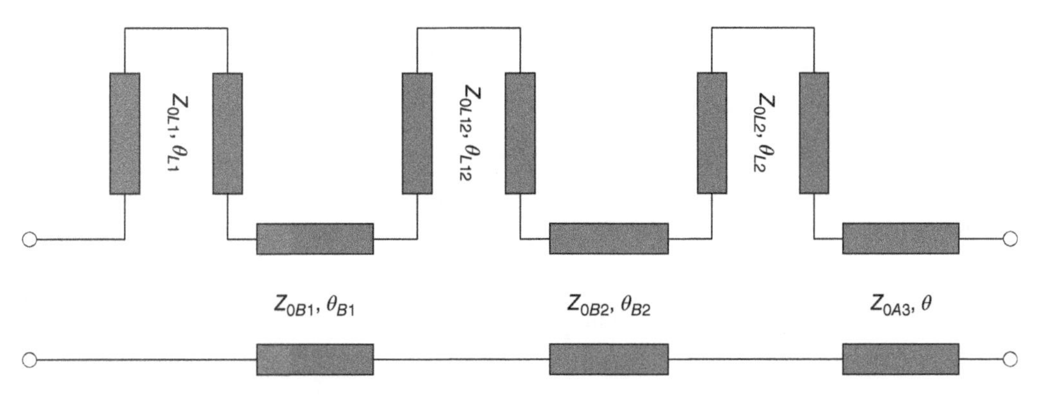

Figure 3.21 Replacing the impedances in series by short-circuited stubs in series.

$$\theta_{B1} = \sin^{-1}\left(\frac{Z_{0B1}}{Z_{0A1}}\right) = \sin^{-1}\left(\frac{140}{180}\right) = 51.1^\circ \quad X_{L1} = Z_{0B1}\sqrt{\left(\frac{Z_{0A1}}{Z_{0B1}}\right)^2 - 1} = 140\sqrt{\left(\frac{180}{140}\right)^2 - 1} = 113.1$$

$$\theta_{B2} = \sin^{-1}\left(\frac{Z_{0B2}}{Z_{0A2}}\right) = \sin^{-1}\left(\frac{140}{165}\right) = 58.0^\circ \quad X_{L2} = Z_{0B2}\sqrt{\left(\frac{Z_{0A2}}{Z_{0B2}}\right)^2 - 1} = 140\sqrt{\left(\frac{165}{140}\right)^2 - 1} = 87.3$$

Then, the impedances in series are replaced by short-circuited stubs in series as shown in Figure 3.21.

In this case, we decide to make a stub length be equal to the length of the UE on its right. This gives

$$Z_{0L1} = \frac{X_{L1}}{\tan\theta_{L1}} = \frac{113.4}{\tan(51.1^\circ)} = 91.5$$

$$\theta_{L1} = \theta_{B1} = 51.1^\circ$$

$$\theta_{L12} = \theta_{B2} = 58.0^\circ \quad Z_{0L12} = \frac{X_{L12}}{\tan\theta_{L12}} = \frac{(113.4 + 87.3)}{\tan(58.0^\circ)} = 125.0$$

$$\theta_{L2} = \theta_{A3} = 45^\circ$$

$$Z_{0L2} = \frac{X_{L2}}{\tan\theta_{L3}} = \frac{87.3}{\tan(90^\circ)} = 87.3$$

Since short-circuited stubs in series are not always desired, one could use the Kuroda's identities of Section 1.8.9 to transform them into open-circuited stubs in parallel.

3.9 Design Procedures

The design of the multistage distributed amplifier starts with the first equalizer–transistor stage terminated on reference impedance Z_0 (e.g., 50 Ω) as shown in Figure 3.22.

The transducer gain for the first stage is given by

$$T_1 = \left(1-|\rho_G|^2\right)\frac{|e_{21,1}|^2|S_{21,1}|^2}{|1-e_{11,1}\rho_G|^2|1-\hat{e}_{22,1}S_{11,1}|^2} \tag{3.46}$$

with

$$\hat{e}_{22,1} = e_{22,1} + \frac{e_{21,1}^2\rho_G}{1-e_{11,1}\rho_G} \tag{3.47}$$

$$\rho_G = \frac{Z_G - Z_0}{Z_G + Z_0} \tag{3.48}$$

and the target transducer gain for the first stage is given by

$$T_{0,1} = \min\left\{\frac{|S_{21,1}(j\omega)|^2}{\left|1-|S_{11,1}(j\omega)|^2\right|}\right\} \tag{3.49}$$

The noise factor for the first stage is given by

$$F_1 = F_{\min,1} + 4R_{n,1}Y_0\frac{|\hat{e}_{22,1}-\rho_{G\min,1}|^2}{\left(1-|\hat{e}_{22,1}|^2\right)|1+\rho_{G\min,1}|^2} \tag{3.50}$$

where $F_{\min,1}$, $R_{n,1}$, and $\rho_{G\min,1}$ are the noise parameters of the first transistor.

The target noise factor for the first stage is given when $\hat{e}_{22,1} = \rho_{G\min,1}$ such that

$$F_{1\ \text{adapted}} = F_{\min,1} \tag{3.51}$$

and we take the maximum value of $F_{1\ \text{adapted}}$ across the band of frequencies used in the optimization [5]:

$$F_{0,1} = \max\left\{F_{1\ \text{adapted}}(j\omega)\right\} \tag{3.52}$$

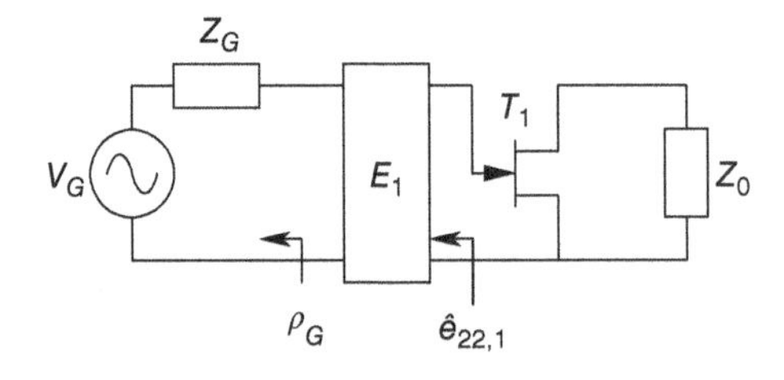

Figure 3.22 Optimizing the first equalizer.

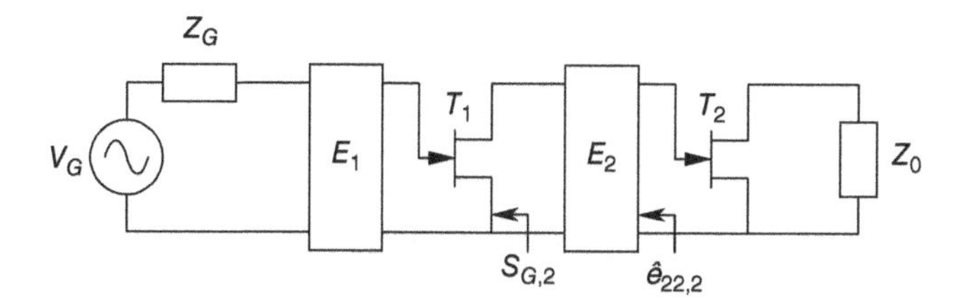

Figure 3.23 Optimizing the second equalizer.

Once the first equalizer has been optimized, the second equalizer–transistor stage is placed between the first transistor and the load Z_0 as shown in Figure 3.23. At this time, the first equalizer is fixed and only $e_{11,2}(t)$, $e_{22,2}(t)$, and $e_{12,2}(t) = e_{21,2}(t)$ of the second equalizer vary between iterations of the optimization.

The transducer gain T_2 for the two first equalizer–transistor stages is

$$T_2 = T_1 \frac{|e_{21,2}|^2 |S_{21,2}|^2}{|1 - e_{11,2} S_{G,2}|^2 |1 - \hat{e}_{22,2} S_{11,2}|^2} \tag{3.53}$$

with

$$\hat{e}_{22,2} = e_{22,2} + \frac{e_{21,2}^2 S_{G,2}}{1 - e_{11,2} S_{G,2}} \tag{3.54}$$

$$S_{G,2} = S_{22,1} + \frac{S_{12,1} S_{21,1} \hat{e}_{22,1}}{1 - S_{11,1} \hat{e}_{22,1}} \tag{3.55}$$

and the target transducer gain for the second stage is given by

$$T_{0,2} = \min\left\{ T_{MAX,1} \frac{|S_{21,2}|^2}{\left|1 - |S_{11,2}|^2\right|} \right\} \tag{3.56}$$

where

$$T_{MAX,1} = \frac{|S_{21,1}|^2}{\left|1 - |S_{11,1}|^2\right| \left|1 - |S_{22,1}|^2\right|} \frac{1}{\left| 1 - \dfrac{S_{12,1} S_{21,1} S_{11,1}^* S_{22,1}^*}{\left(1 - |S_{11,1}|^2\right)\left(1 - |S_{22,1}|^2\right)} \right|^2} \tag{3.57}$$

The noise figure for the first two equalizer–transistor stages is

$$F_2 = F_1 + \frac{1}{T_{A,1}} \left[F_{\min,2} + 4R_{n,2} Y_0 \frac{|\hat{e}_{22,2} - \rho_{G\min,2}|^2}{(1 - |\hat{e}_{22,2}|)^2 |1 + \rho_{G\min,2}|^2} - 1 \right] \tag{3.58}$$

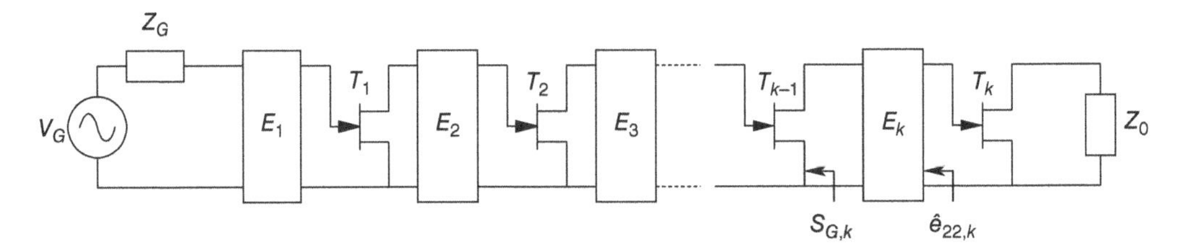

Figure 3.24 Optimization of the kth equalizer.

where $T_{A,1}$ is the available gain given by

$$T_{A,1} = \frac{T_1}{1-|S_{G,2}|^2} \tag{3.59}$$

where $F_{\min,2}$, $R_{n,2}$, and $\rho_{G\min,2}$ are the noise parameters of the second transistor

The target noise factor for the second stage is given when $\hat{e}_{22,2} = \rho_{G\min,2}$ such that

$$F_{2\text{ adapted}} = F_1 + \frac{F_{\min,2}-1}{T_{A,1}} \tag{3.60}$$

We take the maximum value of $F_{2\text{ adapted}}$ across the band of frequencies used in the optimization [5]:

$$F_{0,2} = \max\{F_{2\text{ adapted}}(j\omega)\} \tag{3.61}$$

The procedure is repeated, and Figure 3.24 shows the situation for kth equalizer–transistor stage terminated on Z_0.

The transducer gain T_k for the first k equalizer–transistor stages is

$$T_k = T_{k-1}\frac{|e_{21,k}|^2|S_{21,k}|^2}{|1-e_{11,k}S_{G,k}|^2|1-\hat{e}_{22,k}S_{11,k}|^2} \tag{3.62}$$

with

$$\hat{e}_{22,k} = e_{22,k} + \frac{e_{21,k}^2 S_{G,k}}{1-e_{11,k}S_{G,k}} \tag{3.63}$$

$$S_{G,k} = S_{22,k-1} + \frac{S_{12,k-1}S_{21,k-1}\hat{e}_{22,k-1}}{1-S_{11,k-1}\hat{e}_{22,k-1}} \tag{3.64}$$

and the target transducer gain for the kth stage is given by

$$T_{0,k} = \min\left\{T_{MAX,1}\times T_{MAX,2}\times\cdots\times T_{MAX,k-1}\frac{|S_{21,k}|^2}{|1-|S_{11,k}|^2|}\right\} \tag{3.65}$$

where

$$T_{MAX,i} = \frac{|S_{21,i}|^2}{\left|1-|S_{11,i}|^2\right|\left|1-|S_{22,i}|^2\right|}\frac{1}{\left|1-\frac{S_{12,i}S_{21,i}S_{11,i}^*S_{22,i}^*}{\left(1-|S_{11,i}|^2\right)\left(1-|S_{22,i}|^2\right)}\right|^2} \tag{3.66}$$

The noise figure for the first kth equalizer–transistor stages is

$$F_k = F_{k-1} + \frac{1}{T_{A,k-1}}\left[F_{\min,k} + 4R_{n,k}Y_0\frac{\left|\hat{e}_{22,k} - \rho_{G\min,k}\right|^2}{(1-|\hat{e}_{22,k}|)^2\left|1+\rho_{G\min,k}\right|^2} - 1\right] \tag{3.67}$$

where $T_{A,k-1}$ is the available gain given by

$$T_{A,k-1} = \frac{T_{k-1}}{1-|S_{G,k}|^2} \tag{3.68}$$

where $F_{\min,k}$, $R_{n,k}$, and $\rho_{G\min,k}$ are the noise parameters of the kth transistor.

The target noise factor for the kth stage is given when $\hat{e}_{22,k} = \rho_{G\min,k}$ such that

$$F_{k\ \text{adapted}} = F_{k-1} + \frac{F_{\min,k}-1}{T_{A,k-1}} \tag{3.69}$$

We take the maximum value of $F_{k\ \text{adapted}}$ across the band of frequencies used in the optimization [5]:

$$F_{0,k} = \max\left\{F_{k\ \text{adapted}}(j\omega)\right\} \tag{3.70}$$

Once the kth equalizer has been optimized, a last equalizer E_{k+1} can be added to compensate for an arbitrary load Z_L different than Z_0 as shown in Figure 3.25.

The overall transducer gain T_{GT} in the case of Figure 3.25 is given by

$$T_{GT} = T_k\frac{|e_{21,k+1}|^2\left(1-|\rho_L|^2\right)}{|1-e_{11,k+1}S_{G,k+1}|^2|1-\hat{e}_{22,k+1}\rho_L|^2} \tag{3.71}$$

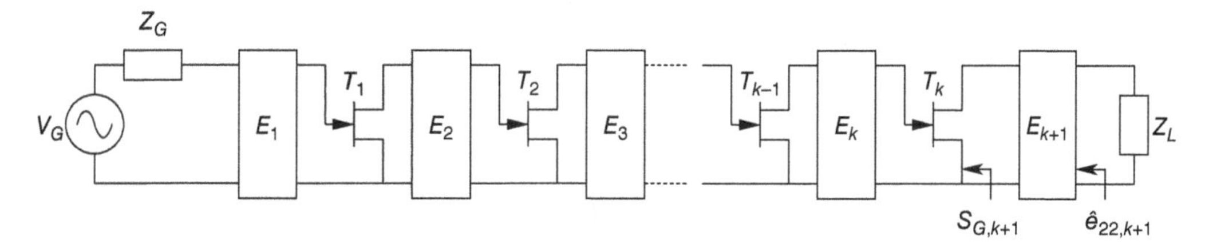

Figure 3.25 Optimizing the $(k+1)$th equalizer.

with

$$\hat{e}_{22,k+1} = e_{22,k+1} + \frac{e_{21,k+1}^2 S_{G,k+1}}{1 - e_{11,k+1} S_{G,k+1}} \tag{3.72}$$

$$S_{G,k+1} = S_{22,k} + \frac{S_{12,k} S_{21,k} \hat{e}_{22,k}}{1 - S_{11,k} \hat{e}_{22,k}} \tag{3.73}$$

$$\rho_L = \frac{Z_L - Z_0}{Z_L + Z_0} \tag{3.74}$$

and the target transducer gain for the $(k+1)^{th}$ equalizer is given by

$$T_{0,k+1} = \min\{T_{MAX,1} \times T_{MAX,2} \times \cdots \times T_{MAX,k-1} \times T_{MAX,k}\} \tag{3.75}$$

Note that the total noise figure F_T does not change with the addition of this lossless equalizer E_{k+1} and remains as

$$F_T = F_k \tag{3.76}$$

3.10 Simulations and Realizations

The procedures described in this chapter were implemented in a program called FREEL.

3.10.1 *Three-Stage 2–8 GHz Distributed Amplifier*

This first example was done for CNET [5, 8]. It consisted of a three-stage amplifier to operate in a 2–8 GHz band. The multistage amplifier was optimized for gain and noise. The equalizers were synthesized using distributed networks.

The transistor was selected as the FHX04X MESFET from Fujitsu. Table 3.1 provides the measured small-signal scattering parameters of the transistor, and Table 3.2 provides the noise parameters of the transistor for few frequency points. Biasing conditions were $V_{ds} = 2\text{V}$, $I_{ds} = 10\text{mA}$.

For wideband amplifiers design, the transistors used additional resistive feedback and resistive adaptation as shown in Figure 3.26. For more information on transistor feedback blocks for

Table 3.1 Measured *S* parameters of the Fujitsu FHX04X MESFET

FREQ (GHz)	$S_{11\ Mag}$	$S_{11\ Ang}$	$S_{21\ Mag}$	$S_{21\ Ang}$	$S_{12\ Mag}$	$S_{12\ Ang}$	$S_{22\ Mag}$	$S_{22\ Ang}$
2	0.983	−18.8	3.658	164.1	0.026	81.0	0.598	−8.3
3	0.964	−28.0	3.585	156.4	0.038	76.7	0.588	−12.3
4	0.938	−37.0	3.439	149.0	0.049	72.8	0.576	−16.0
5	0.909	−45.6	3.377	141.9	0.059	69.2	0.562	−19.6
6	0.877	−54.0	3.255	136.1	0.068	66.0	0.547	−22.9
7	0.844	−62.1	3.128	128.6	0.076	63.2	0.531	−26.0
8	0.811	−69.8	2.999	122.5	0.082	60.8	0.516	−28.9

Table 3.2 Measured noise parameters of the Fujitsu FHX04X MESFET

FREQ (GHz)	F_{min}	$\Gamma_{opt\ Mag}$	$\Gamma_{opt\ Ang}$	R_n
2	0.33	0.80	16	0.50
4	0.35	0.74	31	0.45
6	0.44	0.68	46	0.30
8	0.53	0.63	61	0.30

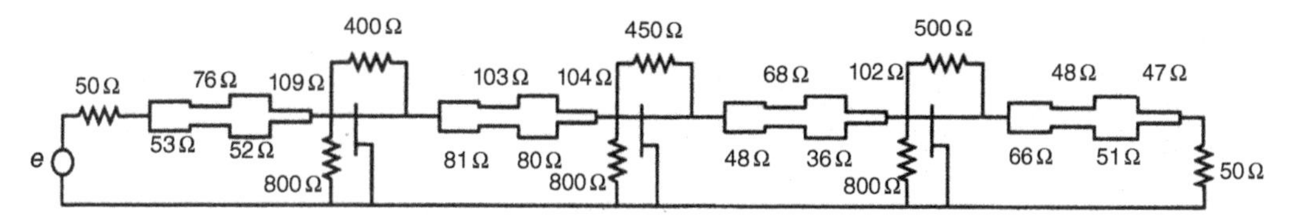

Figure 3.26 Three-stage 2–8 GHz amplifier using distributed equalizers.

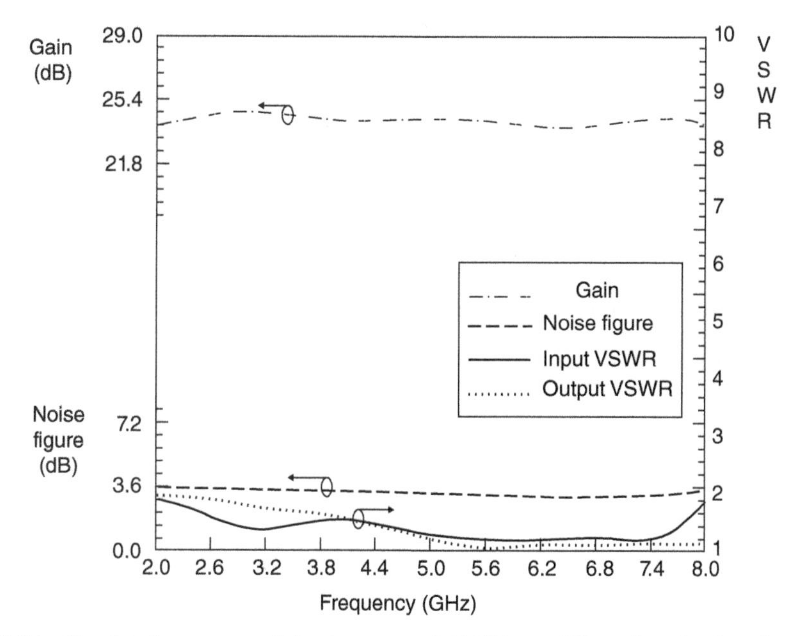

Figure 3.27 Simulated responses of the three-stage 2–8 GHz amplifier using distributed equalizers.

wideband applications, see Section 2.9. As in the case of lumped amplifiers, the value of the parallel resistance is selected for noise figure, VSWR, and gain-bandwidth specifications to meet. The feedback loop, stabilizing shunt gate resistance, and GaAs MESFET transistor form a transistor block whose S parameters are computed using the technique described in Section 3.7. These are the S parameters used for the transistors shown in Figure 3.1. This technique allows the RFT to optimize more accurate equalizers that should provide better actual responses.

The equalizers were made of quarter-wave transmission lines at 18 GHz and had a four-element topology as shown in Figure 3.26. Figure 3.27 shows the simulated transducer gain, VSWR, and noise figure. The optimization has provided a gain of 24 ± 0.3dB, a noise figure less than 3.6 dB, and input and output VSWR less than 1.9.

3.10.2 *Three-Stage 1.15–1.5 GHz Distributed Amplifier*

This second example was also done for CNET [5, 8]. It consisted of a three-stage amplifier to operate in the 1.15–1.5 GHz band. The multistage amplifier was optimized for gain and noise. The equalizers were synthesized using distributed networks.

The transistor was selected as the FSC-10LG HEMT from Fujitsu. Table 3.3 provides the measured small-signal scattering parameters of the transistor, and Table 3.4 provides the noise parameters of the transistor for few frequency points. Biasing conditions were $V_{ds} = 2\text{V}$ and $I_{ds} = 10\text{mA}$.

Figure 3.28 gives the cascaded commensurate line structure topology using FREEL. The normalized quarter wavelength frequencies were fixed at 1.88, 1.813, 2.23, and 4.8 GHz for the first, second, third, and fourth equalizer, respectively. The simulated amplifier responses using FREEL are given in Figure 3.29.

3.10.3 *Three-Stage 1.15–1.5 GHz Distributed Amplifier (Noncommensurate)*

This third example was again done for CNET [5, 8]. It consisted of the same three-stage amplifier to operate in a 1.15–1.5 GHz band but where the equalizers could be realized using microstrip technology.

The transistor was again selected as the FSC-10LG HEMT from Fujitsu, but the biasing conditions were slightly different leading to new scattering and noise parameters used in the RFT. Table 3.5 provides the measured small-signal scattering parameters of the transistor, and Table 3.6 provides the noise parameters of the transistor for few frequency points. Biasing conditions were $V_{ds} = 2\text{V}$ and $I_{ds} = 30\text{mA}$.

Table 3.3 Measured *S* parameters of the Fujitsu FSC-10LG HEMT

FREQ (GHz)	$S_{11\ Mag}$	$S_{11\ Ang}$	$S_{21\ Mag}$	$S_{21\ Ang}$	$S_{12\ Mag}$	$S_{12\ Ang}$	$S_{22\ Mag}$	$S_{22\ Ang}$
1	0.989	–24.1	2.995	155.8	0.039	70.8	0.652	–18.1
2	0.959	–45.7	2.844	135.1	0.068	56.2	0.632	–33.7

Table 3.4 Measured noise parameters of the Fujitsu FSC-10LG HEMT

FREQ (GHz)	$F_{\min}$	$\Gamma_{opt\ Mag}$	$\Gamma_{opt\ Ang}$	R_n
1	0.33	0.80	16.0	0.50
2	0.35	0.74	31.0	0.45

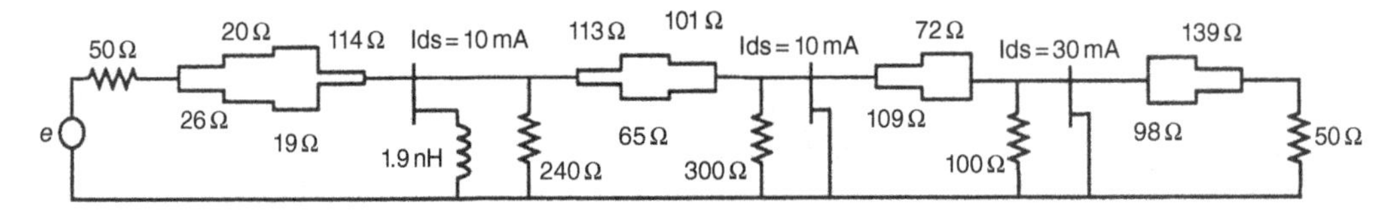

Figure 3.28 Three-stage 1.15–1.5 GHz amplifier using distributed equalizers.

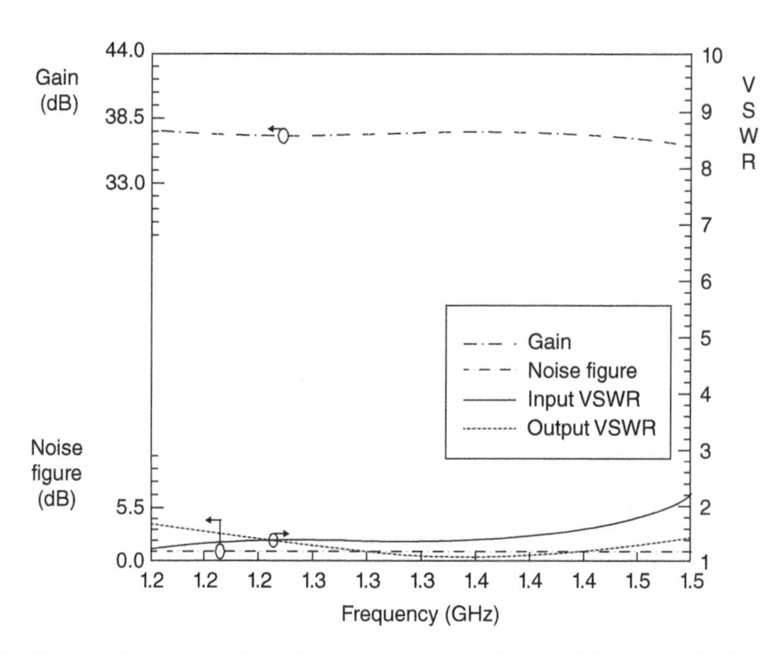

Figure 3.29 Simulated responses of the three-stage 1.15–1.5 GHz amplifier using distributed equalizers.

Table 3.5 Measured S parameters of the Fujitsu FSC-10LG HEMT

FREQ (GHz)	$S_{11\ Mag}$	$S_{11\ Ang}$	$S_{21\ Mag}$	$S_{21\ Ang}$	$S_{12\ Mag}$	$S_{12\ Ang}$	$S_{22\ Mag}$	$S_{22\ Ang}$
1	0.975	−28.1	4.679	154.8	0.026	70.8	0.595	−18.1
2	0.915	−54.7	4.338	130.1	0.047	56.2	0.566	−34.7

Table 3.6 Measured noise parameters of the Fujitsu FSC-10LG HEMT

FREQ (GHz)	$F_{\min}$	$\Gamma_{opt\ Mag}$	$\Gamma_{opt\ Ang}$	R_n
1	0.44	0.68	46.0	0.30
2	0.53	0.63	61.0	0.30

The first equalizer resulted in too low characteristic impedances comparable to those in Figure 3.28. In order to obtain a more compact and easily realizable microstrip circuit, the too low impedance equivalent network technique described in Section 3.8 was used and the results are shown in Figure 3.30. The series transmission lines and shunt stubs are calculated at a center frequency of 1.33 GHz, and the second number next to characteristic impedance value is the electrical length value expressed in degrees. Because these equivalences are narrowband, it was necessary to slightly optimize the circuit using Touchstone® to obtain the same performances as with cascaded UEs. Figure 3.31 gives the final layout of the amplifier. Figure 3.32 shows the measured and simulated gains. The measured gain is quite good at about 30 dB in the passband. The measured VSWR was found to be less than 15 dB. Figure 3.33 shows the measured and simulated noise figures. The measured noise is around–20 dB to about 1.4 GHz and increases to −10 dB at 1.5 GHz.

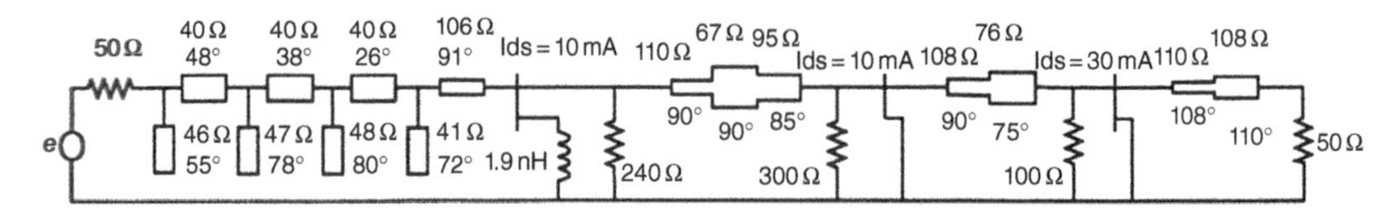

Figure 3.30 Three-stage 1.15–1.5 GHz amplifier using equalizers with noncommensurate distributed elements and open stubs (Touchstone® optimized).

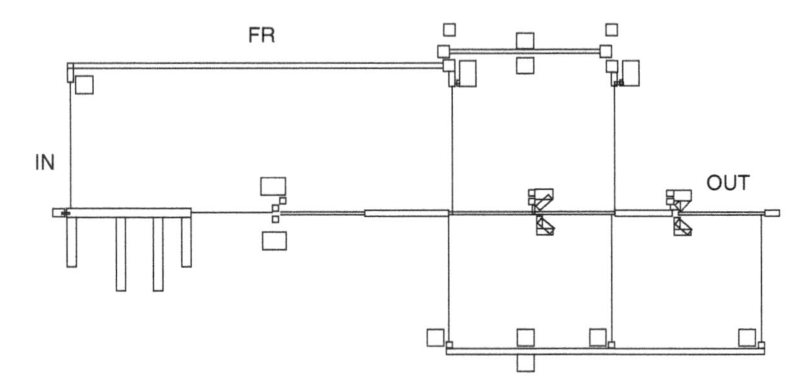

Figure 3.31 Layout of the three-stage 1.15–1.5 GHz amplifier using noncommensurate distributed equalizers.

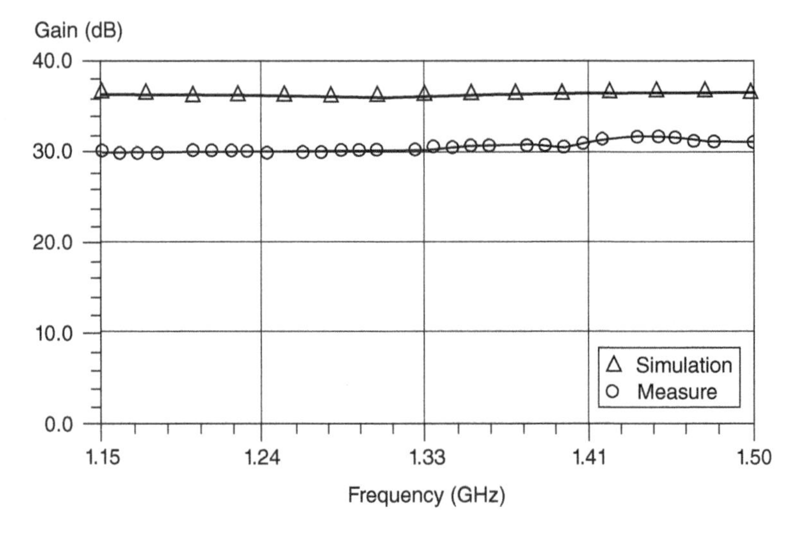

Figure 3.32 Measured and simulated gains for the three-stage 1.15–1.5 GHz amplifier.

3.10.4 Three-Stage 5.925–6.425 GHz Hybrid Amplifier

This example was made for ALCATEL SPACE (THALES SPACE) [6, 9]. It consisted of a three-stage amplifier to operate in the 5.925–6.425 GHz band. The multistage amplifier was optimized for gain and noise. The equalizers were synthesized using distributed networks.

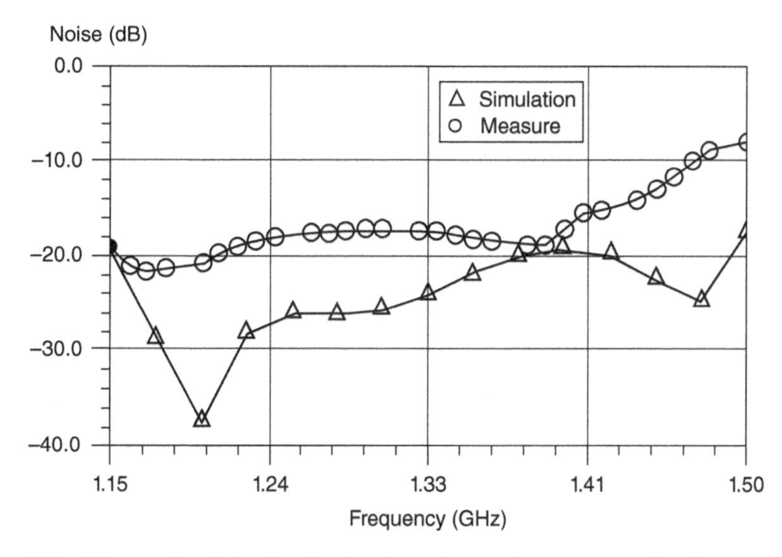

Figure 3.33 Measured and simulated noise figure for the three-stage 1.15–1.5 GHz amplifier.

Table 3.7 Measured S parameters of the NEC NE24283A HJ FET

FREQ (GHz)	$S_{11\ Mag}$	$S_{11\ Ang}$	$S_{21\ Mag}$	$S_{21\ Ang}$	$S_{12\ Mag}$	$S_{12\ Ang}$	$S_{22\ Mag}$	$S_{22\ Ang}$
5	0.829	−86.0	3.912	96.7	0.054	33.9	0.607	−63.5
6	0.785	−100.3	3.646	83.1	0.058	26.5	0.592	−73.8
7	0.751	−113.4	3.375	70.2	0.060	20.1	0.586	−83.3

Table 3.8 Measured noise parameters of the NEC NE24283A HJ FET

FREQ (GHz)	$F_{\min}$	$\Gamma_{opt\ MAG}$	$\Gamma_{opt\ ANG}$	R_n
4	0.32	0.73	57.0	0.26
6	0.37	0.67	83.0	0.20
8	0.43	0.63	105	0.13

The transistor was selected as the ultra-low-noise NE24283A HJ FET from NEC. Table 3.7 provides the measured small-signal scattering parameters of the transistor, and Table 3.8 provides the noise parameters of the transistor for few frequency points. Biasing conditions were $V_{ds} = 2\text{V}$ and $I_{ds} = 10\text{mA}$.

In this example, the bias circuit described in Section 3.7 and Appendix C was included in the scattering parameters representing a transistor during optimization. The equalizers were synthesized as a cascade of UEs as shown in Figure 3.34.

The amplifier was then realized as a microwave integrated circuit (MIC) with discrete capacitors and resistors for the biasing circuit. The MIC itself was enclosed into a brass surrounding box. The layout of the three-stage amplifier is given in Figure 3.35, and the picture of the three-stage amplifier is given in Figure 3.36.

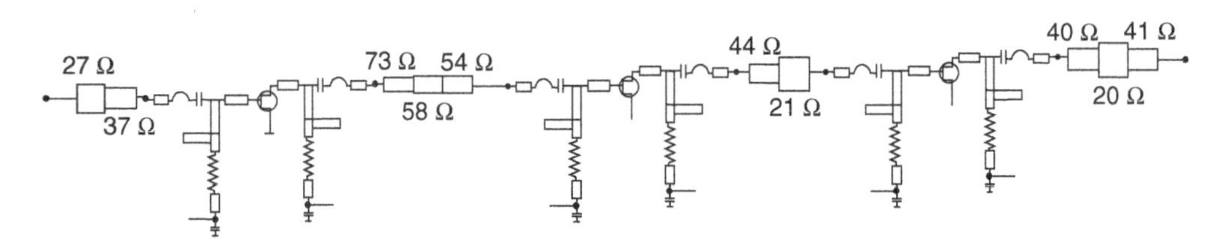

Figure 3.34 Three-stage 5.925–6.425 GHz amplifier using distributed equalizers.

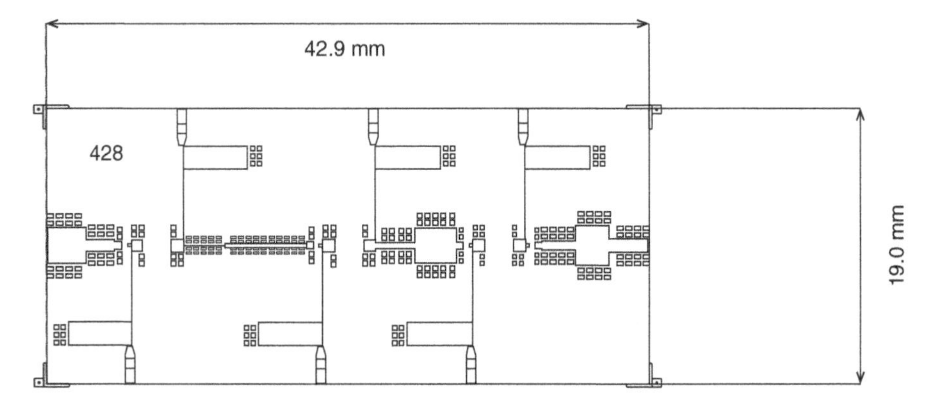

Figure 3.35 Layout of the 5.925–6.425 GHz amplifier.

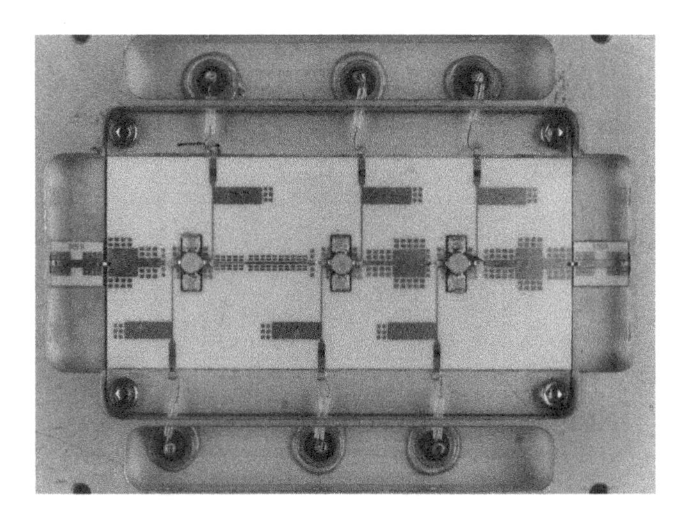

Figure 3.36 Picture of the 5.925–6.425 GHz amplifier.

Figure 3.37 shows the measured gain of the three-stage amplifier over the band 5.925–6.425 GHz.

Table 3.9 summarizes the measured and simulated quantities of this amplifier.

In conclusion, the RFT is also adapted to the design of distributed microwave amplifiers. Using this synthesis theory, we gave the realizations of two three-stage hybrid narrow band amplifiers, one at 1.15 MHz to 1.5 GHz and one at 5.925–6.425 GHz. The theory applies also in the case of

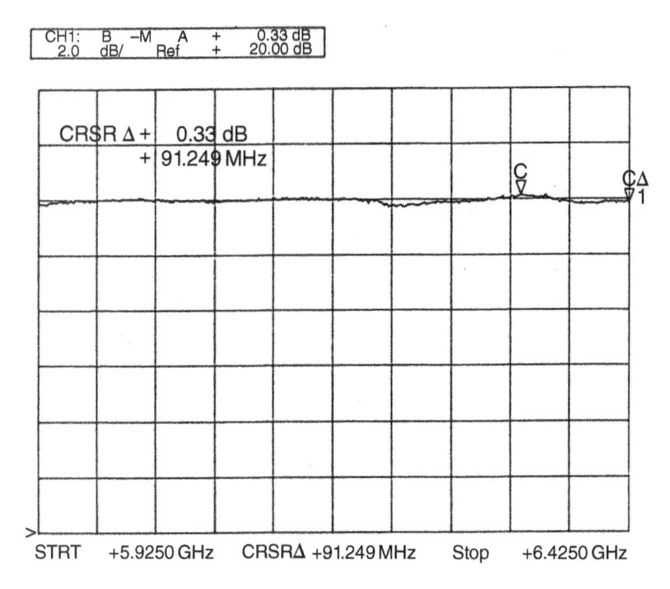

Figure 3.37 Measured gain of the 5.925–6.425 GHz amplifier.

Table 3.9 Measured and simulated characteristics of the three-stage 5.925–6.425 GHz amplifier

	Simulations	Measures
Frequency band	5.925–6.425	5.925–6.425
Gain	34.2	32
Ripple	±0.20 dB	±0.16 dB
Noise	≤0.85 dB	≤1.50 dB
Input VSWR	<–10 dB	<–15 dB
Output VSWR	<–15 dB	<–12 dB

broadband amplifier with an example at 2–8 GHz. The results obtained confirm the value of the synthesis approach used and a computer program FR was created. The amplifiers were made for ALACATEL SPACE and CNET factories with commercial transistors.

References

[1] H. J. Carlin, "Distributed circuit design with transmission elements," Proc. IEEE, vol. 59, pp. 1059–1081, July 1971.

[2] H. J. Carlin, J. J. Komiak, "A new method of broad-band equalization applied to microwave amplifiers," IEEE Trans. Microw. Theory Tech., vol. 27, no. 2, pp. 93–99, December 1982.

[3] B. S. Yarman, H. J. Carlin, "A simplified real frequency technique applied to broadband multistage microwave amplifiers," IEEE Trans. Microw. Theory Tech., vol. 30, pp. 2216–2222, December 1982.

[4] R. Soares, A. Perennec, A. O. Ngongo, P. Jarry, "Application of a simplified real-frequency synthesis method to distributed-element amplifier design," Int. J. Microw. Millimeter-Wave Comput. Aided Eng., Vol. 1, No. 4, pp. 365–378, January 1991.

[5] A. N. Olomo, Synthesis and Realisations of Low Noise and Transimpedance Microwave Amplifiers by the Real Frequency Method, Ph.D. dissertation, Brest University, October 1992.

[6] E. Kerherve, Realisations of distributed low noise microwave amplifiers using the real frequency method, Ph.D. dissertation, Bordeaux University, September 1994.
[7] P. I. Richards, "Resistor transmission line circuits," Proc. IRE, vol. 36, pp. 217–220, February 1948.
[8] A. Perennec, A. N. Olomo, P. Jarry, R. Soares, "Optimization of gain, VSWR and noise of the broadband multistage microwave MMIC amplifier by the real frequency method. Synthesis in lumped and distributed elements," IEEE MTT-S Symposium Digest, DALLAS, TX, pp. 363–366, May 8–10 1990.
[9] E. Kerherve, P. Jarry, P. M. Martin, "Efficient numerical CAD technique for RF and microwave amplifiers," Int. J. Microw. Millimeter-Wave Comput. Aided Eng., http://onlinelibrary.wiley.com/journal/10.1002/%28ISSN%291099-047X.

4

Multistage Transimpedance Amplifiers

4.1 Introduction

In very high data rate fiber-optic systems such as in an optical receiver, it is common place to have a frequency band of interest spanning three decades [1]. The source is no longer a traditional voltage source with series source impedance but rather a current source with high input impedance and a parallel capacitor. This chapter proposes a technique to optimize the transimpedance gain of a multistage amplifier made of transistors and equalizers using the real frequency technique (RFT).

Several circuits employing inductor and capacitor peaking have been reported to increase bandwidth [2, 3]. In this chapter, we will see that inductor peaking can also be used to adjust the input noise of a transimpedance amplifier.

The techniques in this chapter have been implemented in a program called TRANS.

Microwave Amplifier and Active Circuit Design Using the Real Frequency Technique, First Edition.
Pierre Jarry and Jacques N. Beneat.

4.2 Multistage Transimpedance Amplifier Representation

The multistage transimpedance amplifier is represented in Figure 4.1 [4]. It assumes that there are k transistors. The transistor can represent a transistor alone or a transistor with added but known circuitry as long as the S parameters can be provided. An equalizer is associated with each transistor. There is one final equalizer placed after the kth transistor.

In the case of a transimpedance amplifier, the source is no longer represented by a voltage source with series impedance. In this chapter, we are interested in transimpedance amplifiers for optical applications where a photodiode can be modeled by a current source with a parallel capacitor as shown in Figure 4.1.

The equalizer of the kth stage is characterized by the scattering matrix

$$S_{E_k} = \begin{pmatrix} e_{11,k}(s) & e_{12,k}(s) \\ e_{21,k}(s) & e_{22,k}(s) \end{pmatrix} \tag{4.1}$$

The transistor circuit of the kth stage is characterized by the scattering matrix

$$S_{T_k} = \begin{pmatrix} S_{11,k}(s) & S_{12,k}(s) \\ S_{21,k}(s) & S_{22,k}(s) \end{pmatrix} \tag{4.2}$$

In this chapter, the equalizer is assumed to be a lossless and reciprocal lumped two-port so that its scattering matrix can be expressed as

$$S_E = \begin{pmatrix} e_{11}(s) = \dfrac{h(s)}{g(s)} & e_{12}(s) = \dfrac{f(s)}{g(s)} \\ e_{21}(s) = \dfrac{f(s)}{g(s)} & e_{22}(s) = \dfrac{-\varepsilon h(-s)}{g(s)} \end{pmatrix} \tag{4.3}$$

where $f(s)$ is an even or odd function. If $f(s)$ is even then $\varepsilon = 1$, and if $f(s)$ is odd then $\varepsilon = -1$. In addition, the lossless condition $|e_{11}|^2 + |e_{12}|^2 = 1$ leads to

$$g(s)g(-s) = h(s)h(-s) + f(s)f(-s) \tag{4.4}$$

$h(s)$ and $g(s)$ are polynomials of order n with real-valued coefficients:

$$h(s) = h_0 + h_1 s + \cdots + h_{n-1}s^{n-1} + h_n s^n$$

$$g(s) = g_0 + g_1 s + \cdots + g_{n-1}s^{n-1} + g_n s^n$$

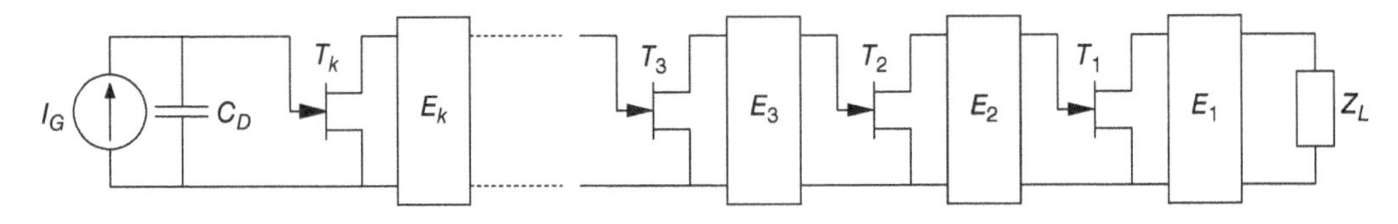

Figure 4.1 Multistage transimpedance amplifier.

and in this chapter $f(s)$ is taken as

$$f(s) = \pm s^k \quad \text{and} \quad \varepsilon = (-1)^k \tag{4.5}$$

where k represents the number of transmission zeros at $\omega = 0$ and with $k \le n$ leading to

$$g(s)g(-s) = h(s)h(-s) + (-1)^k s^{2k} \tag{4.6}$$

For a given equalizer, the problem consists of defining the unknown polynomial coefficients h_0, $h_1, h_2, \ldots, h_n$ that optimize the response of the system. Once $h(s)$ is known, $g(s)g(-s)$ can be computed using

$$G(s^2) = g(s)g(-s) = G_0 + G_1 s^2 + \cdots + G_n s^{2n}$$

where the polynomial coefficients of $G(s^2)$ are given by

$$\begin{aligned}
G_0 &= h_0^2 \\
G_1 &= -h_1^2 + 2h_2h_0 \\
&\vdots \\
G_i &= (-1)^i h_i^2 + 2h_{2i}h_0 + 2\sum_{j=2}^{i}(-1)^{j-1}h_{j-1}h_{2j-i+1} \\
&\vdots \\
G_k &= G_i|_{i=k} + (-1)^k \\
&\vdots \\
G_n &= (-1)^n h_n^2
\end{aligned} \tag{4.7}$$

The left-half roots of $g(s)g(-s)$ are retained to form a Hurwitz polynomial $g(s)$. At this point, the equalizer scattering parameters $e_{11}(s)$, $e_{12}(s)$, $e_{21}(s)$, and $e_{22}(s)$ can be computed. During optimization, they will be needed for computing the error function to be optimized, and after optimization, for example, $e_{11}(s)$ can be used to realize the equalizer.

Note that in this chapter, we will be using a technique based on the admittance matrix Y to optimize the transimpedance gain and voltage standing wave ratio (VSWR).

The admittance matrix Y_e of a lossless and reciprocal two-port such as the equalizer considered in this chapter is given in terms of the polynomials $f(s)$, $g(s)$, and $h(s)$ and their even and odd parts [4]:

$$Y_e = \begin{pmatrix} Y_{e11} & Y_{e12} \\ Y_{e21} & Y_{e22} \end{pmatrix} = \frac{1}{h_o(s) + g_o(s)} \begin{pmatrix} g_e(s) - h_e(s) & -f(s) \\ -f(s) & g_e(s) + h_e(s) \end{pmatrix} \tag{4.8}$$

where $h_e(s)$ and $g_e(s)$ are the even parts, and $h_o(s)$ and $g_o(s)$ are the odd parts of $h(s)$ and $g(s)$, respectively, and

$$\begin{aligned} h(s) &= h_e(s) + h_o(s) \\ g(s) &= g_e(s) + g_o(s) \end{aligned} \tag{4.9}$$

In the case of polynomials, $h_e(s)$ is simply made of the coefficients of $h(s)$ corresponding to the even powers of s, and $h_o(s)$ is made of the coefficients corresponding to the odd power of s. The same applies to $g_e(s)$, $g_o(s)$, and $g(s)$.

Therefore, computing the scattering parameters $e'_{ij}s$ or the admittance matrix coefficients $Y'_{eij}s$ of an equalizer is immediate as soon as $h(s)$ and $g(s)$ have been computed since $f(s) = \pm s^k$ choice is done prior to the optimization process.

For the transistors, one needs to convert the scattering parameters to admittance matrix parameters. The admittance matrix coefficients Y_{ij} are obtained using the following conversion formulas when the scattering parameters S_{ij} are provided for characteristic impedance $Z_0 = 1/Y_0$:

$$Y = \begin{pmatrix} Y_{11} & Y_{12} \\ Y_{21} & Y_{22} \end{pmatrix} = \begin{pmatrix} Y_0 \dfrac{(1-S_{11})(1+S_{22})+S_{12}S_{21}}{(1+S_{11})(1+S_{22})-S_{12}S_{21}} & Y_0 \dfrac{-2S_{12}}{(1+S_{11})(1+S_{22})-S_{12}S_{21}} \\ Y_0 \dfrac{-2S_{21}}{(1+S_{11})(1+S_{22})-S_{12}S_{21}} & Y_0 \dfrac{(1+S_{11})(1-S_{22})+S_{12}S_{21}}{(1+S_{11})(1+S_{22})-S_{12}S_{21}} \end{pmatrix} \tag{4.10}$$

Note that the transistor admittance parameters are computed at the start of the optimization process and do not change afterwards.

4.3 Extension to Distributed Equalizers

The equalizer of the kth stage is characterized by the scattering matrix

$$S_{E_k} = \begin{pmatrix} e_{11,k}(t) & e_{12,k}(t) \\ e_{21,k}(t) & e_{22,k}(t) \end{pmatrix} \tag{4.11}$$

The transistor circuit of the kth stage is characterized by the scattering matrix

$$S_{T_k} = \begin{pmatrix} S_{11,k}(t) & S_{12,k}(t) \\ S_{21,k}(t) & S_{22,k}(t) \end{pmatrix} \tag{4.12}$$

where $t = j\Omega = j\tan\tau_0\omega$ is the Richard's variable defined in Chapter 1.

The equalizer is assumed to be a lossless distributed two-port with a scattering matrix such that

$$S_E = \begin{pmatrix} e_{11}(t) = \dfrac{h(t)}{g(t)} & e_{12}(t) = \dfrac{f(t)}{g(t)} \\ e_{21}(t) = \dfrac{f(t)}{g(t)} & e_{22}(t) = \dfrac{-h(-t)}{g(t)} \end{pmatrix} \tag{4.13}$$

with

$$f(t) = t^r\left(1-t^2\right)^{\frac{n-q}{2}} \tag{4.14}$$

where

q represents the number of possible low-pass stubs
r represents the number of possible high-pass stubs

The lossless condition $|e_{11}|^2 + |e_{12}|^2 = 1$ leads to

$$g(t)g(-t) = f(t)f(-t) + h(t)h(-t) \quad (4.15)$$

so that

$$g(t)g(-t) = h(t)h(-t) + t^{2r}\left(1-t^2\right)^{n-q} \quad (4.16)$$

For a given equalizer, the problem consists of defining the unknown polynomial coefficients h_0, $h_1, h_2, \ldots, h_n$ that optimize the response of the system. Once $h(t)$ is known, $g(t)g(-t)$ can be computed using

$$G(t^2) = g(t)g(-t) = G_0 + G_1 t^2 + \cdots + G_n t^{2n}$$

where

$$\begin{cases} G_0 = h_0^2 + 1 \\ G_1 = -h_1^2 + 2h_2 h_0 - C_{n-q}^1 \\ \ldots \\ G_i = (-1)^i h_i^2 + 2h_{2i} h_0 + 2\sum_{j=2}^{i} (-1)^{j-1} h_{j-1} h_{2i-j+1} + (-1)^i C_{n-q}^i \\ \ldots \\ G_{n-q} = (-1)^{n-q} h_{n-q}^2 + 2h_{2(n-q)} h_0 + 2\sum_{j=2}^{n-q} (-1)^{j-1} h_{j-1} h_{2(n-q)-j+1} + (-1)^{n-q} \\ \ldots \\ G_n = (-1)^n h_n^2 \end{cases} \quad (4.17)$$

with

$$C_n^p = \frac{n!}{p!(n-p)!}$$

The left-half roots of $g(t)g(-t)$ are retained to form a Hurwitz polynomial $g(t)$. At this point, the equalizer scattering parameters $e_{11}(t)$, $e_{12}(t)$, $e_{21}(t)$, and $e_{22}(t)$ can be computed. During optimization, they will be needed for computing the error function to be optimized, and after optimization, for example, $e_{11}(t)$ can be used to realize the equalizer using the synthesis procedures given in Section 3.9.

The admittance matrix Y_e of the lossless and reciprocal distributed equalizer considered in this chapter is given in terms of the polynomials $f(t)$, $g(t)$, and $h(t)$ and their even and odd parts [4]:

$$Y_e = \begin{pmatrix} Y_{e11} & Y_{e12} \\ Y_{e21} & Y_{e22} \end{pmatrix} = \frac{1}{h_o(t)+g_o(t)} \begin{pmatrix} g_e(t)-h_e(t) & -f(t) \\ -f(t) & g_e(t)+h_e(t) \end{pmatrix} \quad (4.18)$$

where $h_e(t)$ and $g_e(t)$ are the even parts, and $h_o(t)$ and $g_o(t)$ are the odd parts of $h(t)$ and $g(t)$, respectively, and

$$\begin{aligned} h(t) &= h_e(t)+h_o(t) \\ g(t) &= g_e(t)+g_o(t) \end{aligned} \quad (4.19)$$

In the case of polynomials, $h_e(t)$ is simply made of the coefficients of $h(t)$ corresponding to the even powers of t, and $h_o(t)$ is made of the coefficients corresponding to the odd power of t. The same applies to $g_e(t)$, $g_o(t)$, and $g(t)$.

The transistor scattering parameters defined at actual frequencies ω_j's are mapped to the scaled frequencies Ω_i's and then converted to admittance matrix coefficients Y_{ij} using (4.10).

4.4 Multistage Transimpedance Gain

First, we recall the properties of the admittance matrix for a typical two-port network as shown in Figure 4.2.

The admittance matrix is defined in terms of the voltages V_1, V_2 and currents I_1, I_2 such that

$$\begin{pmatrix} I_1 \\ I_2 \end{pmatrix} = \begin{pmatrix} Y_{11} & Y_{12} \\ Y_{21} & Y_{22} \end{pmatrix} \begin{pmatrix} V_1 \\ V_2 \end{pmatrix} \quad (4.20)$$

Then, when the two-port is closed on a load impedance Z_L, we have $V_2 = -Z_L I_2$ and

$$\begin{aligned} I_1 &= Y_{11}V_1 + Y_{12}V_2 = Y_{11}V_1 - Z_L Y_{12} I_2 \\ I_2 &= Y_{21}V_1 + Y_{22}V_2 = Y_{21}V_1 - Z_L Y_{22} I_2 \end{aligned}$$

and from the bottom equation

$$I_2 = \frac{Y_{21}V_1}{1+Z_L Y_{22}}$$

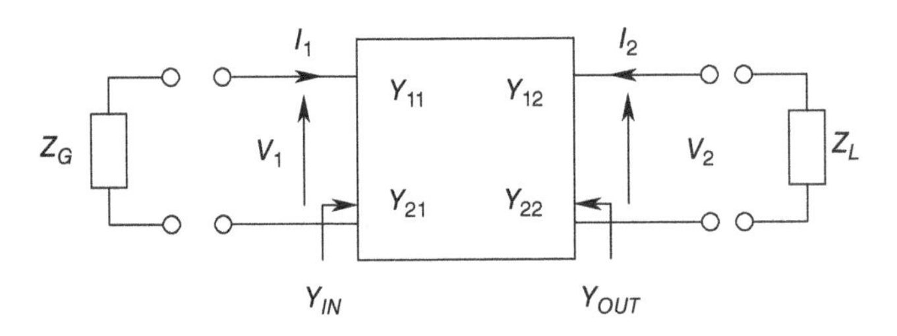

Figure 4.2 Definitions and properties of the admittance matrix.

from which the top equation becomes

$$I_1 = Y_{11}V_1 - Z_L Y_{12}\frac{Y_{21}V_1}{1 + Z_L Y_{22}}$$

Then, the input admittance Y_{IN} when the output of the two-port is connected to load impedance Z_L is given by

$$Y_{IN} = \frac{I_1}{V_1} = Y_{11} - Z_L\frac{Y_{12}Y_{21}}{1 + Z_L Y_{22}}$$

$$Y_{IN} = Y_{11} - \frac{Y_{12}Y_{21}}{Y_{22} + \frac{1}{Z_L}} \tag{4.21}$$

Using a similar approach, one finds that the output admittance Y_{OUT} when the input of the two-port is connected to source impedance Z_G is given by

$$Y_{OUT} = \frac{I_2}{V_2} = Y_{22} - \frac{Y_{12}Y_{21}}{Y_{11} + \frac{1}{Z_G}} \tag{4.22}$$

Note that dividing the first row of (4.20) by V_1 gives

$$\frac{I_1}{V_1} = Y_{11} + Y_{12}\frac{V_2}{V_1}$$

and we define the voltage gain G_v as

$$G_v = \frac{V_2}{V_1} = \frac{1}{Y_{12}}\left(\frac{I_1}{V_1} - Y_{11}\right) = \frac{1}{Y_{12}}\left(Y_{11} - \frac{Y_{12}Y_{21}}{Y_{22} + \frac{1}{Z_L}} - Y_{11}\right) = -\frac{Y_{21}}{Y_{22} + \frac{1}{Z_L}}$$

$$G_v = -\frac{Y_{21}}{Y_{22} + \frac{1}{Z_L}} \tag{4.23}$$

and we define the transimpedance gain Z_t as [4]

$$Z_t = \frac{|G_v|}{|Y_{IN}|} = \frac{|Y_{21}|}{\left|Y_{22} + \frac{1}{Z_L}\right|}\frac{1}{|Y_{IN}|} \tag{4.24}$$

To define the multistage transimpedance gain, we start with the transistor stage terminated on load Z_L as shown in Figure 4.3.

The voltage gain of the first stage is given by the product of the voltage gain of the transistor stage and the voltage gain of the equalizer stage:

$$G_{v,1} = \frac{Y_{21,1}}{Y_{22,1} + Y_{eIN,1}}\frac{Y_{e21,1}}{Y_{e22,1} + \frac{1}{Z_L}} \tag{4.25}$$

The transimpedance gain of this first stage is then given by

$$Z_{t,1} = \left|\frac{G_{v,1}}{Y_{IN,1}}\right| = \left|\frac{Y_{21,1}}{Y_{22,1}+Y_{eIN,1}}\frac{Y_{e21,1}}{Y_{e22,1}+\dfrac{1}{Z_L}}\right|\left|\frac{1}{Y_{IN,1}}\right| \tag{4.26}$$

where $Y_{eIN,1}$ is the input admittance of equalizer 1 loaded on impedance Z_L as seen in Figure 4.3 and given by

$$Y_{eIN,1} = Y_{e11,1} - \frac{Y_{e12,1}Y_{e21,1}}{Y_{e22,1}+\dfrac{1}{Z_L}} \tag{4.27}$$

and where $Y_{IN,1}$ is the input admittance of transistor 1 on impedance $1/Y_{eIN,1}$ as seen in Figure 4.3 and given by

$$Y_{IN,1} = Y_{11,1} - \frac{Y_{12,1}Y_{21,1}}{Y_{22,1}+Y_{eIN,1}} \tag{4.28}$$

and where $Y'_{eij,1}s$ are the admittance matrix coefficients of equalizer 1 and $Y'_{ij,1}s$ are the admittance matrix coefficients of transistor 1.

Figure 4.4 shows the generalization of the transimpedance gain to k transistor–equalizer stages. In this case, the gain after k transistor–equalizer stages is given by

$$G_{v,k} = G_{v,k-1}\frac{Y_{21,k}}{Y_{22,k}+Y_{eIN,k}}\frac{Y_{e21,k}}{Y_{e22,k}+Y_{IN,k-1}} \tag{4.29}$$

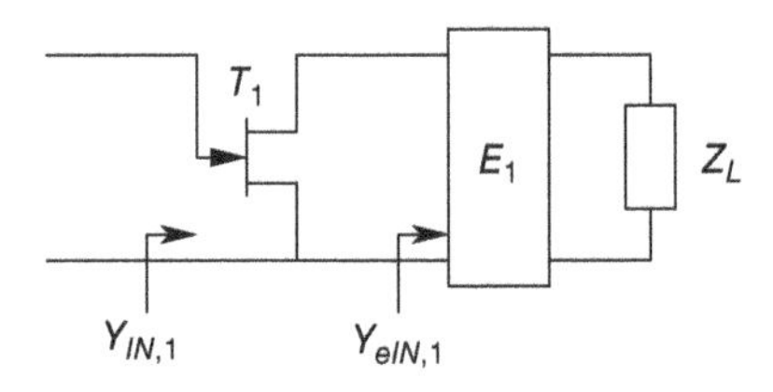

Figure 4.3 First transistor–equalizer stage loaded by Z_L.

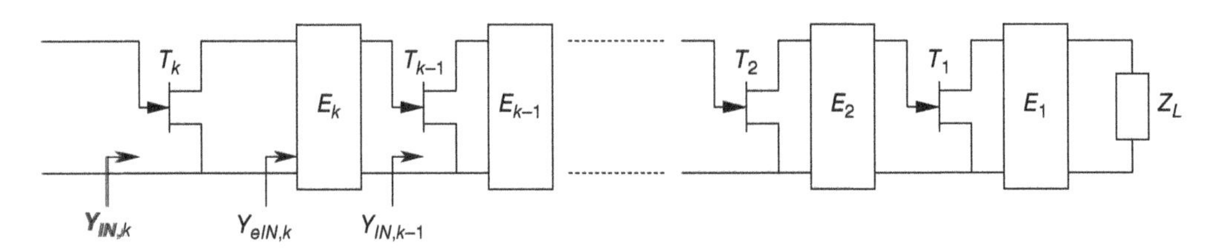

Figure 4.4 Transimpedance gain of k transistor–equalizer stages.

where $Y_{eIN,k}$ is the input admittance of equalizer k loaded on impedance $1/Y_{IN,k-1}$ as seen in Figure 4.4 and given by

$$Y_{eIN,k} = Y_{e11,k} - \frac{Y_{e12,k}Y_{e21,k}}{Y_{e22,k} + Y_{IN,k-1}} \tag{4.30}$$

and where $Y_{IN,k}$ is the input admittance of transistor k on impedance $1/Y_{eIN,k}$ as seen in Figure 4.4 and given by

$$Y_{IN,k} = Y_{11,k} - \frac{Y_{12,k}Y_{21,k}}{Y_{22,k} + Y_{eIN,k}} \tag{4.31}$$

where

$Y'_{eij,k}s$ are the admittance matrix coefficients of equalizer k
$Y'_{ij,k}s$ are the admittance matrix coefficients of transistor k

and the transimpedance gain after k transistor–equalizer stages is given by

$$Z_{t,k} = \left|\frac{G_{v,k}}{Y_{IN,k}}\right| \tag{4.32}$$

4.5 Multistage VSWR

The output VSWR when the input is terminated on some impedance Z_G as shown in Figure 4.5 is computed recursively from the source [4]

The output VSWR is given by

$$\text{VSWR}_{OUT} = \frac{1+|\Gamma_{OUT}|}{1-|\Gamma_{OUT}|} \tag{4.33}$$

where

$$\Gamma_{OUT} = \frac{Y_0 - Y_{eOUT,1}}{Y_0 + Y_{eOUT,1}} \tag{4.34}$$

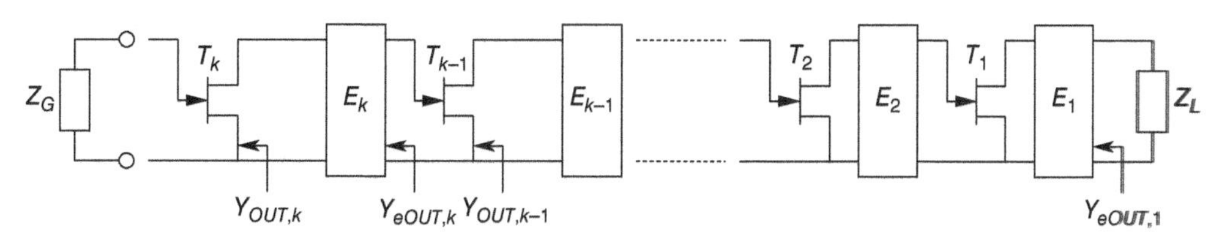

Figure 4.5 VSWR for k transistor–equalizer stages.

with

$$
\begin{aligned}
Y_{OUT,k} &= Y_{22,k} - \frac{Y_{12,k}Y_{21,k}}{Y_{11,k} + \dfrac{1}{Z_G}} \\
Y_{eOUT,k} &= Y_{e22,k} - \frac{Y_{e12,k}Y_{e21,k}}{Y_{e11,k} + Y_{OUT,k}} \\
Y_{OUT,k-1} &= Y_{22,k-1} - \frac{Y_{12,k-1}Y_{21,k-1}}{Y_{11,k-1} + Y_{eOUT,k}} \\
&\vdots \\
Y_{OUT,1} &= Y_{22,1} - \frac{Y_{12,1}Y_{21,1}}{Y_{11,1} + Y_{eOUT,2}} \\
Y_{eOUT,1} &= Y_{e22,1} - \frac{Y_{e12,1}Y_{e21,1}}{Y_{e11,1} + Y_{OUT,1}}
\end{aligned}
\tag{4.35}
$$

4.6 Optimization Process

The optimization algorithm is used to define polynomial coefficients $h_0, h_1, \ldots, h_n$ of $h(t)$ of an equalizer that minimize a least mean square error function that includes desired target functions with different weight factors.

For the transimpedance amplifiers of this chapter, the error function consists mainly of the transimpedance gain:

$$E = \sum_{i=1}^{m} \left[W_1 \left(\frac{Z_{t,k}(\omega_i, \boldsymbol{H})}{Z_{t0,k}} - 1 \right)^2 \right] \tag{4.36}$$

where ω_i are the frequencies used for optimization.

$Z_{t,k}(\omega_i, \boldsymbol{H})$ is the transimpedance gain for k equalizer–transistor stages given by the equations in Section **4.4** and is used to optimize the kth equalizer. The transimpedance gain depends on frequency through the scattering parameters $e_{ij}(j\omega)$ of the equalizer and the scattering parameters $S_{ij}(j\omega)$ of the transistor. The target transimpedance gain $Z_{t0,k}$ for optimizing the kth equalizer is taken as a constant in the frequency band being optimized. The expressions for the target transimpedance gain are provided in Section **4.7**.

The vector $\boldsymbol{H}$ includes the polynomial coefficients $h_0, h_1, \ldots, h_n$ and is used as the $\boldsymbol{X}$ vector in the Levenberg–More optimization algorithms of Chapter 2 and Appendix B.

The main computation in the optimization routine is to define the step $\Delta\boldsymbol{H}$ to add to the elements of vector $\boldsymbol{H}$ at the next iteration given by

$$\Delta\boldsymbol{H} = -\left[\boldsymbol{J}^T\boldsymbol{J} + \lambda\boldsymbol{D}^T\boldsymbol{D}\right]^{-1}\boldsymbol{J}^T e(\boldsymbol{H}) \tag{4.37}$$

where

$e(\boldsymbol{H})$ is the error vector
J is the Jacobian matrix of the error with elements $\partial e_j / \partial h_i$

D is a diagonal matrix
λ is the Levenberg–Marquardt–More parameter described in Appendix B

4.7 Design Procedures

The design procedure starts with the design of equalizer 1 next to the load Z_L. We are interested in using a target transimpedance gain that is constant. However, rather than using some arbitrary value, this section provides a technique that gives a target transimpedance gain that depends on the transistor scattering parameters and gain of the already optimized stages.

We start by defining the transimpedance gain of the first transistor and load without any equalizer as shown in Figure 4.6.

Using the results of Section **4.4**, the transimpedance gain of transistor 1 loaded on Z_L with no equalizer 1 is given by

$$Z_{T\ t,1} = \left|\frac{Y_{21,1}}{Y_{22,1} + \frac{1}{Z_L}}\right| \left|\frac{1}{Y_{T\ IN,1}}\right|$$

where $Y_{T\ IN,1}$ is the input admittance of transistor 1 when loaded on impedance Z_L:

$$Y_{T\ IN,1} = Y_{11,1} - \frac{Y_{12,1} Y_{21,1}}{Y_{22,1} + \frac{1}{Z_L}} \tag{4.38}$$

where $Y_{ij,1}$ are the admittance matrix coefficients of transistor 1.

Since the admittance matrix coefficients of transistor 1 vary with frequency, the target constant value for the transimpedance gain is taken as [4]

$$Z_{t0,1} = \max\{Z_{T\ t,1}(j\omega)\}$$

To optimize the first equalizer, one uses Figure 4.7.

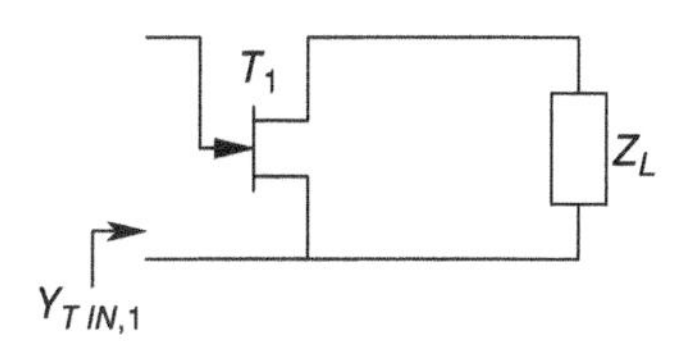

Figure 4.6 Defining the target transimpedance gain for the first stage.

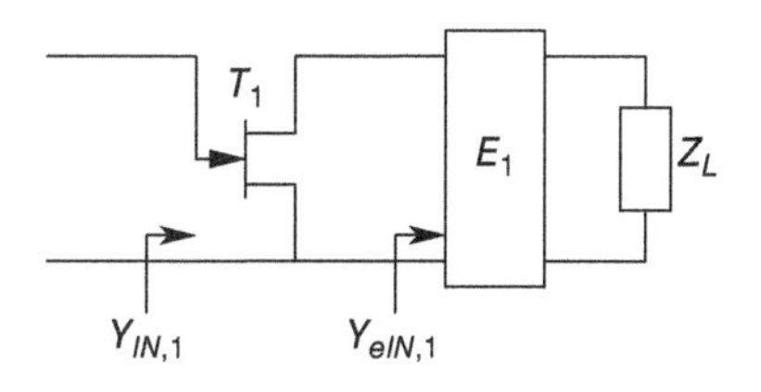

Figure 4.7 Optimizing the first equalizer.

For the first equalizer, the error function to optimize is given by

$$E=\sum_{i=1}^{m}\left(\frac{Z_{t,1}(\omega_i,\boldsymbol{H})}{Z_{T\ t0,1}}-1\right)^2 \tag{4.39}$$

where

$$Z_{t,1}=\left|\frac{G_{v,1}}{Y_{IN,1}}\right| \tag{4.40}$$

$$G_{v,1}=\frac{Y_{21,1}}{Y_{22,1}+Y_{eIN,1}}\frac{Y_{e21,1}}{Y_{e22,1}+\frac{1}{Z_L}} \tag{4.41}$$

$$Z_{t0,1}=\max\{Z_{T\ t,1}(j\omega)\} \tag{4.42}$$

The process is repeated for the second equalizer. First, transistor 2 is added without equalizer 2 to define the target transimpedance gain for the second stage as shown in Figure 4.8. Note that at this point equalizer 1 is totally defined since it was optimized in the aforementioned step, and the admittance matrix coefficients for all transistors are assumed to have been computed prior to the start of the optimization process.

The transimpedance gain of transistor 2 loaded on impedance $1/Y_{IN,1}$ and first transistor–equalizer stage with no equalizer 2 is given by

$$Z_{T\ t,2}=|G_{v,1}|\left|\frac{Y_{21,2}}{Y_{22,2}+Y_{IN,1}}\right|\left|\frac{1}{Y_{T\ IN,2}}\right|$$

where $G_{v,1}$ is the voltage gain of the first transistor–equalizer stage and where $Y_{T\ IN,2}$ is the input admittance of transistor 2 when loaded on impedance $1/Y_{IN,1}$:

$$Y_{T\ IN,2}=Y_{11,2}-\frac{Y_{12,2}Y_{21,2}}{Y_{22,2}+Y_{IN,1}} \tag{4.43}$$

with $Y_{ij,2}$ the admittance matrix coefficients of transistor 2.

Since the admittance matrix coefficients of transistor 2 vary with frequency, the target constant value for the transimpedance gain is taken as

$$Z_{t0,2}=\max\{Z_{T\ t,2}(j\omega)\}$$

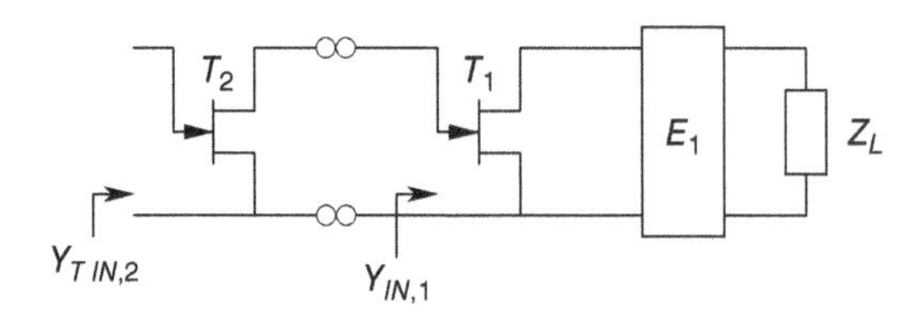

Figure 4.8 Defining the target transimpedance gain for the second stage.

To optimize the second equalizer, one uses Figure 4.9.

For the second equalizer, the error function to optimize is given by

$$E = \sum_{i=1}^{m} \left(\frac{Z_{t,2}(\omega_i, \boldsymbol{H})}{Z_{T\ t0,2}} - 1 \right)^2 \tag{4.44}$$

where

$$Z_{t,2} = \left| \frac{G_{v,2}}{Y_{IN,2}} \right| \tag{4.45}$$

$$G_{v,2} = G_{v,1} \frac{Y_{21,2}}{Y_{22,2} + Y_{eIN,2}} \frac{Y_{e21,2}}{Y_{e22,2} + Y_{IN,1}} \tag{4.46}$$

$$Z_{t0,2} = \max\{Z_{T\ t,2}(j\omega)\} \tag{4.47}$$

The process is repeated for the other stages until we reach the last transistor. First, transistor k and the photodiode are added without equalizer k to define the target transimpedance gain for the last stage as shown in Figure 4.10.

In this case, the photodiode modeled here as a parallel capacitor C_D and the transistor k are thought of as a transistor block. One computes the admittance matrix coefficients $Y'_{ij,k}$ of this transistor block using the model of the photo diode and measured scattering parameters of transistor k. This can be done, for example, using cascade, series, and parallel combinations of $ABCD$ matrices as was the case in Section 3.7 to define the scattering parameters of a transistor block made of the transistor with its biasing circuit.

The transimpedance gain of transistor block k loaded on impedance $1/Y_{IN,k-1}$ and the $k-1$ transistor–equalizer stages is given by

$$Z_{T\ t,k} = |G_{v,k-1}| \left| \frac{Y'_{21,k}}{Y'_{22,k} + Y_{IN,k-1}} \right| \left| \frac{1}{Y_{T\ IN,k}} \right|$$

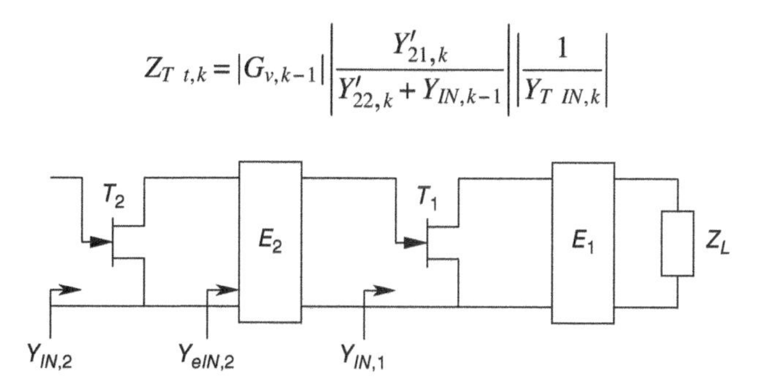

Figure 4.9 Optimizing the second equalizer.

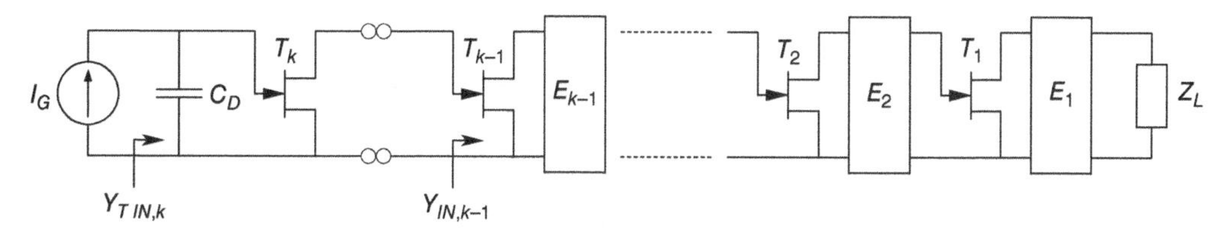

Figure 4.10 Defining the target transimpedance gain for the last stage k.

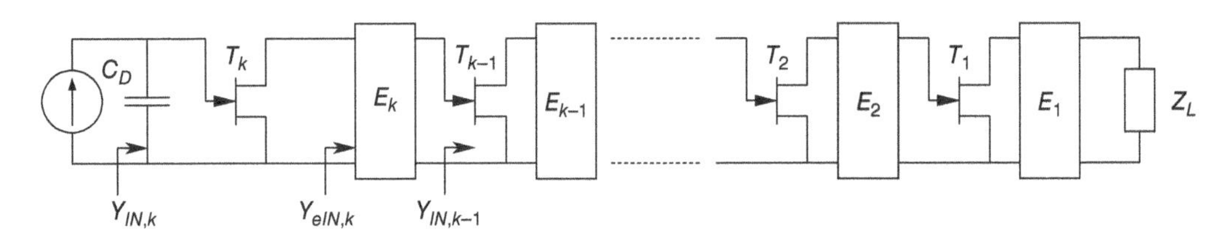

Figure 4.11 Optimizing the last equalizer k.

where $G_{v,k-1}$ is the voltage gain of the $k-1$ transistor–equalizer stages and where $Y_{T\ IN,k}$ is the input admittance of transistor block k when loaded on impedance $1/Y_{IN,k-1}$:

$$Y_{T\ IN,k} = Y'_{11,k} - \frac{Y'_{12,k}Y'_{21,k}}{Y'_{22,k} + Y_{IN,k-1}} \tag{4.48}$$

where $Y'_{ij,k}$ are the admittance matrix coefficients of transistor block k made of the photodiode and transistor k.

Since the admittance matrix coefficients of transistor k vary with frequency, the target constant value for the transimpedance gain is taken as

$$Z_{t0,k} = \max\{Z_{T\ t,k}(j\omega)\}$$

To optimize the last equalizer k, one uses Figure 4.11.

For the last equalizer k, the error function to optimize is given by

$$E = \sum_{i=1}^{m}\left(\frac{Z_{t,k}(\omega_i,\boldsymbol{H})}{Z_{T\ t0,k}} - 1\right)^2 \tag{4.49}$$

$$Z_{t,k} = \left|\frac{G_{v,k}}{Y_{IN,k}}\right| \tag{4.50}$$

$$G_{v,k} = G_{v,k-1}\frac{Y'_{21,k}}{Y'_{22,k} + Y_{eIN,k}}\frac{Y_{e21,k}}{Y_{e22,k} + Y_{IN,k-1}} \tag{4.51}$$

$$Z_{t0,k} = \max\{Z_{T\ t,k}(j\omega)\} \tag{4.52}$$

where $Y'_{ij,k}$ are the admittance matrix coefficients of transistor block k made of the photodiode and transistor k.

4.8 Noise Model of the Receiver Front End

The equivalent circuit of a receiver front end that takes into account various noise sources is given in Figure 4.12 [5, 6]. In this particular receiver front-end example, the parameters were such that $R_D = 10\Omega$, $C_D = 0.4\text{pF}$, $L_D = 2\text{nH}$, and $L_f = 0.5\text{nH}$.

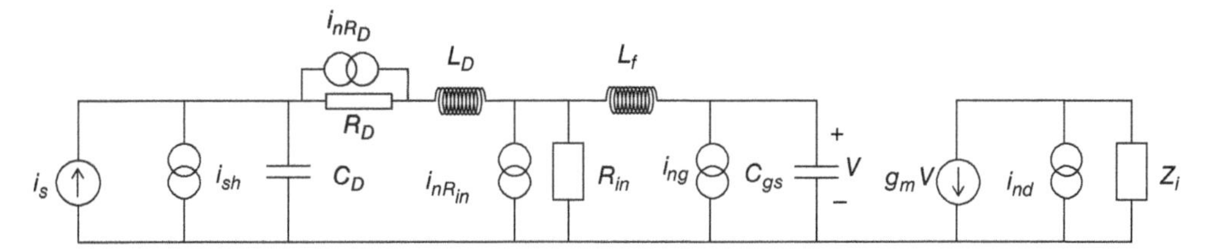

Figure 4.12 Receiver front-end equivalent circuit with noises sources.

The noise current sources inherent in the transistor are characterized by a drain noise, a gate noise, and a noise correlation coefficient given by

$$\begin{aligned}\overline{i_{nd}^2} &= 4KTPg_m\Delta f\\ \overline{i_{ng}^2} &= 4KTR\frac{(\omega C_{gs})^2}{g_m}\Delta f\\ \overline{i_{ng}^* i_{nd}} &= jC_{cor}\sqrt{PR}4KT(\omega C_{gs})\Delta f\end{aligned} \tag{4.53}$$

In this particular receiver front-end example, $P = 1$, $R = 0.5$, and $C_{cor} = 0.9$.

The noise current sources inherent in the pin-diode are characterized by a shot noise and a series thermal resistor noise given by

$$\begin{aligned}\overline{i_{sh}^2} &= 2qI_{dc}\Delta f\\ \overline{i_{nR_D}^2} &= \frac{4KT}{R_D}\Delta f\end{aligned} \tag{4.54}$$

where I_{dc} is the average detected photocurrent.

The noise current sources at the input of the circuit are characterized by

$$\begin{aligned}\overline{i_{inR_D}^2} &= 4KTR_D|Y_D|^2\\ \overline{i_{inR_{in}}^2} &= \frac{4KT}{R_{in}}\left(1 + |Y_D|^2|Z_1|^2 + 2\mathrm{Re}\{Y_D Z_1\}\right)\end{aligned} \tag{4.55}$$

$$\begin{aligned}\overline{i_{inFET}^2} = &\frac{4KT}{g_m}P\left(|C|^2 + |Y_D|^2|A|^2 + 2\mathrm{Re}\{Y_D Z_1\}\right)\\ &+ \frac{4KT}{g_m}R\omega^2 C_{gs}{}^2\left(|D|^2 + |Y_D|^2|D|^2 + 2\mathrm{Re}\{Y_D BD^*\}\right)\\ &+ \frac{4KT}{g_m}2\mathrm{Re}\left\{\left[CD^* + |Y_D|^2 AB^* + Y_D(AD^* - BC^*)\right]jC_{cor}\sqrt{PR}(\omega C_{gs})\right\}\end{aligned} \tag{4.56}$$

with

$$\begin{aligned}Y_D &= j\omega C_D\\ Z_1 &= R_D + j\omega L_D\\ R_i &= R_c // R_p\end{aligned} \tag{4.57}$$

where R_c and R_p are the feedback and biasing resistors, respectively, and where A, B, C, D are the $ABCD$ matrix coefficients obtained by placing Z_1, R_{in}, L_f, and C_{gs} and C_{gd} in cascade.

The total noise current source at the input of the circuit is then given by [6, 7]

$$\overline{i_{in}^2} = \overline{i_{inFET}^2} + \overline{i_{inR_D}^2} + \overline{i_{inR_{in}}^2} + \overline{i_{sh}^2} \tag{4.58}$$

It will be shown in the next section that $\overline{i_{inFET}^2}$ can be adjusted by inductor L_f.

It is then advantageous to optimize the transistor input noise by placing an equalizer between the photodiode and the first transistor [7] in the transimpedance amplifier.

4.9 Two-Stage Transimpedance Amplifier Example

The procedures described in this chapter were implemented in a program called TRANS. It was used to optimize a two-stage transimpedance amplifier for transimpedance gain and to explore the effect of peaking inductor L_f placed at the gate of the first transistor on the noise at the input of the circuit.

In this example, the transistors are the Futjitsu HEMT FHX04X chip [4]. In the results, the device model values used were $C_{gs} = 0.2\text{pF}$, $C_{gd} = 0.025\text{pF}$, $g_m = 50\text{ms}$, $g_d = 5.3\text{ms}$, $C_{ds} = 0.049\text{pF}$, $R_i = 2.5\Omega$, $\tau = 0.85\text{ps}$, $R_G = R_S = R_D = 1.3\Omega$, $L_G = L_D = 0.1\text{nH}$, and $L_S = 0.08\text{nH}$.

The transimpedance gain was optimized in the 3–7 GHz frequency band. The optimized equalizers were synthesized either using the procedures in Chapter 2 for a lumped element realization as shown in Figure 4.13 and/or using the procedures in Chapter 3 for a distributed network realization as shown in Figure 4.14. In Figures 4.13 and 4.14, the inductors are given in nH, the capacitors in pF, and the resistors or UE characteristic impedances in Ω.

Figure 4.15 shows that the simulated transimpedance gain was about $44.4 \pm 0.3\text{dB}\Omega$ for the lumped realization. Figure 4.16 shows a similar transimpedance gain for the distributed realization. This corresponds to a preamplifier gain S_{21} of $8.5 \pm 0.3\text{dB}$. Figures 4.15 and 4.16 also show that the output return loss S_{22} is less than -13dB.

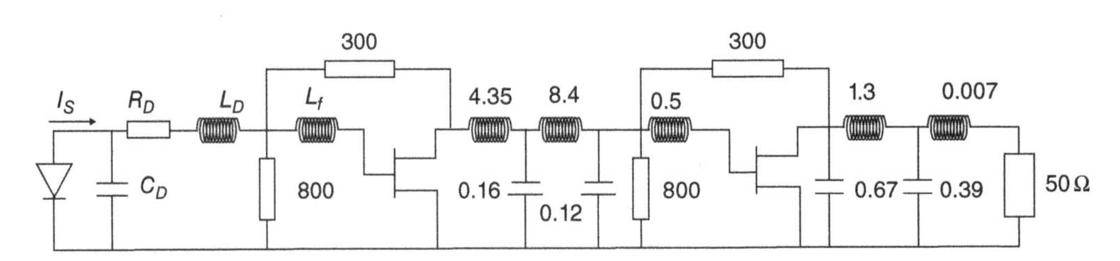

Figure 4.13 Lumped two-stage transimpedance amplifier.

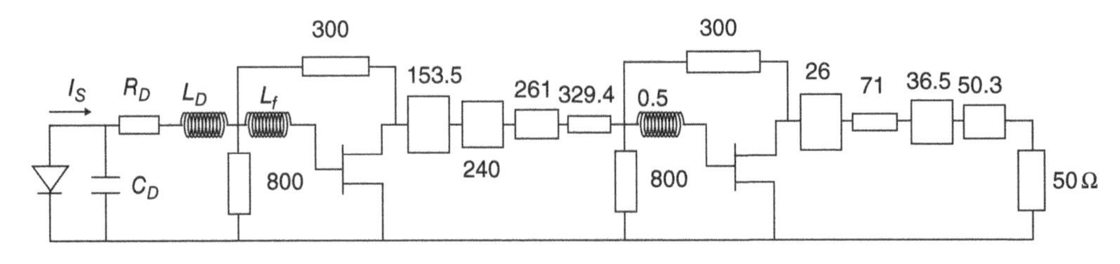

Figure 4.14 Distributed two-stage transimpedance amplifier.

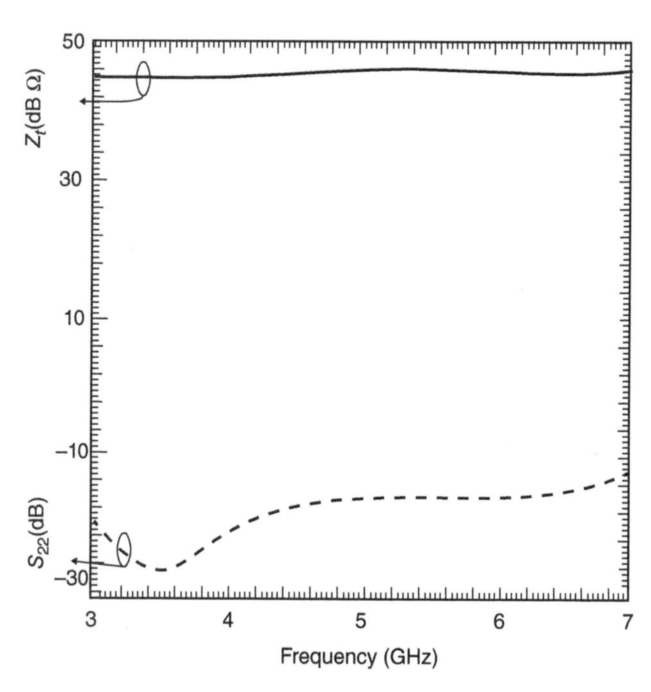

Figure 4.15 Simulated transimpedance gain and output return loss for the lumped realization. Z_t is the trans-impedance gain and S_{22} is the return loss.

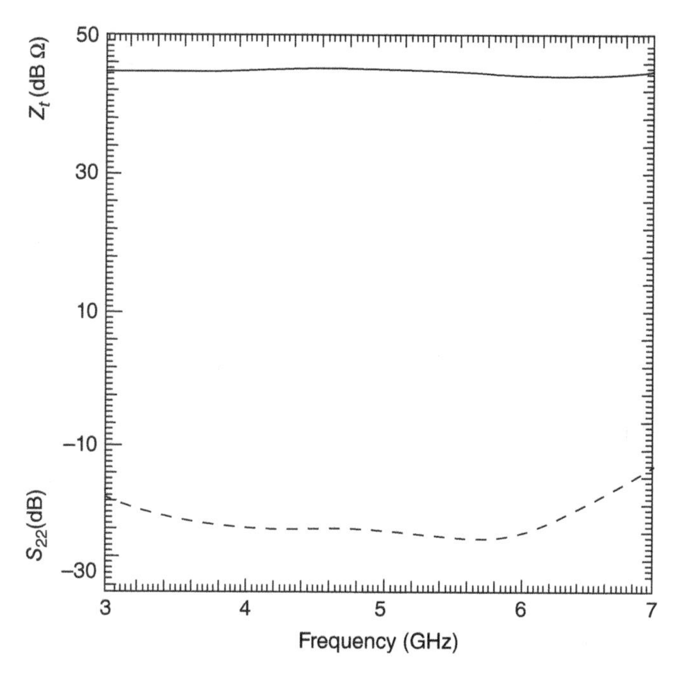

Figure 4.16 Simulated transimpedance gain and output return loss for the distributed realization. Z_t is the trans-impedance gain and S_{22} is the return loss (same idea as Fig. 4.15).

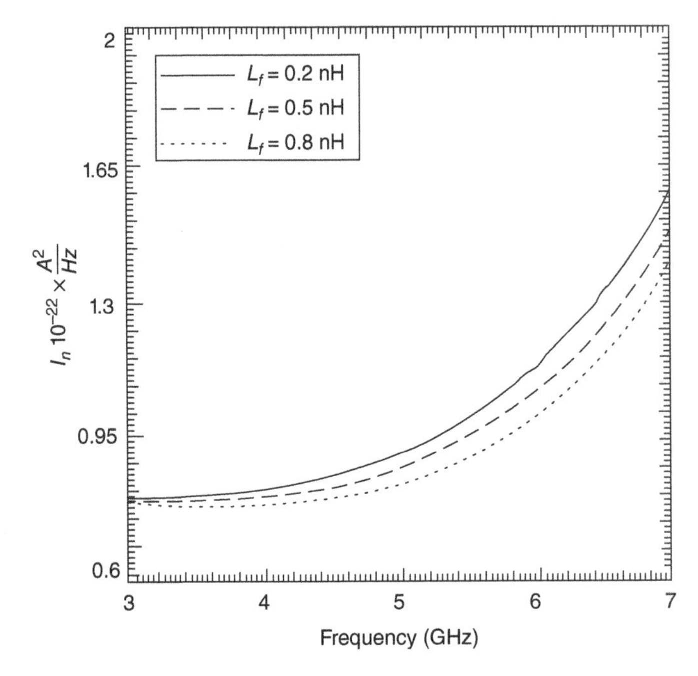

Figure 4.17 Simulated input noise current density $\overline{I_{in}^2}$ when varying inductor L_f.

The input noise current density $\overline{I_{in}^2}$ was calculated for the first stage of the transimpedance amplifier. It was found to be less than $1.51 \times 10^{-22}\ \mathrm{A^2/Hz}$.

Figure 4.17 shows the input noise current density over the 3–7 GHz band when varying peaking inductor L_f. It is seen that the input noise density decreases as peaking inductor L_f at the gate of the first transistor increases.

Therefore, it is possible to improve the input noise current density by using a peaking inductor L_f at the gate of the first transistor.

In conclusion, it is shown how a technique based on the RFT can be used to synthesize a multistage transimpedance network. The noise is improved by using inductor peaking in the method. The output load only needs to be specified by empirical data, and an a priori choice of an equalizer topology is not assumed. The results show that this technique has wide flexibility in its range of application. Different than classical multistage amplifier design, this method designs the multistage amplifier from the load to the source. The computer program TRANS developed on this method enables the optimization of gain, noise, and VSWR. The validity and advantages of this new formalism are demonstrated by a lumped and a distributed example. This study was made for CNET.

References

[1] A. Perennec, R. Soares, P. Jarry, P. Legaud, M. Goloubkoff, "Computer-aided-design of hybrid and monolithic broadband amplifiers for optoelectronic receivers," IEEE Trans. Microw. Theory, vol. MTT-37, no. 9, pp. 1475–1478, September 1989.

[2] N. Okawa, "Fiber-optic multi-gigabit GaAs MIC front-end circuit with inductor peaking," J. Lightwave Technol., vol. 6, pp.1665–1671, 1988.

[3] H. Kikuchi, Y. Miyagawa, T. Kimura, "Broad-band GaAs monolithic equalizing amplifiers for multi-gigabit-per-second optical receivers," IEEE Trans. Microw. Theory, vol. MTT 38, no. 12, pp. 1916–1923, December 1990.
[4] A. N. Olomo, Synthesis and Realisations of Low Noise and Transimpedance Microwave Amplifiers by the Real Frequency Method, Ph.D. dissertation, Brest University, October 1992.
[5] A. Olomo, A. Perennec, R. Soares, P. Jarry, "Computer-aided-design for broad-band multistage microwave trans-impedance amplifiers, synthesis in lumped and distributed elements," IEEE Microwave and Millimeter-Wave Monolithic Circuits Symposium Digest, Albuquerque, NM, June 1–5 1992.
[6] A. Olomo, A. Perennec, R. Soares, P. Jarry, "CAD for broad-band multistage microwave trans-impedance amplifiers," Gallium Arsenide Applications Symposium, GAAS' 1992, Noordwijk, the Netherlands, pp. 27–29, April 1992.
[7] K. E. Alameh, R. A. Minasian, "Tuned optical receivers for microwave sub-carrier multiplexed lightwave systems," IEEE Trans. Microw. Theory, vol. MTT 38, No. 5, pp. 546–551, May 1990.

5

Multistage Lossy Distributed Amplifiers

5.1 Introduction

To achieve broader bandwidths, the equalizer network topology is being modified to introduce lossy junctions in the cascade of distributed networks introduced in Chapter 3. This chapter presents a new idea that is to optimize the lossy junctions as part of the optimization of the equalizers used in a multistage amplifier.

The lossy junctions are introduced between lossless unit elements (UEs) composing an equalizer so that the equalizers can be used to offset power gain issues at low frequencies. In this resistive matching case, the real frequency technique (RFT) is used to directly synthesize the resistors and the UEs to satisfy both gain flatness and stability. The distributed multistage amplifier RFT of Chapter 3 is successfully modified to introduce the lossy elements and incorporated into a CAD program called SYNTARD.

Microwave Amplifier and Active Circuit Design Using the Real Frequency Technique, First Edition.
Pierre Jarry and Jacques N. Beneat.

The new concept is first validated by designing a single-stage 0.1–5 GHz microwave amplifier. Then, the modified multistage RFT is used to design a two-stage 0.1–9 GHz broadband amplifier. In order to realize the equalizers, a lossy UE network synthesis method is given.

5.2 Lossy Distributed Network

In this chapter, the equalizers are composed of the lossy distributed structure shown in Figure 5.1. This structure consists of a cascade of lossless distributed UEs with characteristic impedances Z_{0i}. Each UE has a parallel resistor R_i on its input and a parallel resistor R_{i+1} on its output.

The RFT technique relies on the scattering matrices of the equalizers and transistors, so the first step is to characterize the scattering matrix of the lossy distributed network.

From Chapter 1, a UE with a length of $l=\lambda/4$ with characteristic impedance Z_{0i} has an *ABCD* matrix that can be expressed in terms of the Richard's variable t by

$$\begin{pmatrix} A & B \\ C & D \end{pmatrix} = \frac{1}{\sqrt{1-t^2}} \begin{pmatrix} 1 & tZ_{0i} \\ \frac{t}{Z_{0i}} & 1 \end{pmatrix}$$

where

$$t = j\Omega = j\tan\left(\frac{\pi}{2}\frac{\omega}{\omega_0}\right) = j\tan\theta$$

Also from Chapter 1, the *ABCD* matrix of a parallel resistor R_i is given by

$$\begin{pmatrix} A & B \\ C & D \end{pmatrix} = \begin{pmatrix} 1 & 0 \\ \frac{1}{R_i} & 1 \end{pmatrix}$$

According to (1.54), when terminations are equal to $R_G = R_L = 1$, the S matrix is given in terms of the *ABCD* parameters by

$$\begin{pmatrix} S_{11} & S_{12} \\ S_{21} & S_{22} \end{pmatrix} = \begin{pmatrix} \frac{A+B-C-D}{A+B+C+D} & \frac{2(AD-BC)}{A+B+C+D} \\ \frac{2}{A+B+C+D} & \frac{-A+B-C+D}{A+B+C+D} \end{pmatrix}$$

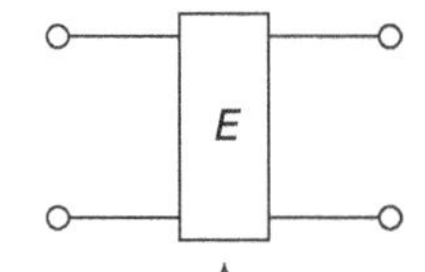

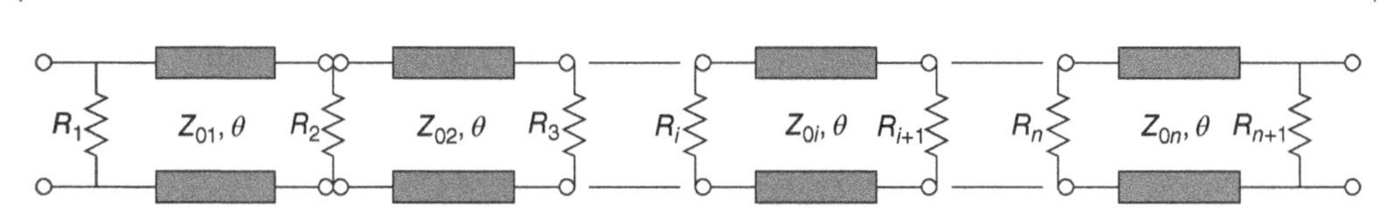

Figure 5.1 An equalizer is made of a lossy distributed network.

For the rest of this section, we will use the normalized characteristic impedance z_{0i} given by

$$z_{0i} = \frac{Z_{0i}}{Z_0} \quad (5.1)$$

where Z_0 is the characteristic impedance of reference (e.g., 50 Ω) to represent the UEs. We will also use the normalized resistance r_i to represent the parallel resistors:

$$r_i = \frac{R_i}{Z_0} \quad (5.2)$$

Using the *ABCD* matrix of a UE given in (1.74) with the conversion from *ABCD* to *S* parameters shows that a UE of length $l = \lambda/4$ with characteristic impedance Z_{0i} will have the scattering matrix:

$$\begin{pmatrix} S_{11} & S_{12} \\ S_{21} & S_{22} \end{pmatrix} = \frac{1}{2 + t\left(z_{0i} + \frac{1}{z_{0i}}\right)} \begin{pmatrix} t\left(z_{0i} - \frac{1}{z_{0i}}\right) & 2\sqrt{1-t^2} \\ 2\sqrt{1-t^2} & t\left(z_{0i} - \frac{1}{z_{0i}}\right) \end{pmatrix} \quad (5.3)$$

Note the reciprocity and symmetry of the UE:

$$S_{11} = S_{22}$$
$$S_{21} = S_{12}$$

Using the *ABCD* to *S* parameter conversion, the scattering matrix of parallel resistor R_i is given by

$$\begin{pmatrix} S_{11} & S_{12} \\ S_{21} & S_{22} \end{pmatrix} = \frac{1}{2 + \frac{1}{r_i}} \begin{pmatrix} -\frac{1}{r_i} & 2 \\ 2 & -\frac{1}{r_i} \end{pmatrix} \quad (5.4)$$

To simplify the notations, we will use the normalized admittance u_i to describe the scattering matrix of a parallel resistor:

$$\begin{pmatrix} S_{11} & S_{12} \\ S_{21} & S_{22} \end{pmatrix} = \frac{1}{2 + u_i} \begin{pmatrix} -u_i & 2 \\ 2 & -u_i \end{pmatrix} \quad (5.5)$$

where

$$u_i = \frac{1}{r_i} \quad (5.6)$$

We are now interested in defining the scattering matrix of a building block of the lossy distributed network shown in Figure 5.2. It is made of the single UE of characteristic impedance Z_{0i} with two parallel resistors R_i and R_{i+1}.

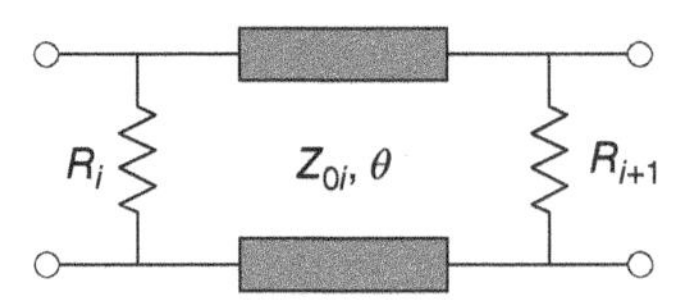

Figure 5.2 Building block of the lossy distributed network.

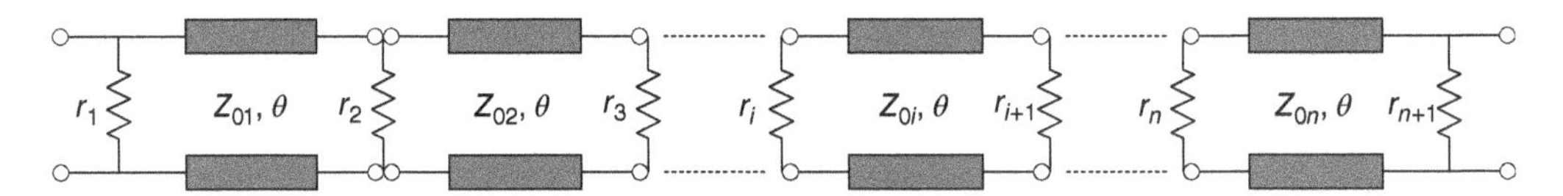

Figure 5.3 Real network N.

Using the multiplication property of the *ABCD* matrices, we find that the *ABCD* matrix of the building block of the lossy distributed network is given by

$$\begin{pmatrix} A & B \\ C & D \end{pmatrix} = \frac{1}{\sqrt{1-t^2}} \begin{pmatrix} 1+t\dfrac{Z_{0i}}{R_{i+1}} & tZ_{0i} \\ \dfrac{1}{R_i}+\dfrac{1}{R_{i+1}}+t\left(\dfrac{1}{Z_{0i}}+\dfrac{Z_{0i}}{R_iR_{i+1}}\right) & 1+t\dfrac{Z_{0i}}{R_i} \end{pmatrix} \tag{5.7}$$

Then converting from *ABCD* to the scattering parameters gives the following *S* parameters for the building block of the lossy distributed network:

$$S_{11} = \frac{-u_i - u_{i+1} + t\left[\left(z_{0i} - \dfrac{1}{z_{0i}}\right) + z_{0i}\left(-u_i + u_{i+1} - u_i u_{i+1}\right)\right]}{2 + u_i + u_{i+1} + t\left[\left(z_{0i} + \dfrac{1}{z_{0i}}\right) + z_{0i}\left(u_i + u_{i+1} + u_i u_{i+1}\right)\right]} \tag{5.8}$$

$$S_{12} = S_{21} = \frac{2\sqrt{1-t^2}}{2 + u_i + u_{i+1} + t\left[\left(z_{0i} + \dfrac{1}{z_{0i}}\right) + z_{0i}\left(u_i + u_{i+1} + u_i u_{i+1}\right)\right]} \tag{5.9}$$

$$S_{22} = \frac{-u_i - u_{i+1} + t\left[\left(z_{0i} - \dfrac{1}{z_{0i}}\right) + z_{0i}\left(u_i - u_{i+1} - u_i u_{i+1}\right)\right]}{2 + u_i + u_{i+1} + t\left[\left(z_{0i} + \dfrac{1}{z_{0i}}\right) + z_{0i}\left(u_i + u_{i+1} + u_i u_{i+1}\right)\right]} \tag{5.10}$$

Note that a key point will be to decompose the scattering parameters in terms of even and odd parts around the admittances $u_i, u_{i+1}, \ldots$.

Next, we define the general properties of the scattering matrix of the lossy distributed network using the concept of image network established by Pang [1].

Starting with a real network N that is made of lossless UEs with normalized characteristic impedances z_{0i} and normalized parallel resistors r_i as shown in Figure 5.3, the image network

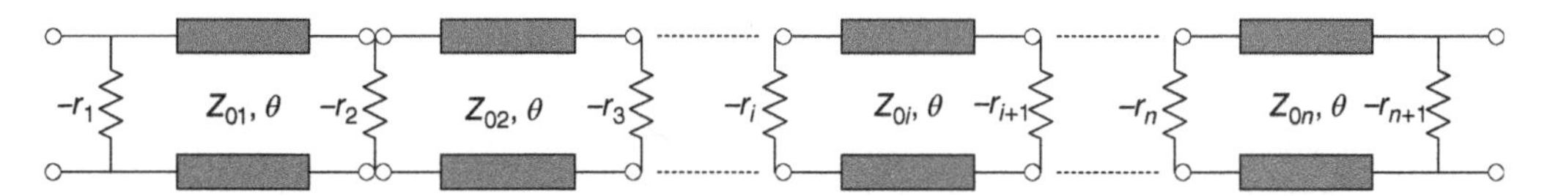

Figure 5.4 Image network $\bar{N}$.

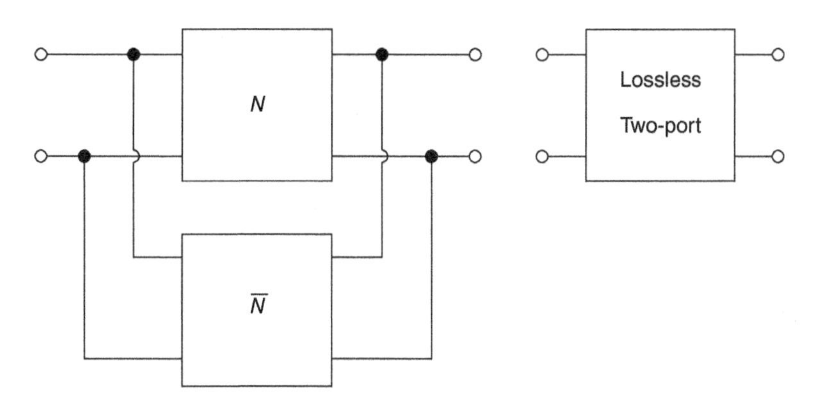

Figure 5.5 Connecting the real and image networks in parallel.

$\bar{N}$ is formed by the same lossless UEs with normalized characteristic impedances z_{0i} but with negative normalized parallel resistors $-r_i$ as shown in Figure 5.4.

If we now connect the real network N and the image network $\bar{N}$ in parallel as shown in Figure 5.5, then the result is a lossless two-port.

In addition, if the real network N is characterized by its scattering matrix S and the image network $\bar{N}$ is characterized by its scattering matrix $\bar{S}$, then the scattering matrices have the property [1]:

$$I - S\bar{S}^* = 0 \tag{5.11}$$

From this property, the S parameters of the real and image networks satisfy the relation:

$$S_{11}\bar{S}_{11}^* + S_{21}\bar{S}_{21}^* = 1 \tag{5.12}$$

from which it is found that the scattering parameters of a lossy distributed network are such that [2]

$$S_{11}(t) = \frac{h(t)}{g(t)} \tag{5.13}$$

$$S_{12}(t) = S_{21}(t) = \frac{(1-t^2)^{n/2}}{g(t)} \tag{5.14}$$

$$S_{22}(t) = -\frac{\overline{h(-t)}}{g(t)} \tag{5.15}$$

where $\overline{h(-t)}$ is obtained by replacing t by $-t$ and by replacing all the normalized resistors r_i by $-r_i$ in the expression of $h(t)$.

It is advantageous to decompose $h(t)$ into an even part $\alpha(t)$ and an odd part $\beta(t)$ with respect to the $r_i's$. In this case, we have

$$h(t)=\alpha(t)+\beta(t)$$

and

$$\overline{h(-t)}=\alpha(-t)-\beta(-t)$$

To better understand what the even and odd parts $\alpha(t)$ and $\beta(t)$ correspond to, we use the example of the S_{11} parameter of Equation (5.8):

$$S_{11}=\frac{h(t)}{g(t)}=\frac{-u_i-u_{i+1}+t\left[\left(z_{0i}-\frac{1}{z_{0i}}\right)+z_{0i}\left(-u_i+u_{i+1}-u_iu_{i+1}\right)\right]}{2+u_i+u_{i+1}+t\left[\left(z_{0i}+\frac{1}{z_{0i}}\right)+z_{0i}\left(u_i+u_{i+1}+u_iu_{i+1}\right)\right]}$$

so that

$$h(t)=-u_i-u_{i+1}+t\left[\left(z_{0i}-\frac{1}{z_{0i}}\right)+z_{0i}\left(-u_i+u_{i+1}-u_iu_{i+1}\right)\right]$$

We note that since $u_i=1/r_i$, $\alpha(t)$ and $\beta(t)$ are also the even and odd parts with respect to the $u_i's$.

In order to define $\alpha(t)$ and $\beta(t)$ corresponding to the aforementioned example, we can rewrite $h(t)$ to regroup the terms that are dependent of the $u_i's$ and the terms that are not dependent of the $u_i's$:

$$h(t)=t\left(z_{0i}-\frac{1}{z_{0i}}\right)-u_i-u_{i+1}+tz_{0i}\left(-u_i+u_{i+1}-u_iu_{i+1}\right)$$

from which we find that the part that remains the same after making $u_i's=-u_i's$ is

$$\alpha(t)=t\left(z_{0i}-\frac{1}{z_{0i}}\right)-tz_{0i}u_iu_{i+1}$$

and what is left is

$$\beta(t)=-u_i-u_{i+1}+tz_{0i}\left(-u_i+u_{i+1}\right)$$

We can check that $\alpha(t)=\alpha(t)$ and $\beta(t)=-\beta(t)$ when making $u_i's=-u_i's$.

If we now compute $\overline{h(-t)}$, we find that

$$\begin{aligned}\overline{h(-t)}&=\alpha(-t)-\beta(-t)\\&=-t\left(z_{0i}-\frac{1}{z_{0i}}\right)+tz_{0i}u_iu_{i+1}-\left(-u_i-u_{i+1}-tz_{0i}\left(-u_i+u_{i+1}\right)\right)\end{aligned}$$

$$= -t\left(z_{0i}-\frac{1}{z_{0i}}\right)-(-u_i-u_{i+1}-tz_{0i}(-u_i+u_{i+1}+u_iu_{i+1}))$$

and

$$-\overline{h(-t)} = t\left(z_{0i}-\frac{1}{z_{0i}}\right)+(-u_i-u_{i+1}-tz_{0i}(-u_i+u_{i+1}+u_iu_{i+1}))$$

and we can check that this is the same as the numerator of the S_{22} parameter in (5.10).

Finally, the unknown coefficients to be optimized in the RFT will be the polynomial coefficients $\alpha_0, \alpha_1, \alpha_2, \ldots, \alpha_n$ and $\beta_0, \beta_1, \beta_2, \ldots, \beta_n$ of $\alpha(t)$ and $\beta(t)$.

For example, in the aforementioned case, we have

$$\alpha(t) = \alpha_1 t + \alpha_0$$
$$\beta(t) = \beta_1 t + \beta_0$$

with

$$\begin{cases} \alpha_1 = z_{0i}-\dfrac{1}{z_{0i}}-z_{0i}u_iu_{i+1} \\ \alpha_0 = 0 \end{cases} \quad \text{and} \quad \begin{matrix} \beta_1 = z_{0i}(-u_i+u_{i+1}) \\ \beta_0 = -u_i-u_{i+1} \end{matrix}$$

5.3 Multistage Lossy Distributed Amplifier Representation

The multistage lossy distributed amplifier is represented in Figure 5.6 [2]. It assumes that there are k transistors. An equalizer is associated with each transistor and is assumed to be made of a lossy distributed network as described in Section **5.2**. There is one final equalizer placed after the kth transistor. The source and load impedances are assumed arbitrary, complex, and frequency dependent. The transistor can represent a transistor alone or a transistor with added but known circuitry as long as the S parameters can be provided.

The equalizer of the kth stage is characterized by the scattering matrix

$$S_{E_k} = \begin{pmatrix} e_{11,k}(t) & e_{12,k}(t) \\ e_{21,k}(t) & e_{22,k}(t) \end{pmatrix} \tag{5.16}$$

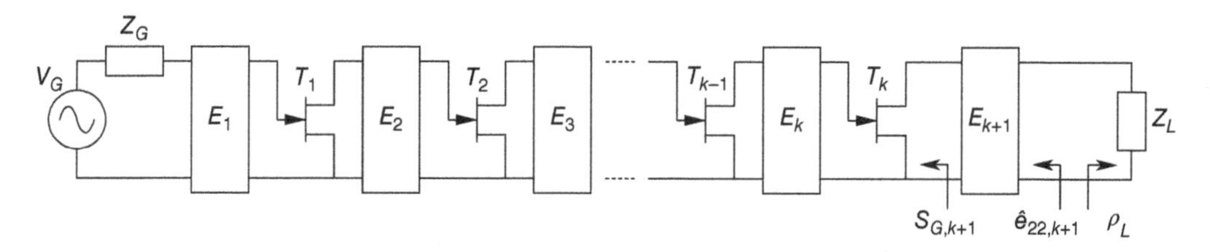

Figure 5.6 Multistage lossy distributed amplifier representation.

The transistor circuit of the kth stage is characterized by the scattering matrix

$$S_{T_k} = \begin{pmatrix} S_{11,k}(t) & S_{12,k}(t) \\ S_{21,k}(t) & S_{22,k}(t) \end{pmatrix} \tag{5.17}$$

where t is the Richard's variable defined in Chapter 1.

The equalizer is made of a lossy distributed network described in Section **5.2** and has the scattering matrix:

$$S_E = \begin{pmatrix} e_{11}(t) = \dfrac{h(t)}{g(t)} & e_{12}(t) = \dfrac{f(t)}{g(t)} \\ e_{21}(t) = \dfrac{f(t)}{g(t)} & e_{22}(t) = -\dfrac{\overline{h(-t)}}{g(t)} \end{pmatrix} \tag{5.18}$$

where $h(t)$ and $g(t)$ are polynomials of order n with real-valued coefficients:

$$h(s) = h_0 + h_1 t + \cdots + h_{n-1} t^{n-1} + h_n t^n$$
$$g(s) = g_0 + g_1 t + \cdots + g_{n-1} t^{n-1} + g_n t^n$$

and where $f(t)$ is a function of the form

$$f(t) = \left(1 - t^2\right)^{\frac{n}{2}} \tag{5.19}$$

According to Section **5.2**, the polynomial $h(t)$ is decomposed into an even part polynomial $\alpha(t)$ and an odd part polynomial $\beta(t)$ such that

$$h(t) = \alpha(t) + \beta(t) \tag{5.20}$$

and $\overline{h(-t)}$ is defined as

$$\overline{h(-t)} = \alpha(-t) - \beta(-t) \tag{5.21}$$

In addition, the scattering parameters of the equalizer made of a lossy distributed network satisfy the relation

$$e_{11}\overline{e_{11}}^* + e_{21}\overline{e_{21}}^* = 1$$

so that

$$g(t)\overline{g(-t)} = h(t)\overline{h(-t)} + \left(1 - t^2\right)^n \tag{5.22}$$

The problem then consists of defining the unknown polynomial coefficients $\alpha_0, \alpha_1, \alpha_2, \ldots, \alpha_n$ and $\beta_0, \beta_1, \beta_2, \ldots, \beta_n$ of the equalizer that optimize the response of the system.

Once $\alpha_0, \alpha_1, \alpha_2, \ldots, \alpha_n$ and $\beta_0, \beta_1, \beta_2, \ldots, \beta_n$ are known, then $h(t)$ is computed using $h(t) = \alpha(t) + \beta(t)$ and $\overline{h(-t)}$ is computed using $\overline{h(-t)} = \alpha(-t) - \beta(-t)$ so that $g(t)\overline{g(-t)}$ can be defined from

$$G(t) = g(t)\overline{g(-t)} = h(t)\overline{h(-t)} + \left(1-t^2\right)^n = G_0 + G_1 t + G_1 t^2 + \cdots + G_{2n} t^{2n}$$

with

$$\begin{aligned}
G_0 &= \alpha_0^2 - \beta_0^2 + 1 \\
G_1 &= 2\alpha_0\beta_1 - 2\alpha_1\beta_0 \\
&\vdots \\
G_k = G_{2i} &= (-1)^i\left(\alpha_i^2 - \beta_i^2 + C_n^i\right) + 2\sum_{j=1}^{i}\left[(-1)^{j-1}\left(\alpha_{j-1}\alpha_{2i-j+1} - \beta_{j-1}\beta_{2i-j+1}\right)\right] \quad (k \text{ even}) \\
G_{k+1} = G_{2i+1} &= 2\sum_{j=1}^{i+1}\left[(-1)^{j-1}\left(\alpha_{j-1}\beta_{2i-j+2} - \alpha_{2i-j+2}\beta_{j-1}\right)\right] \quad (k+1 \text{ odd}) \\
&\vdots \\
G_{2n} &= (-1)^n\left(1 + \alpha_n^2 - \beta_n^2\right)
\end{aligned} \tag{5.23}$$

where

$$C_n^p = \frac{n!}{p!(n-p)!}$$

The left-half roots of $g(t)\overline{g(-t)}$ are retained to form a Hurwitz polynomial $g(t)$. At this point, all the equalizer scattering parameters $e_{11}(t)$, $e_{12}(t)$, $e_{21}(t)$, and $e_{22}(t)$ can be computed. During optimization, they will be needed for computing the error function to be optimized, and after optimization, for example, $e_{11}(t)$ can be used to realize the equalizer using the synthesis procedures given in Section **5.7**.

One should note that for lossless distributed equalizers the roots of $g(t)g(-t)$ are complex conjugate and symmetrical with respect to the imaginary axis as shown in Figure 5.7. However, for lossy distributed equalizers, the roots of $g(t)\overline{g(-t)}$ are complex conjugate but no longer symmetrical with respect to the imaginary axis as shown in Figure 5.8.

One should select the roots in the left-half plane to construct the $g(t)$ polynomial.

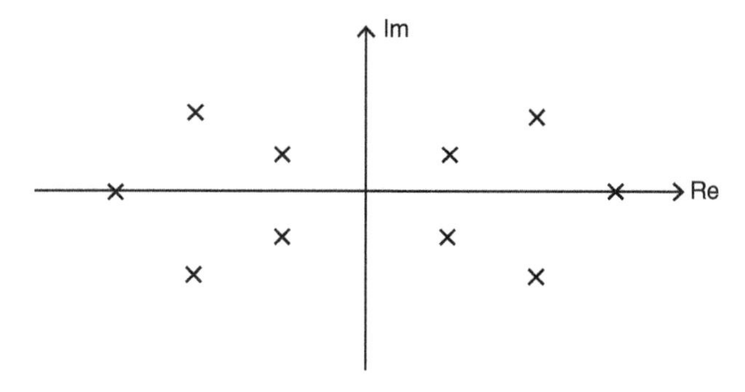

Figure 5.7 Roots of $g(t)g(-t)$ are complex conjugate and symmetrical around imaginary axis for lossless distributed equalizers.

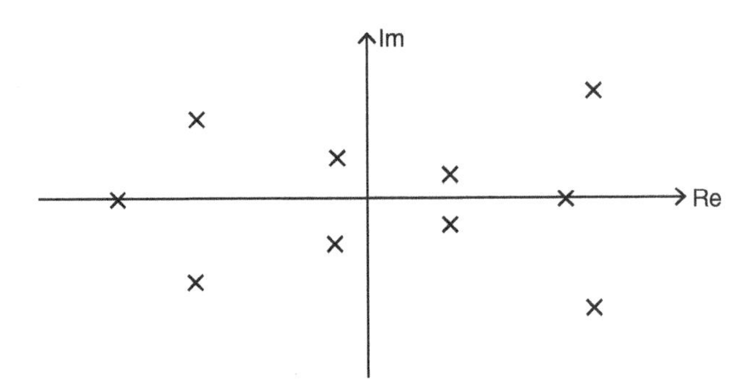

Figure 5.8 Roots of $g(t)\overline{g(-t)}$ are complex conjugate and but no longer symmetrical around imaginary axis for lossy distributed equalizers.

5.4 Multistage Transducer Gain

The overall transducer gain T_{GT} of the multistage lossy distributed amplifier when terminated on arbitrary terminations Z_G and Z_L as shown in Figure 5.6 is given by

$$T_{GT} = T_k \frac{|e_{21,k+1}|^2\left(1-|\rho_L|^2\right)}{|1-e_{11,k+1}S_{G,k+1}|^2|1-\hat{e}_{22,k+1}\rho_L|^2} \tag{5.24}$$

with

$$\hat{e}_{22,k+1} = e_{22,k+1} + \frac{e_{21,k+1}^2 S_{G,k+1}}{1-e_{11,k+1}S_{G,k+1}} \tag{5.25}$$

$$S_{G,k+1} = S_{22,k} + \frac{S_{12,k}S_{21,k}\hat{e}_{22,k}}{1-S_{11,k}\hat{e}_{22,k}} \tag{5.26}$$

$$\rho_L = \frac{Z_L - Z_0}{Z_L + Z_0} \tag{5.27}$$

where

$S_{G,k+1}$ is the output reflection coefficient of transistor k when loaded on input
$\hat{e}_{22,k+1}$ is the output reflection of equalizer $k+1$ when loaded on input
ρ_L is the reflection coefficient of the load impedance Z_L

and where T_k is the transducer gain of the first k equalizer–transistor stages terminated on Z_0 as shown in Figure 5.9. Z_0 is the reference used for the scattering parameters. For example, $Z_0 = 50\Omega$ is often used to provide the measured scattering parameters of a transistor.

The transducer gain T_k of the first k equalizer–transistor stages is computed recursively by

$$T_k = T_{k-1} \frac{|e_{21,k}|^2|S_{21,k}|^2}{|1-e_{11,k}S_{G,k}|^2|1-\hat{e}_{22,k}S_{11,k}|^2} \quad \text{for } k>1 \tag{5.28}$$

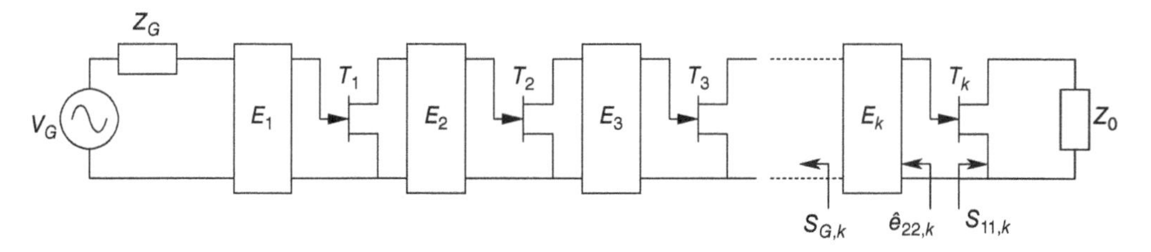

Figure 5.9 Transducer gain for k equalizer–transistor stages terminated on Z_0.

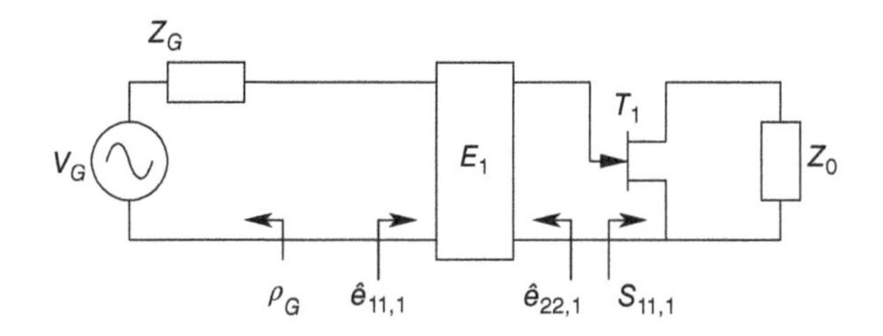

Figure 5.10 Transducer gain for the first stage terminated on Z_0.

with

$$\hat{e}_{22,k} = e_{22,k} + \frac{e_{21,k}^2 S_{G,k}}{1 - e_{11,k} S_{G,k}} \tag{5.29}$$

$$S_{G,k} = S_{22,k-1} + \frac{S_{12,k-1} S_{21,k-1} \hat{e}_{22,k-1}}{1 - S_{11,k-1} \hat{e}_{22,k-1}} \tag{5.30}$$

where

T_{k-1} is the transducer gain of the first $k-1$ equalizer–transistor stages
$e_{ij,k}$ are the scattering parameters of equalizer k
$S_{ij,k}$ are the scattering parameters of transistor k
$\hat{e}_{22,k}$ is the output reflection coefficient of equalizer k when loaded on input
$S_{G,k}$ is the output reflection coefficient of transistor $k-1$ when loaded on input
$S_{G,k}$ is the reflection coefficient of the k − 1 prior stages

The recursion starts with the first stage ($k = 1$) as shown in Figure 5.10.
The transducer gain of the first stage is given by

$$T_1 = \left(1 - |\rho_G|^2\right) \frac{|e_{21,1}|^2 |S_{21,1}|^2}{|1 - e_{11,1} S_{G,1}|^2 |1 - \hat{e}_{22,1} S_{11,1}|^2} \tag{5.31}$$

with

$$\hat{e}_{22,1} = e_{22,1} + \frac{e_{21,1}^2 S_{G,1}}{1 - e_{11,1} S_{G,1}} \tag{5.32}$$

$$S_{G,1}=\rho_G \tag{5.33}$$

$$\rho_G=\frac{Z_G-Z_0}{Z_G+Z_0} \tag{5.34}$$

where

$S_{G,1}$ would be the output reflection coefficient of a transistor 0, but since there is no transistor 0, it is equal to the reflection coefficient of the source ρ_G
$\hat{e}_{22,1}$ is the output reflection of equalizer 1 when loaded on input
ρ_G is the reflection coefficient of the source impedance Z_G

The transducer gain equations are demonstrated in Appendix A.

5.5 Multistage VSWR

The input VSWR corresponding to the k equalizer–transistor stages terminated on Z_0 as shown in Figure 5.10 is computed using

$$\text{VSWR}_{IN}=\frac{1+|\hat{e}_{11,1}|}{1-|\hat{e}_{11,1}|} \tag{5.35}$$

As transistor–equalizer stages are added to form the situation of Figure 5.9, the input VSWR is calculated recursively using

$$\begin{aligned}
&S_{L,k}=S_{11,k}\\
&\hat{e}_{11,k}=e_{11,k}+\frac{e_{21,k}^2S_{L,k}}{1-e_{22,k}S_{L,k}}\\
&S_{L,k-1}=S_{11,k-1}+\frac{S_{12,k-1}S_{21,k-1}\hat{e}_{11,k}}{1-S_{22,k-1}\hat{e}_{11,k}}\\
&\hat{e}_{11,k-1}=e_{11,k-1}+\frac{e_{21,k-1}^2S_{L,k-1}}{1-e_{22,k-1}S_{L,k-1}}\\
&\vdots\\
&\hat{e}_{11,1}=e_{11,1}+\frac{e_{21,1}^2S_{L,1}}{1-e_{22,1}S_{L,1}}
\end{aligned} \tag{5.36}$$

where

$\hat{e}_{11,k}$ is the input reflection coefficient of equalizer k when loaded on output
$S_{L,k}$ is the input reflection coefficient of transistor k when loaded on output

After the last $(k+1)$th equalizer has been added for compensating arbitrary load impedance Z_L as shown in Figure 5.6, the input VSWR is computed using the recursive formula starting with

$$\hat{e}_{11,k+1} = e_{11,k+1} + \frac{e_{21,k+1}^2 \rho_L}{1 - e_{22,k+1}\rho_L}$$
$$S_{L,k} = S_{11,k} + \frac{S_{12,k}S_{21,k}\hat{e}_{11,k+1}}{1 - S_{22,k}\hat{e}_{11,k+1}} \tag{5.37}$$
$$\rho_L = \frac{Z_L - Z_0}{Z_L + Z_0}$$

The output VSWR including the $(k+1)$th equalizer and terminated on Z_L as shown in Figure 5.6 is given by

$$\mathrm{VSWR}_{OUT} = \frac{1 + |\hat{e}_{22,k+1}|}{1 - |\hat{e}_{22,k+1}|} \tag{5.38}$$

with

$$\hat{e}_{22,k+1} = e_{22,k+1} + \frac{e_{21,k+1}^2 S_{G,k+1}}{1 - e_{11,k+1}S_{G,k+1}}$$
$$S_{G,k+1} = S_{22,k} + \frac{S_{12,k}S_{21,k}\hat{e}_{22,k}}{1 - S_{11,k}\hat{e}_{22,k}} \tag{5.39}$$

$\hat{e}_{22,k+1}$ is the output reflection coefficient of equalizer $k+1$ when loaded on input
$S_{G,k+1}$ is the output reflection coefficient of transistor k when loaded on input

and one uses (5.29) and (5.30) for recursively computing the output VSWR.

5.6 Optimization Process

The optimization algorithm is used to define the polynomial coefficients $h_0, h_1, \ldots, h_n$ of $h(t)$ of an equalizer that minimizes a least mean square error function that includes desired target functions with different weight factors.

For the lossy distributed amplifiers of this chapter, the error function consists of the transducer gain and the VSWR:

$$E = \sum_{i=1}^{m}\left[W_1\left(\frac{G_{T,k}(\Omega_i,\boldsymbol{H})}{G_{T0,k}} - 1\right)^2 + W_2\left(\frac{\mathrm{VSWR}_k(\Omega_i,\boldsymbol{H})}{\mathrm{VSWR}_{0,k}} - 1\right)^2\right] \tag{5.40}$$

where the functions are given in terms of Richard's frequency $\Omega_i = \tan\left(\frac{\pi}{2}\frac{\omega_i}{\omega_0}\right)$ and where ω_i are the actual frequencies and ω_0 is the normalization frequency used in the Richard's transform.

$G_{T,k}(\Omega_i, \boldsymbol{H})$ is the transducer power gain for k equalizer–transistor stages given by the equations in Section **5.4** and used to optimize the kth equalizer. The transducer power gain depends on frequency through the scattering parameters $e_{ij}(j\Omega)$ of the equalizer and the scattering parameters $S_{ij}(j\Omega)$ of the transistor. The target transducer power gain $G_{T0,k}$ for optimizing the kth equalizer is taken as a constant in the frequency band being optimized. The expressions for the target transducer gain are provided in Section **5.8**.

$\text{VSWR}_{IN,k}(\Omega_i, \boldsymbol{H})$ is the input VSWR at stage k given by the equations in Section **5.5**. The target $\text{VSWR}_{IN0,k}$ is constant over the frequency band being optimized. The value is often between 1 and 2. Note that the input VSWR is generally optimized using the first equalizer.

$\text{VSWR}_{OUT,k}(\Omega_i, \boldsymbol{H})$ is the output VSWR at stage k given by the equations in Section **5.5**. The target $\text{VSWR}_{OUT0,k}$ is a constant over the frequency band being optimized. The value is often between 1 and 2. Note that the output VSWR is generally optimized using the last equalizer.

The vector $\boldsymbol{H}$ includes the polynomial coefficients $h_0, h_1, \ldots, h_n$ and is used as the $\boldsymbol{X}$ vector in the Levenberg–More optimization algorithms of Chapter 2 and Appendix B.

The main computation in the optimization routine is to define the step $\Delta\boldsymbol{H}$ to add to the elements of vector $\boldsymbol{H}$ at the next iteration given by

$$\Delta\boldsymbol{H} = -\left[\boldsymbol{J}^T\boldsymbol{J} + \lambda\boldsymbol{D}^T\boldsymbol{D}\right]^{-1}\boldsymbol{J}^T e(\boldsymbol{H})$$

where $e(\boldsymbol{H})$ is the error vector, J is the Jacobian matrix of the error with elements $\partial e_j/\partial h_i$, D is a diagonal matrix, and λ is the Levenberg–Marquardt–More parameter described in Appendix B.

An important element is the Jacobian matrix. It is based on derivatives of the error functions. For the transducer power gain, analytical expressions for the derivatives that can be determined are rather tedious and lengthy to express and are written into a program. In the case of the input VSWR, the analytical expressions become very cumbersome and it is preferred to use a numerical derivative such that

$$\frac{\partial \text{VSWR}_{IN}(\Omega_j, \boldsymbol{H})}{\partial h_i} = \frac{\text{VSWR}_{IN}(\Omega_j, \boldsymbol{H} + \Delta h_i) - \text{VSWR}_{IN}(\Omega_j, \boldsymbol{H})}{\Delta h_i} \tag{5.41}$$

where a step of $\Delta h_i = 10^{-5}$ was found satisfactory when starting with $h(t)$ polynomial coefficients initialized at $h_i = \pm 1$.

Note that in the case of lossy distributed amplifiers one must keep track of the even part $\alpha(t)$ and the odd part $\beta(t)$ of $h(t)$ as defined in Sections **5.2** and **5.3**. Theoretically this will involve a Jacobian matrix such that

$$\boldsymbol{J} = \begin{bmatrix} \frac{\partial e_1(\boldsymbol{X})}{\partial \alpha_1} & \frac{\partial e_1(\boldsymbol{X})}{\partial \alpha_2} & \cdots & \frac{\partial e_1(\boldsymbol{X})}{\partial \alpha_n} & \frac{\partial e_1(\boldsymbol{X})}{\partial \beta_1} & \frac{\partial e_1(\boldsymbol{X})}{\partial \beta_2} & \cdots & \frac{\partial e_1(\boldsymbol{X})}{\partial \beta_n} \\ \frac{\partial e_2(\boldsymbol{X})}{\partial \alpha_1} & \frac{\partial e_2(\boldsymbol{X})}{\partial \alpha_2} & \cdots & \frac{\partial e_2(\boldsymbol{X})}{\partial \alpha_n} & \frac{\partial e_2(\boldsymbol{X})}{\partial \beta_1} & \frac{\partial e_2(\boldsymbol{X})}{\partial \beta_2} & \cdots & \frac{\partial e_2(\boldsymbol{X})}{\partial \beta_n} \\ \frac{\partial e_3(\boldsymbol{X})}{\partial \alpha_1} & \frac{\partial e_3(\boldsymbol{X})}{\partial \alpha_2} & \cdots & \frac{\partial e_3(\boldsymbol{X})}{\partial \alpha_n} & \frac{\partial e_3(\boldsymbol{X})}{\partial \beta_1} & \frac{\partial e_3(\boldsymbol{X})}{\partial \beta_2} & \cdots & \frac{\partial e_3(\boldsymbol{X})}{\partial \beta_n} \\ \frac{\partial e_4(\boldsymbol{X})}{\partial \alpha_1} & \frac{\partial e_4(\boldsymbol{X})}{\partial \alpha_2} & \cdots & \frac{\partial e_4(\boldsymbol{X})}{\partial \alpha_n} & \frac{\partial e_4(\boldsymbol{X})}{\partial \beta_1} & \frac{\partial e_4(\boldsymbol{X})}{\partial \beta_2} & \cdots & \frac{\partial e_4(\boldsymbol{X})}{\partial \beta_n} \\ \vdots & \vdots & \vdots\ \vdots & & \vdots & \vdots & \vdots\ \vdots & \\ \frac{\partial e_m(\boldsymbol{X})}{\partial \alpha_1} & \frac{\partial e_m(\boldsymbol{X})}{\partial \alpha_2} & \cdots & \frac{\partial e_m(\boldsymbol{X})}{\partial \alpha_n} & \frac{\partial e_m(\boldsymbol{X})}{\partial \beta_1} & \frac{\partial e_m(\boldsymbol{X})}{\partial \beta_2} & \cdots & \frac{\partial e_m(\boldsymbol{X})}{\partial \beta_n} \end{bmatrix}$$

Since this presents a serious algorithmic change to the optimization routine and since it nearly doubles the amount of simultaneous unknown parameters to optimize, the idea was to optimize the $\alpha_0, \alpha_1, \ldots, \alpha_n$ parameters first by taking the $\beta_0, \beta_1, \ldots, \beta_n$ parameters all equal to zero. This gives the optimized step $\Delta\boldsymbol{\alpha}$ such that

$$\Delta\boldsymbol{\alpha} = -\left[\boldsymbol{J}^T\boldsymbol{J} + \lambda\boldsymbol{D}^T\boldsymbol{D}\right]^{-1}\boldsymbol{J}^T e(\boldsymbol{\alpha}) \tag{5.42}$$

Note that this optimizes the UE characteristic impedances in the lossy distributed equalizer.

The optimized step $\Delta\boldsymbol{\alpha}$ is used to compute new values for the $\alpha_0, \alpha_1, \ldots, \alpha_n$ parameters that are then used to optimize the $\beta_0, \beta_1, \ldots, \beta_n$ parameters and the entire $h(t)$ polynomial using

$$\Delta\boldsymbol{H} = -\left[\boldsymbol{J}^T\boldsymbol{J} + \lambda\boldsymbol{D}^T\boldsymbol{D}\right]^{-1}\boldsymbol{J}^T e(\boldsymbol{H}) \tag{5.43}$$

and where $h(t) = \alpha(t) + \beta(t)$.

Note that to start the optimization process taking $\alpha_i = \pm 1$ and $\beta_i = \pm 0.5$ was found to work well.

5.7 Synthesis of the Lossy Distributed Network

The synthesis of a lossy distributed network with parallel resistors is based on extracting a UE using Richard's theorem combined with extracting a parallel resistor [2–5].

In Richard's theorem, a lossless UE with characteristic impedance Z_{01} is extracted from the input impedance $Z(t)$ and the remaining input impedance is $Z_1(t)$ as shown in Figure 5.11.

In this case, the characteristic impedance of the extracted UE is given by

$$Z_{01} = Z(1)$$

and the remaining input impedance is given by

$$Z_1(t) = Z(1)\frac{Z(t) - tZ(1)}{Z(1) - tZ(t)}$$

For the synthesis of a lossy distributed network with parallel resistors, we are interested in extracting a parallel resistor R_1 and a lossless UE with characteristic impedance Z_{01} from the input impedance $Z(t)$ as shown in Figure 5.12. We are also interested in finding the remaining input impedance $Z_2(t)$.

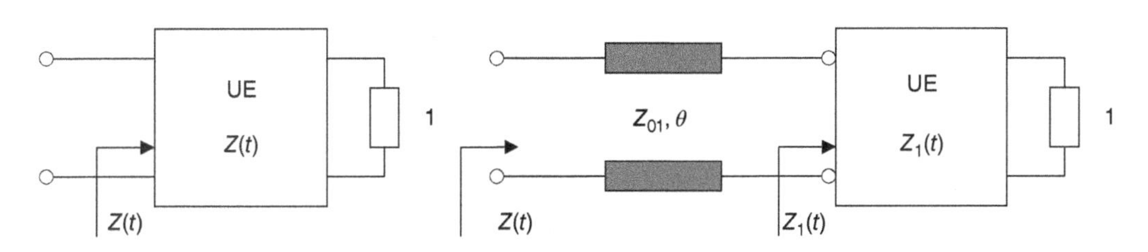

Figure 5.11 Richard's theorem for cascade of lossless UEs.

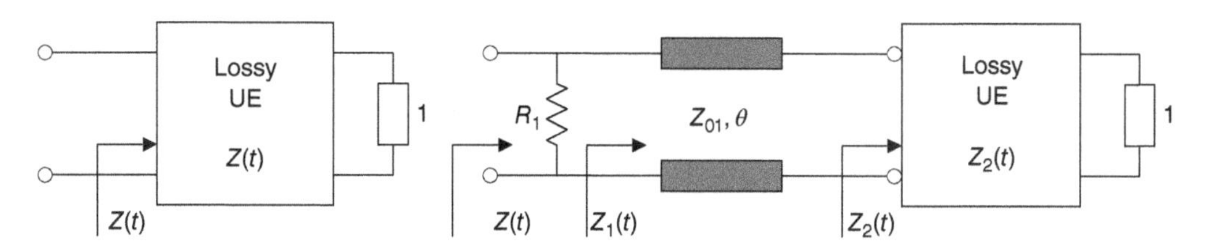

Figure 5.12 Extracting a parallel resistor and lossless UE from the input impedance of a lossy distributed network.

After extracting the parallel resistor and computing the remaining input impedance $Z_1(t)$, one can extract a lossy UE with characteristic impedance Z_{01} and also define the remaining input impedance $Z_2(t)$ using Richard's theorem such that

$$Z_2(t)=Z_1(1)\frac{Z_1(t)-tZ_1(1)}{Z_1(1)-tZ_1(t)} \tag{5.44}$$

$$Z_{01}=Z_1(1) \tag{5.45}$$

We can rewrite (5.44) such that

$$Z_1(t)=Z_1(1)\frac{Z_2(t)+tZ_1(1)}{Z_1(1)+tZ_2(t)} \tag{5.46}$$

Since the starting input impedance $Z(t)$ is equal to the resistor R_1 in parallel with $Z_1(t)$, we have

$$Z(t)=\frac{R_1Z_1(t)}{R_1+Z_1(t)} \tag{5.47}$$

If we evaluate $Z(t)$ at $t=1$, we have

$$Z(1)=\frac{R_1Z_1(1)}{R_1+Z_1(1)} \tag{5.48}$$

If we evaluate $Z(t)$ at $t=-1$ and since $Z_1(-1)=-Z_1(1)$, we have

$$Z(-1)=\frac{-R_1Z_1(1)}{R_1-Z_1(1)} \tag{5.49}$$

Rearranging Equation (5.47), we can express $Z_1(1)$ such that

$$Z_1(1)=\frac{R_1Z(-1)}{Z(-1)-R_1} \tag{5.50}$$

and if we combine Equation (5.47) into Equation (5.48), we have

$$Z(1) = \frac{R_1 \dfrac{R_1 Z(-1)}{Z(-1) - R_1}}{R_1 + \dfrac{R_1 Z(-1)}{Z(-1) - R_1}} = \frac{R_1^2 Z(-1)}{2R_1 Z(-1) - R_1^2} = \frac{R_1 Z(-1)}{2Z(-1) - R_1}$$

Rewriting this equation gives

$$2Z(-1)Z(1) - R_1 Z(1) = R_1 Z(-1)$$

from which we get

$$R_1 = \frac{2Z(1)Z(-1)}{Z(1) + Z(-1)} \tag{5.51}$$

If we rewrite Equation (5.47) such that

$$Z_1(t) = \frac{R_1 Z(t)}{R_1 - Z(t)} \tag{5.52}$$

then combining (5.51) into (5.52), the UE characteristic impedance Z_{01} is obtained by

$$Z_{01} = Z_1(1) = \frac{R_1 Z(1)}{R_1 - Z(1)} = \frac{\dfrac{2Z(1)Z(-1)}{Z(1) + Z(-1)} Z(1)}{\dfrac{2Z(1)Z(-1)}{Z(1) + Z(-1)} - Z(1)}$$

$$Z_{01} = \frac{2Z(1)Z(-1)}{Z(-1) - Z(1)} \tag{5.53}$$

and the remaining lossy UE impedance $Z_2(t)$ is then given by (5.44).

We can further reduce the equations so that we do not need to explicitly compute $Z_1(t)$ by combining Equation (5.52) into Equation (5.44) and using the fact that $Z_{01} = Z_1(1)$:

$$Z_2(t) = Z_1(1) \frac{Z_1(t) - tZ_1(1)}{Z_1(1) - tZ_1(t)} = Z_{01} \frac{\dfrac{R_1 Z(t)}{R_1 - Z(t)} - tZ_{01}}{Z_{01} - t\dfrac{R_1 Z(t)}{R_1 - Z(t)}} = Z_{01} \frac{R_1 Z(t) - tZ_{01}(R_1 - Z(t))}{Z_{01}(R_1 - Z(t)) - tR_1 Z(t)}$$

and

$$Z_2(t) = Z_{01} \frac{(R_1 + tZ_{01})Z(t) - tZ_{01} R_1}{Z_{01} R_1 - (tR_1 + Z_{01})Z(t)} \tag{5.54}$$

In summary,

$$R_1 = \frac{2Z(1)Z(-1)}{Z(1) + Z(-1)}$$

$$Z_{01} = \frac{2Z(1)Z(-1)}{Z(-1)-Z(1)}$$

$$Z_2(t) = Z_{01}\frac{(R_1 + tZ_{01})Z(t) - tZ_{01}R_1}{Z_{01}R_1 - (tR_1 + Z_{01})Z(t)}$$

To follow is an example that illustrates the technique.

First, the lossy UE input impedance is computed from the optimized equalizer input reflection parameter $e_{11}(t)$ such that

$$Z(t) = \frac{1 + e_{11}(t)}{1 - e_{11}(t)}$$

For example, the aforementioned step provided the following lossy distributed network input impedance:

$$Z(t) = \frac{3924t^3 + 6072t^2 + 3348t + 180}{1361t^3 + 3366t^2 + 2541t + 621}$$

First, we extract a parallel resistor R_1 and lossless UE with characteristic impedance Z_{01} such that

$$Z(1) = \frac{3924 + 6072 + 3348 + 180}{1361 + 3366 + 2541 + 621} = \frac{13524}{7889} = 1.7143$$

and

$$Z(-1) = \frac{-3924 + 6072 - 3348 + 180}{-1361 + 3366 - 2541 + 621} = \frac{-1020}{85} = -12$$

so that

$$R_1 = \frac{2Z(1)Z(-1)}{Z(1) + Z(-1)} = \frac{2(1.7143)(-12)}{(1.7143) + (-12)} = 4.0$$

and

$$Z_{01} = \frac{2Z(1)Z(-1)}{Z(-1) - Z(1)} = \frac{2(1.7143)(-12)}{(-12) - (1.7143)} = 3.0$$

and the remaining lossy UE input impedance is given by

$$Z_2(t) = Z_{01}\frac{(R_1 + tZ_{01})Z(t) - tZ_{01}R_1}{Z_{01}R_1 - (tR_1 + Z_{01})Z(t)} = 3\frac{(4+3t)Z(t) - t3\times 4}{3\times 4 - (4t+3)Z(t)}$$

$$Z_2(t) = 3\frac{(4+3t)\dfrac{3924t^3 + 6072t^2 + 3348t + 180}{1361t^3 + 3366t^2 + 2541t + 621} - 12t}{12 - (4t+3)\dfrac{3924t^3 + 6072t^2 + 3348t + 180}{1361t^3 + 3366t^2 + 2541t + 621}} = \frac{285t^2 + 405t + 45}{327t^2 + 411t + 144}$$

Then, at this point, the new lossy distributed network input impedance is

$$Z(t)=\frac{285t^2+405t+45}{327t^2+411t+144}$$

and the procedure is repeated to extract R_2 and Z_{02}:

$$Z(1)=\frac{285+405+45}{327+411+144}=\frac{735}{882}=0.8333$$

and

$$Z(-1)=\frac{285-405+45}{327-411+144}=\frac{-75}{60}=-1.25$$

$$R_2=\frac{2Z(1)Z(-1)}{Z(1)+Z(-1)}=\frac{2(0.8333)(-1.25)}{(0.8333)+(-1.25)}=5.0$$

$$Z_{02}=\frac{2Z(1)Z(-1)}{Z(-1)-Z(1)}=\frac{2(0.8333)(-1.25)}{(-1.25)-(0.8333)}=1.0$$

$$Z_2(t)=Z_{02}\frac{(R_2+tZ_{02})Z(t)-tZ_{02}R_2}{Z_{02}R_2-(tR_2+Z_{02})Z(t)}=1\frac{(5+t)Z(t)-t1\times 5}{1\times 5-(5t+1)Z(t)}$$

$$Z_2(t)=\frac{(5+t)\dfrac{285t^2+405t+45}{327t^2+411t+144}-5t}{5-(5t+1)\dfrac{285t^2+405t+45}{327t^2+411t+144}}=\frac{18t+3}{19t+9}$$

Then, at this point, the new lossy distributed network input impedance is

$$Z(t)=\frac{18t+3}{19t+9}$$

and the procedure is repeated to extract R_3 and Z_{03}:

$$Z(-1)=\frac{18+3}{19+9}=\frac{21}{28}=0.75$$

$$Z(-1)=\frac{-18+3}{-19+9}=\frac{-15}{-10}=1.5$$

$$R_3=\frac{2Z(1)Z(-1)}{Z(1)+Z(-1)}=\frac{2(0.75)(1.5)}{(0.75)+(1.5)}=1.0$$

$$Z_{03}=\frac{2Z(1)Z(-1)}{Z(-1)-Z(1)}=\frac{2(0.75)(1.5)}{(1.5)-(0.75)}=3.0$$

$$Z_2(t)=Z_{03}\frac{(R_3+tZ_{03})Z(t)-tZ_{03}R_3}{Z_{03}R_3-(tR_3+Z_{03})Z(t)}=3\frac{(1+3t)Z(t)-t3\times 1}{3\times 1-(t+3)Z(t)}$$

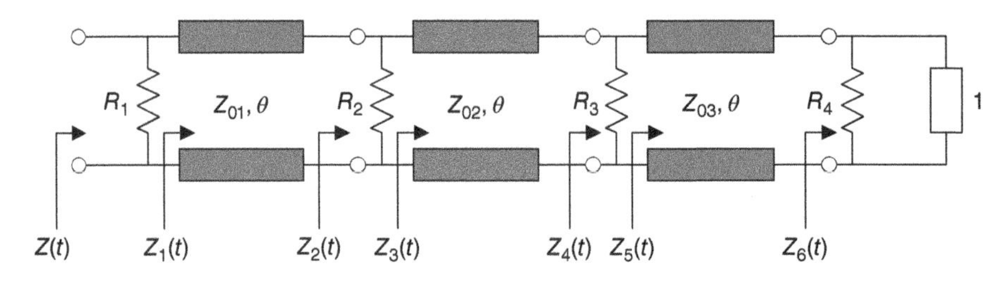

Figure 5.13 Lossy UE synthesis example.

$$Z_2(t) = 3\frac{(1+3t)\dfrac{18t+3}{19t+9} - 3t}{3-(t+3)\dfrac{18t+3}{19t+9}} = \frac{1}{2}$$

At this point, $Z_2(t)$ represents the input impedance after the third UE called $Z_6(t)$ in Figure 5.13. This means that $Z_6(t)$ is the last resistor R_4 in parallel with the 1 Ω normalized termination, so that

$$Z_6(t) = \frac{R_4 \times 1\Omega}{R_4 + 1\Omega}$$

and

$$R_4 = \frac{Z_6(t) \times 1\Omega}{1\Omega - Z_6(t)} = \frac{0.5 \times 1}{1-0.5} = 1.0$$

What follows is a MATLAB program that does the synthesis of this example:

```
clc;
clear;
close all;

syms t;
Z = (3924*t^3 + 6072*t^2 + 3348*t + 180)/(1361*t^3 + 3366*t^2 + 2541*t
+ 621);
% ==================================================
t = 1;
a = eval(Z);
t = -1;
b = eval(Z);
R1 = 2*(a*b)/(a+b);
Z01 = 2*(a*b)/(b-a);
fprintf('R1 = %6.4f, Z01 = %6.4f \n', R1, Z01);

syms t;
Z2 = Z01*simplify(((R1+t*Z01)*Z - R1*Z01*t)/(R1*Z01 - (t*R1+Z01)*Z));
% ==================================================
```

```
t = 1;
a = eval(Z2);
t = -1;
b = eval(Z2);
R2 = 2*(a*b)/(a+b);
Z02 = 2*(a*b)/(b-a);
fprintf('R2 = %6.4f, Z02 = %6.4f \n', R2, Z02);

syms t;
Z4 = Z02*simplify( ((R2+t*Z02)*Z2 - R2*Z02*t)/(R2*Z02 -(t*R2+Z02)
*Z2) );
% =====================================================
t = 1;
a = eval(Z4);
t = -1;
b = eval(Z4);
R3 = 2*(a*b)/(a+b);
Z03 = 2*(a*b)/(b-a);
fprintf('R3 = %6.4f, Z03 = %6.4f \n', R3, Z03);

syms t;
Z6 = Z03*simplify( ((R3+t*Z03)*Z4 - R3*Z03*t)/(R3*Z03 -(t*R3+Z03)
*Z4) );
```

Running the MATLAB program gives

```
R1 = 4.0000, Z01 = 3.0000
R2 = 5.0000, Z02 = 1.0000
R3 = 1.0000, Z03 = 3.0000
Z6 = 1/2
```

5.8 Design Procedures

The design of the multistage lossy distributed amplifier starts with the first equalizer–transistor stage terminated on reference impedance Z_0 (e.g., 50 Ω) as shown in Figure 5.14.

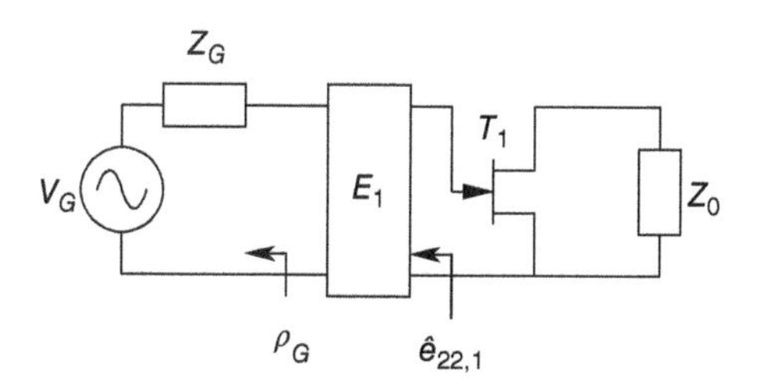

Figure 5.14 Optimizing the first equalizer.

The transducer gain for the first stage is given by

$$T_1 = \left(1-|\rho_G|^2\right)\frac{|e_{21,1}|^2|S_{21,1}|^2}{|1-e_{11,1}\rho_G|^2|1-\hat{e}_{22,1}S_{11,1}|^2} \tag{5.55}$$

with

$$\hat{e}_{22,1} = e_{22,1} + \frac{e_{21,1}^2\rho_G}{1-e_{11,1}\rho_G} \tag{5.56}$$

$$\rho_G = \frac{Z_G-Z_0}{Z_G+Z_0} \tag{5.57}$$

and the target transducer gain for the first stage is given by

$$T_{0,1} = \min\left\{\frac{|S_{21,1}(j\omega)|^2}{\left|1-|S_{11,1}(j\omega)|^2\right|}\right\} \tag{5.58}$$

Once the first equalizer has been optimized, the second equalizer–transistor stage is placed between the first transistor and the load Z_0 as shown in Figure 5.15. At this time, the first equalizer is fixed and only $e_{11,2}(t)$, $e_{22,2}(t)$, and $e_{12,2}(t)=e_{21,2}(t)$ of the second equalizer vary between iterations of the optimization.

The transducer gain T_2 for the two first equalizer–transistor stages is

$$T_2 = T_1\frac{|e_{21,2}|^2|S_{21,2}|^2}{|1-e_{11,2}S_{G,2}|^2|1-\hat{e}_{22,2}S_{11,2}|^2} \tag{5.59}$$

with

$$\hat{e}_{22,2} = e_{22,2} + \frac{e_{21,2}^2S_{G,2}}{1-e_{11,2}S_{G,2}} \tag{5.60}$$

$$S_{G,2} = S_{22,1} + \frac{S_{12,1}S_{21,1}\hat{e}_{22,1}}{1-S_{11,1}\hat{e}_{22,1}} \tag{5.61}$$

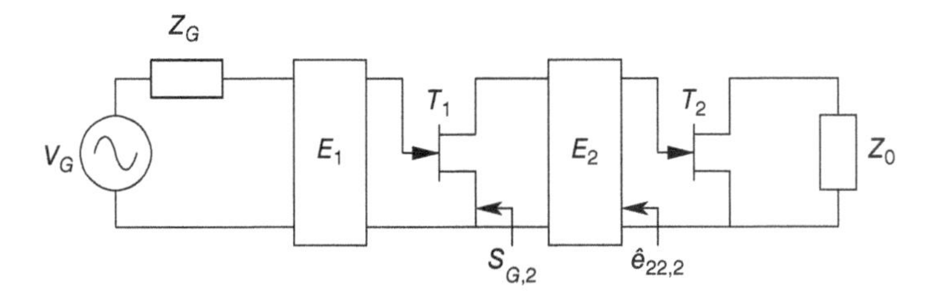

Figure 5.15 Optimizing the second equalizer.

and the target transducer gain for the second stage is given by

$$T_{0,2} = \min\left\{ T_{MAX,1} \frac{|S_{21,2}|^2}{\left|1-|S_{11,2}|^2\right|} \right\} \tag{5.62}$$

where

$$T_{MAX,1} = \frac{|S_{21,1}|^2}{\left|1-|S_{11,1}|^2\right|\left|1-|S_{22,1}|^2\right|} \frac{1}{\left|1-\frac{S_{12,1}S_{21,1}S^*_{11,1}S^*_{22,1}}{\left(1-|S_{11,1}|^2\right)\left(1-|S_{22,1}|^2\right)}\right|^2} \tag{5.63}$$

The procedure is repeated, and Figure 5.16 shows the situation for *k*th equalizer–transistor terminated on Z_0.

The transducer gain T_k for the first *k* equalizer–transistor stages is

$$T_k = T_{k-1} \frac{|e_{21,k}|^2 |S_{21,k}|^2}{|1-e_{11,k}S_{G,k}|^2 |1-\hat{e}_{22,k}S_{11,k}|^2} \tag{5.64}$$

$$\hat{e}_{22,k} = e_{22,k} + \frac{e^2_{21,k}S_{G,k}}{1-e_{11,k}S_{G,k}} \tag{5.65}$$

$$S_{G,k} = S_{22,k-1} + \frac{S_{12,k-1}S_{21,k-1}\hat{e}_{22,k-1}}{1-S_{11,k-1}\hat{e}_{22,k-1}} \tag{5.66}$$

and the target transducer gain for the *k*th stage is given by

$$T_{0,k} = \min\left\{ T_{MAX,1} \times T_{MAX,2} \times \cdots \times T_{MAX,k-1} \frac{|S_{21,k}|^2}{\left|1-|S_{11,k}|^2\right|} \right\} \tag{5.67}$$

where

$$T_{MAX,i} = \frac{|S_{21,i}|^2}{\left|1-|S_{11,i}|^2\right|\left|1-|S_{22,i}|^2\right|} \frac{1}{\left|1-\frac{S_{12,i}S_{21,i}S^*_{11,i}S^*_{22,i}}{\left(1-|S_{11,i}|^2\right)\left(1-|S_{22,i}|^2\right)}\right|^2} \tag{5.68}$$

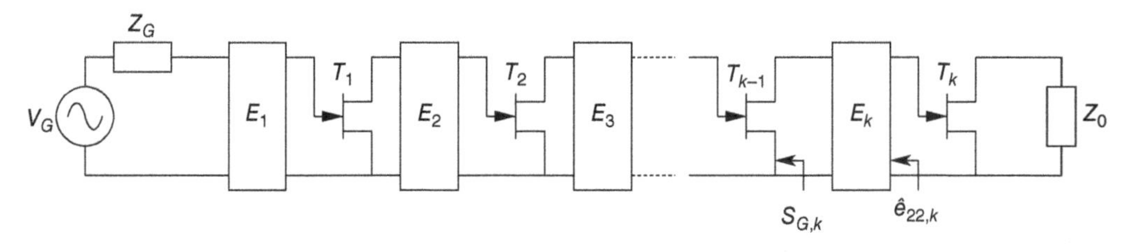

Figure 5.16 Optimization of the *k*th equalizer.

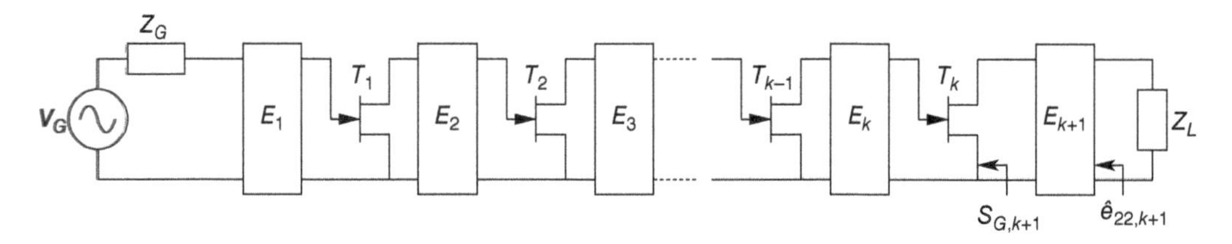

Figure 5.17 Optimizing the $(k+1)$th equalizer.

Once the kth equalizer has been optimized, a last equalizer E_{k+1} can be added to compensate for an arbitrary load Z_L different than Z_0 as shown in Figure 5.17.

The overall transducer gain T_{GT} in the case of Figure 5.17 is given by

$$T_{GT} = T_k \frac{|e_{21,k+1}|^2 \left(1 - |\rho_L|^2\right)}{|1 - e_{11,k+1} S_{G,k+1}|^2 |1 - \hat{e}_{22,k+1} \rho_L|^2} \tag{5.69}$$

with

$$\hat{e}_{22,k+1} = e_{22,k+1} + \frac{e_{21,k+1}^2 S_{G,k+1}}{1 - e_{11,k+1} S_{G,k+1}} \tag{5.70}$$

$$S_{G,k+1} = S_{22,k} + \frac{S_{12,k} S_{21,k} \hat{e}_{22,k}}{1 - S_{11,k} \hat{e}_{22,k}} \tag{5.71}$$

$$\rho_L = \frac{Z_L - Z_0}{Z_L + Z_0} \tag{5.72}$$

and the target transducer gain for the $(k+1)$th equalizer is given by

$$T_{0,k+1} = \min\{T_{MAX,1} \times T_{MAX,2} \times \cdots \times T_{MAX,k-1} \times T_{MAX,k}\} \tag{5.73}$$

5.9 Realizations

5.9.1 *Single-Stage Broadband Hybrid Realization*

The procedures described in this chapter were implemented in a program called SYNTARD. The first realization example was to design a 100 MHz to 5 GHz broadband amplifier [4, 5]. The amplifier used the NEC 72084A GaAs FET transistor that was biased with $V_{GS} = -1.25$ V, $V_{DS} = 3$ V, and $I_{DS} = 3.12$ mA. Table 5.1 provides the scattering parameters used to characterize the transistor.

The equalizers were optimized for gain and input and output VSWR and are shown in Figure 5.18. Note that some of the parallel resistors from optimization were very large and were omitted from the final circuit.

The amplifier was realized on a 0.635-mm-thick alumina substrate with $\lambda/4$ lines and surface mount resistors as shown in Figure 5.19.

A picture of the single-stage amplifier and the picture of a two-transistor design described later are given in Figure 5.20.

Table 5.1 Scattering parameters for the NEC 72084A GaAs FET transistor

	S_{11}		S_{21}		S_{12}		S_{22}	
FREQ (GHz)	Mag	Ang	Mag	Ang	Mag	Ang	Mag	Ang
0.1	0.997	−2.1	3.963	177.3	0.003	105	0.638	−2.1
0.95	0.966	−26.9	3.83	154.2	0.033	73.2	0.620	−18.5
2	0.882	−55.7	3.493	128.1	0.063	57.3	0.586	−36.8
3	0.797	−82.1	3.110	105.9	0.082	42.7	0.539	−51.6
4	0.723	−104.4	2.807	86.3	0.095	34.2	0.501	−65.9
5	0.668	−124.9	2.530	68.1	0.102	25.9	0.473	−79.2

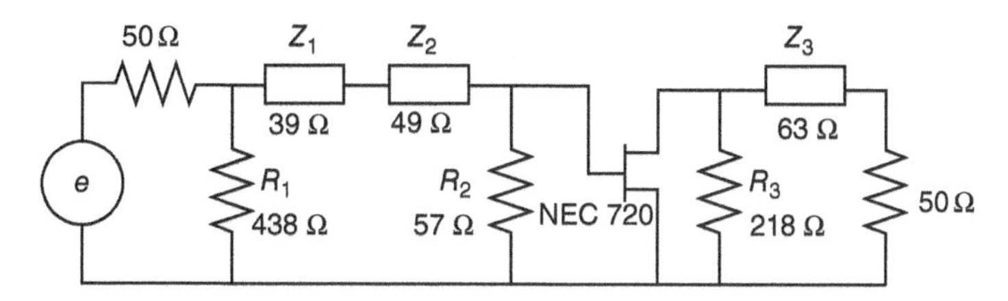

Figure 5.18 Single-stage amplifier with input and output equalizers.

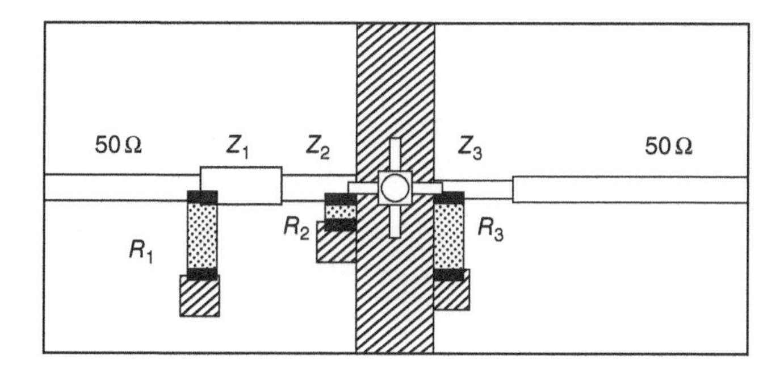

Figure 5.19 Layout of the single-stage 100 MHz to 5 GHz amplifier.

The measured and simulated gains of the single-stage amplifier are shown in Figure 5.21. The small-signal measured gain is about 4.5 dB across the 100 MHz to 5 GHz frequency range. The measured and simulated gains are in very good agreement over this band.

Figure 5.22 shows the measured and simulated input and output return loss. The measured return losses are less than −5 dB.

The simulated and measured input return losses are in very good agreement across the entire 100 MHz to 5 GHz frequency band. The simulated and measured output return losses get further apart at higher frequencies.

5.9.2 *Two-Stage Broadband Hybrid Realization*

For the second example, a two-stage amplifier has been optimized and simulated over the 100 MHz to 9 GHz frequency band using the SYNTARD program [4, 5]. The amplifier used the NEC 71083

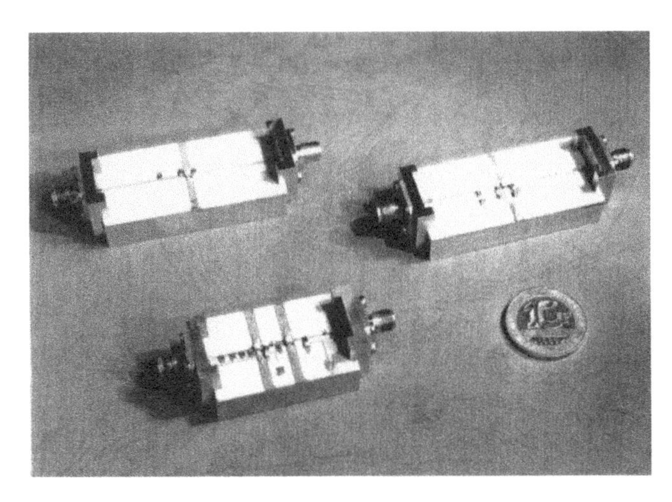

Figure 5.20 Single-stage and two-stage broadband amplifiers using lossy distributed equalizers.

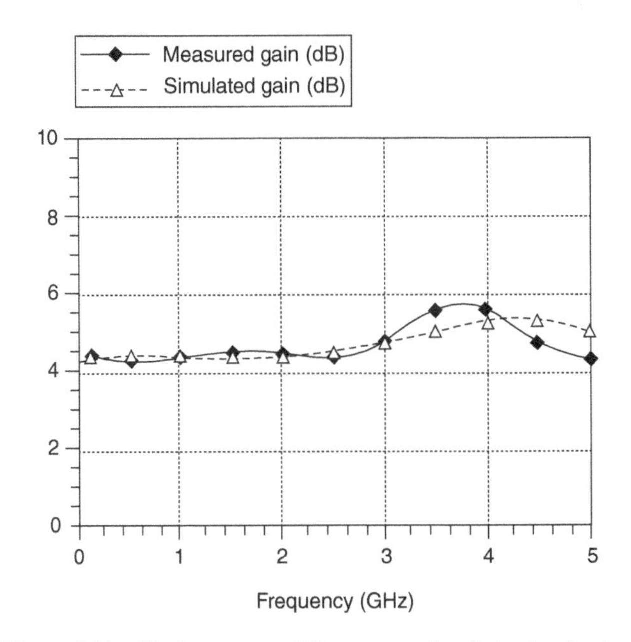

Figure 5.21 Single-stage amplifier measured and simulated gains.

GaAs FET transistor biased with $V_{GS}=0\,\text{V}$, $V_{DS}=3\,\text{V}$, and $I_{DS}=40.0$ mA. Table 5.2 provides the scattering parameters used to characterize the transistor.

The equalizers were optimized for gain, input, and output VSWR and are shown in Figure 5.23. Note that some of the parallel resistors from optimization were very large and were omitted from the final circuit.

The amplifier was realized on a 0.635-mm-thick alumina substrate with $\lambda/4$ lines and surface mount resistors. It is shown along with the picture of the single-stage amplifier in Figure 5.20.

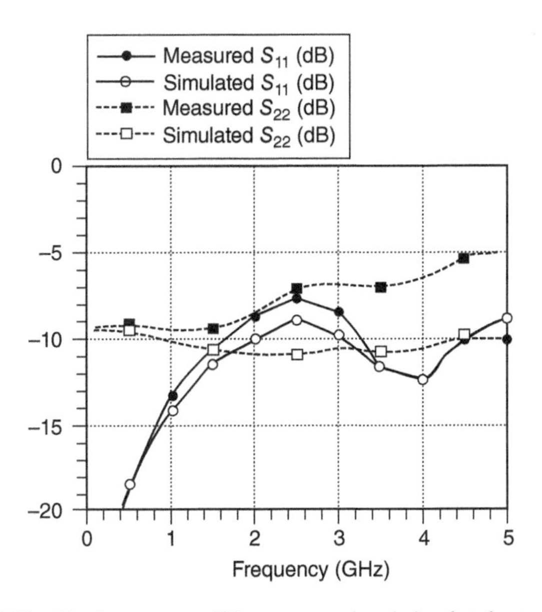

Figure 5.22 Single-stage amplifier measured and simulated return losses.

Table 5.2 Scattering parameters for the NEC 71083 GaAs FET transistor

	S_{11}		S_{21}		S_{12}		S_{22}	
FREQ (GHz)	Mag	Ang	Mag	Ang	Mag	Ang	Mag	Ang
0.1	0.999	−2.511	4.48	177.8	0.0017	88.15	0.588	−1.45
1	0.981	−24.4	4.409	157.6	0.017	75.7	0.581	−15.0
3	0.866	−67.1	3.865	117.8	0.043	51.5	0.545	−43.6
5	0.772	−105.1	3.317	82.8	0.057	34.1	0.502	−69.7
7	0.686	−139.9	2.817	51.7	0.065	23.1	0.473	−94.9
9	0.626	−170.5	2.476	23.0	0.073	15.7	0.472	−118.7

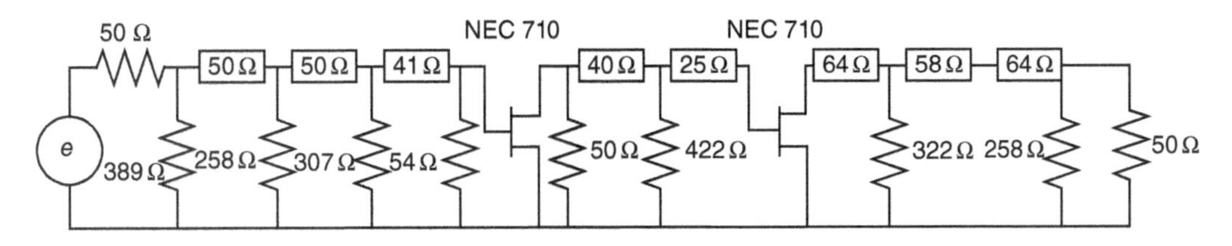

Figure 5.23 Two-stage amplifier using three lossy distributed equalizers.

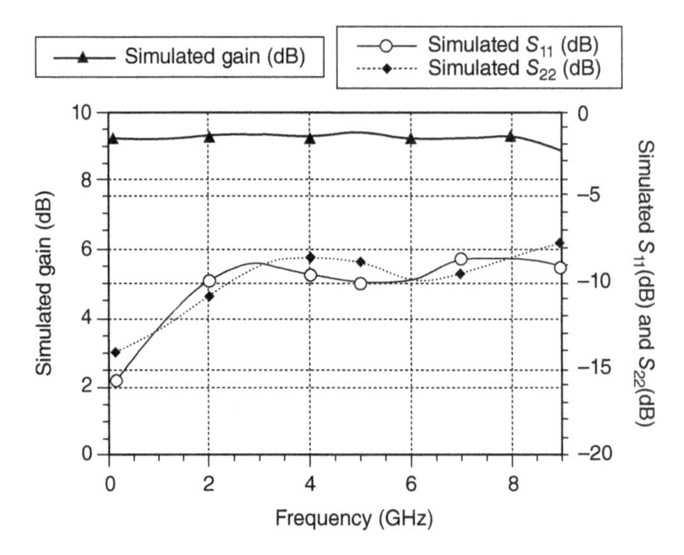

Figure 5.24 Two-stage amplifier simulated gain and return losses.

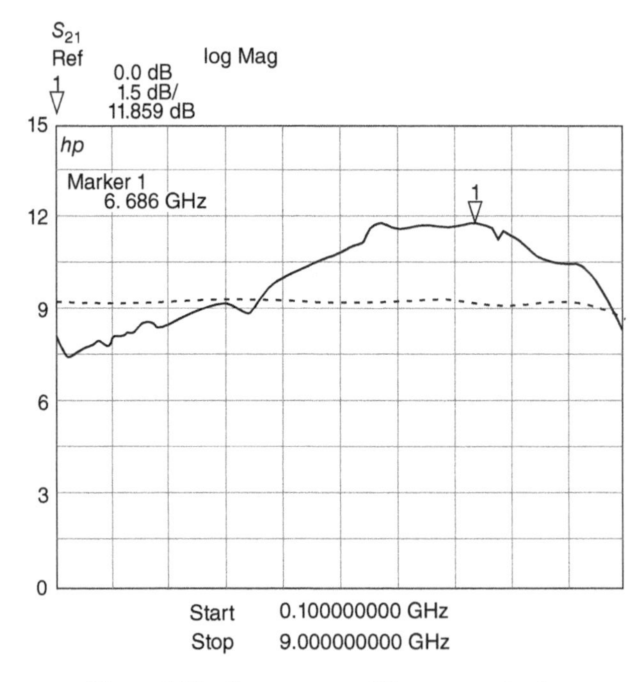

Figure 5.25 Two-stage amplifier measured gain.

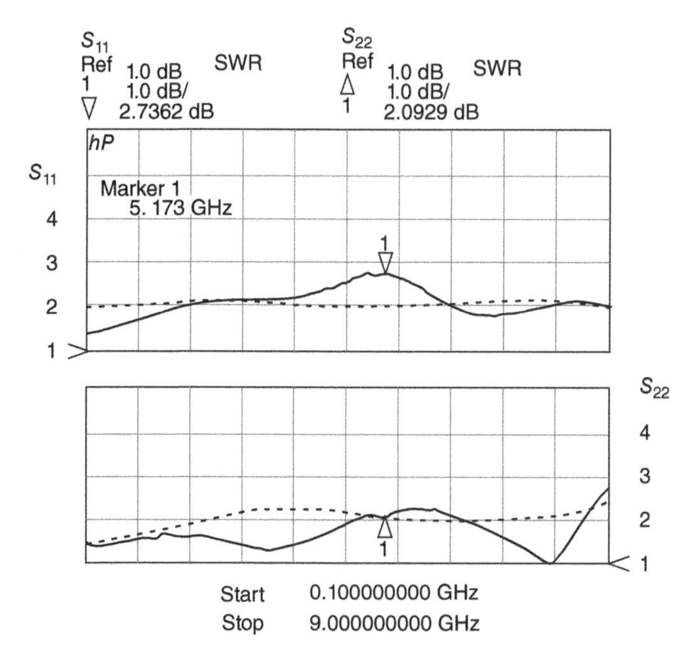

Figure 5.26 Two-stage amplifier measured VSWRs.

Figure 5.24 shows the simulated gain and return losses. The simulated gain is about 9.1 dB with ±0.2 dB flatness across the 100 MHz to 9 GHz frequency band. The simulated return losses are less than −7.8 dB over the prescribed band.

Figure 5.25 shows the measured gain. It is about 9 dB with a peak value of 11.5 dB at 8 GHz. The measured VSWR are given in Figure 5.26. They are in good agreement with the simulations and are less than 2.5 in the prescribed band.

In conclusion, the proposed method can be used to design microwave lossy distributed amplifiers. As in the case of FREEL, it does not require any two-port active models because it directly accepts scattering parameter data. Moreover, the stability and the feasibility of compute equalizers are ensured by the Hurwitz factorization. The validity and advantages of the proposed method have been demonstrated by two design examples. The amplifiers were made for the CNET using commercial transistors.

References

[1] K. K. Pang, Y. Nemoto, "Cascade transmission line network with lossy junctions," IEEE Trans. Circ. Syst., vol. CAS 26, pp. 52–58, January 1979.

[2] P. M. Martin, "Synthesis of Distributed and Lossy Matching Networks Using the Modified Real Frequency Method," Ph.D. dissertation, Brest University, December 1993.

[3] P. M. Martin, A. Perennec, P. Jarry, "Optimization and synthesis of distributed and lossy matching networks by the modified real frequency method," PIERS, Cambridge, MA, pp. 310, July 1991.

[4] E. Kerherve, P. Jarry, P. M. Martin, "Design of broad-band matching network with lossy junctions using the real-frequency technique," IEEE Trans. Microw. Theory Tech., vol. MTT 46, no. 3, pp. 242–249, March 1998.

[5] E. Kerherve, P. Jarry, P. M. Martin, "Efficient numerical CAD technique for RF and Microwave Amplifiers," Int. J. RF Microw. Comput. Aided Eng., vol. 8, no. 2, 1998, pp. 131–141, March 1998.

6

Multistage Power Amplifiers

6.1 Introduction

Solid-state power amplifiers are widely used in wireless and satellite communications. With the evolving new generations of communication networks, microwave circuit and system requirements will continue to increase. For power amplifiers, this means linearity and efficiency features need to improve constantly. There are no systematic procedures for designing optimum power amplifier currently available. A main drawback of the coupled large-signal model-harmonic-balance technique is the requirement of expensive equipment and prohibitive time since the number of measurements increases with the accuracy of the large signal model without any guarantee of convergence. In this

Microwave Amplifier and Active Circuit Design Using the Real Frequency Technique, First Edition.
Pierre Jarry and Jacques N. Beneat.

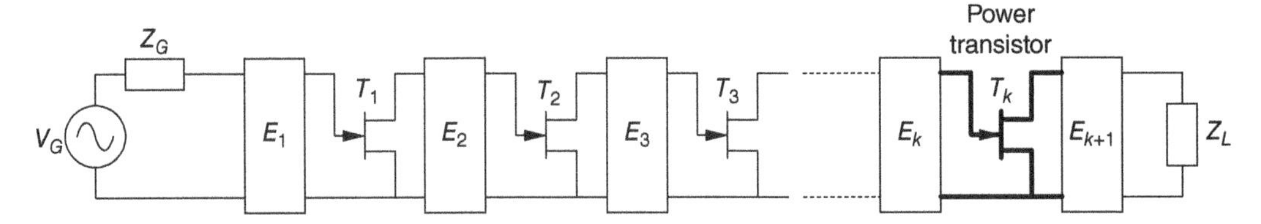

Figure 6.1 Multistage power amplifier representation.

chapter, we propose to design multistage power amplifiers, based on the real frequency technique (RFT) and the use of large-signal *S* parameter data generated from a few reference points. This simple numerical approach does not require algebraic expressions of a transfer function nor a circuit topology.

6.2 Multistage Power Amplifier Representation

The multistage power amplifier is represented in Figure 6.1 [1]. It assumes that there are *k* transistors, but the last transistor is assumed to be a power transistor. An equalizer is associated with each transistor. There is one final equalizer placed after the *k*th transistor. The source and load impedances can be arbitrary, complex, and frequency dependent. A transistor can represent a transistor alone or a transistor with added but known circuitry as long as the *S* parameters can be provided.

The equalizer of the *k*th stage is characterized by the scattering matrix

$$S_{E_k} = \begin{pmatrix} e_{11,k}(t) & e_{12,k}(t) \\ e_{21,k}(t) & e_{22,k}(t) \end{pmatrix} \tag{6.1}$$

The transistor circuit of the *k*th stage is characterized by the scattering matrix

$$S_{T_k} = \begin{pmatrix} S_{11,k}(t) & S_{12,k}(t) \\ S_{21,k}(t) & S_{22,k}(t) \end{pmatrix} \tag{6.2}$$

where *t* is the Richard's variable defined in Chapter 1.

An equalizer is assumed to be a lossless distributed two-port with a scattering matrix such that

$$S_E = \begin{pmatrix} e_{11}(t) = \dfrac{h(t)}{g(t)} & e_{12}(t) = \dfrac{f(t)}{g(t)} \\ e_{21}(t) = \dfrac{f(t)}{g(t)} & e_{22}(t) = \dfrac{-h(-t)}{g(t)} \end{pmatrix} \tag{6.3}$$

where $h(t)$ and $g(t)$ are polynomials of order *n* with real-valued coefficients:

$$h(s) = h_0 + h_1 t + \cdots + h_{n-1}t^{n-1} + h_n t^n$$

$$g(s) = g_0 + g_1 t + \cdots + g_{n-1}t^{n-1} + g_n t^n$$

and where $f(t)$ is a function of the form

$$f(t) = t^r\left(1-t^2\right)^{\frac{n-q}{2}} \tag{6.4}$$

where q represents the number of possible low-pass stubs and r represents the number of possible high-pass stubs.

The lossless condition $|e_{11}|^2 + |e_{12}|^2 = 1$ leads to

$$g(t)g(-t) = f(t)f(-t) + h(t)h(-t) \tag{6.5}$$

so that

$$g(t)g(-t) = h(t)h(-t) + t^{2r}\left(1 - t^2\right)^{n-q} \tag{6.6}$$

For a given equalizer, the problem consists of defining the unknown polynomial coefficients $h_0, h_1, h_2, \ldots, h_n$ that optimize the response of the system. Once $h(t)$ is known $g(t)g(-t)$ can be computed using

$$G\left(t^2\right) = g(t)g(-t) = G_0 + G_1 t^2 + \cdots + G_n t^{2n}$$

where

$$\begin{cases} G_0 = h_0^2 + 1 \\ G_1 = -h_1^2 + 2h_2 h_0 - C_{n-q}^1 \\ \ldots \\ G_i = (-1)^i h_i^2 + 2h_{2i} h_0 + 2\sum_{j=2}^{i} (-1)^{j-1} h_{j-1} h_{2i-j+1} + (-1)^i C_{n-q}^i \\ \ldots \\ G_{n-q} = (-1)^{n-q} h_{n-q}^2 + 2h_{2(n-q)} h_0 + 2\sum_{j=2}^{n-q} (-1)^{j-1} h_{j-1} h_{2(n-q)-j+1} + (-1)^{n-q} \\ \ldots \\ G_n = (-1)^n h_n^2 \end{cases} \tag{6.7}$$

with

$$C_n^p = \frac{n!}{p!(n-p)!}$$

The left-half roots of $g(t)g(-t)$ are retained to form a Hurwitz polynomial $g(t)$. At this point, the equalizer scattering parameters $e_{11}(t)$, $e_{12}(t)$, $e_{21}(t)$, and $e_{22}(t)$ can be computed. During optimization, they will be needed for computing the error function to be optimized, and after optimization, for example, $e_{11}(t)$ can be used to realize the equalizer using the synthesis procedures given in Section 3.9.

In the case of power amplifiers, the last equalizer E_{k+1} will be optimized differently than all other equalizers. In addition, the last transistor (power transistor) will require knowledge of both frequency behavior and large signal power behavior of its scattering parameters. All other transistors are assumed to be working in their linear region and only small-signal scattering parameters showing frequency behavior are needed.

This means that the scattering parameter data of the power transistor will need to be two-dimensional where one needs to provide the frequency behavior of a scattering parameter for several input power levels. Clearly, this will be extremely lengthy to measure so that a two-dimensional interpolation technique is provided in Section 6.3.2.

Once the last equalizer E_{k+1} has been optimized for what is called added power, the other equalizers are optimized using the traditional RFT techniques of Chapter 3. After optimization, all equalizers are realized using the distributed synthesis procedures given in Section 3.9.

6.3 Added Power Optimization

The first objective is to take into account the nonlinear behavior of the power transistor. Several reasons lead us to choose a transistor description based on large-signal S parameters:

1. It avoids defining a specific large-signal transistor model
2. The RFT is based on using the scattering parameters of equalizers and transistors
3. Large-signal S parameters can be used to optimize the added power of an active circuit leading to optimum performance for power amplifiers

In addition, rather than relying on cumbersome load-pull measurements, a two-dimensional interpolation technique is used to provide scattering parameter frequency behavior for any input power level using about 9 reference points.

The optimization of added power usually used in oscillator design is found to be well suited for power amplifier design because their nonlinear properties have a common nature and common description. In addition, large-signal S parameters have been successfully used in oscillators and power amplifier design [2]. Gilmore and Rosenbaum presented an approach to obtain maximum power delivered to the load when designing microwave oscillators using large-signal S parameters [3].

6.3.1 *Requirements for Maximum Added Power*

An active two-port device can be described by its scattering parameters in terms of incident and reflected waves a_i and b_i, where $i = 1$ or 2 as shown in Figure 6.2.

The incident and reflected waves are related by the scattering parameters such that

$$\begin{aligned} b_1 &= S_{11} \cdot a_1 + S_{12} \cdot a_2 \\ b_2 &= S_{21} \cdot a_1 + S_{22} \cdot a_2 \end{aligned}$$

We define the input and output powers as

$$\begin{aligned} P_{in} &= |a_1|^2 - |b_1|^2 \\ P_{out} &= |b_2|^2 - |a_2|^2 \end{aligned} \quad (6.8)$$

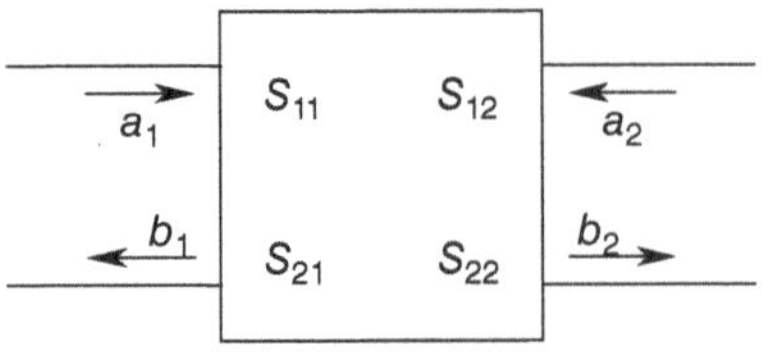

Figure 6.2 Scattering parameter representation for an active two-port device.

and the added power as the output power minus the input power such that

$$P_{add} = P_{out} - P_{in} \tag{6.9}$$

or

$$P_{add} = |b_2|^2 - |a_2|^2 + |b_1|^2 - |a_1|^2 \tag{6.10}$$

Then the goal is to define the input reflection coefficient Γ_{in} that maximizes the added power delivered to the load. The input reflection coefficient Γ_{in} is defined as the ratio of the input reflected wave over the input incident wave

$$\Gamma_{in} = \frac{b_1}{a_1} = A + jB \tag{6.11}$$

The added power delivered to the load can be expressed in terms of the input reflection Γ_{in} and the scattering parameters of the active device such that

$$P_{add} = |a_1|^2 \left[\frac{1}{|S_{12}|^2} \left[|S_{22}\Gamma_{in}|^2 + |\Delta|^2 - 2\mathrm{Re}\left\{ S_{22}^* \Gamma_{in}^* \Delta \right) \right\} - |\Gamma_{in}|^2 - |S_{11}|^2 + 2\mathrm{Re}\left\{ \Gamma_{in}^* S_{11} \right\} \right] + |\Gamma_{in}|^2 - 1] \tag{6.12}$$

with

$$\Delta = S_{11}S_{22} - S_{12}S_{21}$$

To define the maximum added power, the derivatives with respect to real and imaginary parts of the input reflection coefficient Γ_{in} are defined and made equal to zero:

$$\frac{\partial P_{add}}{\partial A} = \frac{\partial P_{add}}{\partial B} = 0$$

This leads to the two equations:

$$2A\left[1 + \frac{|S_{22}|^2 - 1}{|S_{12}|^2}\right] - \frac{1}{|S_{12}|^2}\left[2\mathrm{Re}\left\{S_{22}^*\Delta\right\} + 2\mathrm{Re}\left\{S_{11}\right\}\right] = 0 \tag{6.13}$$

$$2B\left[1 + \frac{|S_{22}|^2 - 1}{|S_{12}|^2}\right] - \frac{1}{|S_{12}|^2}\left[2\mathrm{Im}\left\{S_{22}^*\Delta\right\} + 2\mathrm{Im}\left\{S_{11}\right\}\right] = 0 \tag{6.14}$$

The values of A and B satisfying these two equations provide an optimum input reflection coefficient such that

$$\Gamma_{in\ (opt)} = \frac{S_{11} - S_{22}^*\Delta}{1 - |S_{22}|^2 - |S_{12}|^2} \tag{6.15}$$

The optimum added power in terms of incident wave power and scattering parameters of the active device is then given by

$$P_{add\ (opt)} = |a_1|^2 \left[\frac{|S_{11}|^2 + |S_{12}|^2 + |S_{21}|^2 + |S_{22}|^2 - |\Delta|^2 - 1}{1 - |S_{22}|^2 - |S_{12}|^2} \right] \tag{6.16}$$

In theory, the approach described earlier is only valid for small-signal S parameters. It was, however, found that this approach is also valid for weakly nonlinear amplifier design.

The technique to define the optimal added power to the load is to

- First define the large-signal S parameters for different input powers using, for example, a harmonic balance procedure [4] with a classical commercial simulator if direct S parameter measurements are prohibitive
- Compute $P_{add\,(opt)}$ using different input powers $|a_1|^2$ and their corresponding large-signal S parameters
- Plot the curve of $P_{add\,(opt)}$ versus $|a_1|^2$ and select the value of $|a_1|^2$ that provides the maximum value of $P_{add\,(opt)}$

In the aforementioned procedure, the large-signal S parameters are provided for different input powers but at a single-frequency point like the center frequency of the passband.

The optimal input power is then used to generate the large-signal S parameter data over the frequency points used in the classical RFT technique for multistage amplifier design. Only the power transistor requires the use of large-signal S parameter frequency data, the other transistors are assumed to be operating in their linear region, and small-signal S parameter frequency data are sufficient.

Note that this might not always be the best assumption for the transistor before the power transistor since if the optimum input power level is high then this transistor might be operating in its nonlinear region, and therefore, the small-signal S parameter frequency data are inaccurate.

A technique for generating the large signal S parameter frequency data at any input power value is the subject of Section 6.3.2.

6.3.2 Two-Dimensional Interpolation

The generation of the large-signal S parameter data over a given frequency band and at different input power levels is based on a two-dimensional interpolation technique that uses cubic spline functions [1, 5].

In the one-dimensional case, cubic spline interpolation enables one to control the continuity of successive derivative functions unlike the Lagrange, Newton, or Hermite methods.

For example, in Figure 6.3, for a function whose values $y_i = y(x_i)$ are known at N different points x_i, one can use cubic spline interpolation to define its value at any point x within the interval $[x_i, x_{i+1}]$.

The general expression of the interpolation function is determined from a spline cubic interpolant in each interval. For the $(i-1)$th and ith intervals, the polynomials are given by

$$P_{i-1}(x) = a_{i-1}(x - x_{i-1})^3 + b_{i-1}(x - x_{i-1})^2 + c_{i-1}(x - x_{i-1}) + d_{i-1} \tag{6.17}$$

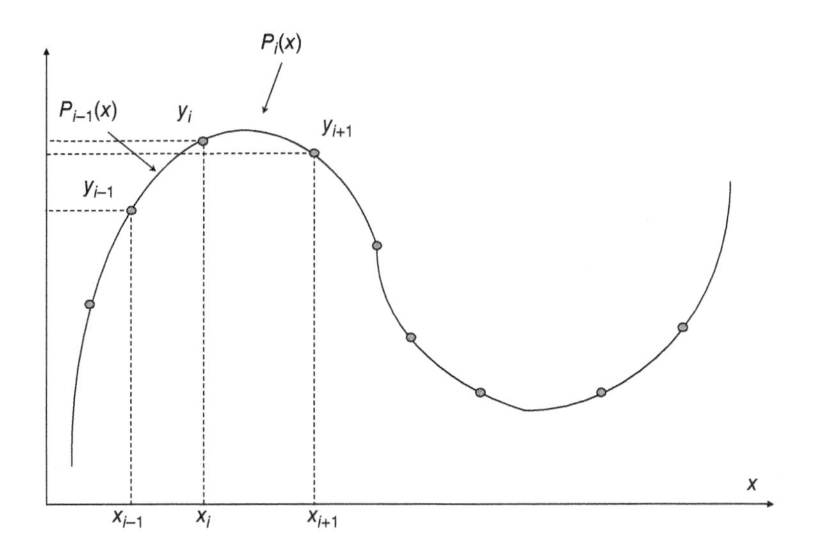

Figure 6.3 One-dimensional cubic spline interpolation.

$$P_i(x) = a_i(x-x_i)^3 + b_i(x-x_i)^2 + c_i(x-x_i) + d_i \tag{6.18}$$

where the coefficients $a_{i-1}, b_{i-1}, c_{i-1}, d_{i-1}, a_i, b_i, c_i$, and d_i are the unknowns. Using the continuity property of the first and second derivatives:

$$a_i = \frac{P_i''(x_{i+1}) - P_i''(x_i)}{6\Delta_{i+1}} \tag{6.19}$$

$$b_i = \frac{P_i''(x_i)}{2} \tag{6.20}$$

$$c_i = \frac{y_{i+1}-y_i}{\Delta_{i+1}} - \frac{P_i''(x_{i+1}) - 2P_i''(x_i)}{6}\Delta_{i+1} \tag{6.21}$$

$$d_i = y_i \tag{6.22}$$

with $\Delta_i = x_i - x_{i-1}$.

We need to resolve the following system:

$$\frac{y_{i+1}-y_i}{\Delta_{i+1}} = \frac{1}{6}\left[2P_i''(x_i)\Delta_i + 2P_i''(x_i)\Delta_{i+1} + P_{i-1}''(x_{i+1})\Delta_i + P_i''(x_{i+1})\Delta_{i+1}\right] \tag{6.23}$$

If we call S_i the second derivative at junction point i, then the problem consists of solving a set of $n-1$ equations of the type:

$$\Delta_i S_{i-1} + 2(\Delta_i + \Delta_{i+1})S_i + \Delta_{i+1}S_{i+1} = 6\left[\frac{y_{i+1}-y_i}{\Delta_{i+1}} - \frac{y_i - y_{i-1}}{\Delta_i}\right] \tag{6.24}$$

but where S_0 and S_n must be provided.

This system of $n-1$ equations can be expressed in matrix form such that

$$\begin{bmatrix} 2(\Delta_1+\Delta_2) & \Delta_2 & 0 & \cdots & \cdots & 0 \\ \Delta_2 & 2(\Delta_2+\Delta_3) & \Delta_3 & \cdots & \cdots & \\ \cdots & \cdots & \cdots & \cdots & \cdots & \cdots \\ 0 & 0 & 0 & \cdots & & \\ 0 & 0 & 0 & \cdots & \Delta_{n-1} & 2(\Delta_{n-1}+\Delta_n) \end{bmatrix} \begin{bmatrix} S_1 \\ S_2 \\ \cdots \\ \cdots \\ S_{n-1} \end{bmatrix} = 6 \begin{bmatrix} (y_2-y_1)/\Delta_2-(y_1-y_0)/\Delta_1 \\ (y_3-y_2)/\Delta_3-(y_2-y_1)/\Delta_2 \\ \cdots \\ \cdots \\ (y_n-y_{n-1})/\Delta_n-(y_{n-1}-y_{n-2})/\Delta_{n-1} \end{bmatrix} \tag{6.25}$$

Solving this system provides the coefficients a_i, b_i, c_i, and d_i of the cubic spine polynomial for each of the $[x_i, x_{i+1}]$ intervals.

In the two-dimensional case, a vector x_{1a} contains m reference values for the first dimension and a second vector x_{2a} contains n reference values for the second dimension. The reference function values for y are then tabulated in a matrix y_a that has m rows and n columns such that the jth row, kth column element is given by

$$y_a[j,k] = y(x_{1a}[j], x_{2a}[k]) \tag{6.26}$$

For example, x_{1a} contains the reference input power levels, x_{2a} contains the reference frequency values, and y is the S_{11} parameter value at one reference input power level and at one reference frequency value. The y_a matrix contains the S_{11} parameter values for all reference power levels specified in x_{1a} and all reference frequency values specified in x_{2a}.

The matrix y_a is then used to find the interpolating function to estimate $y(x_1, x_2)$, where x_1 is not a reference value found in x_{1a} and x_2 is not a reference value found in x_{2a}. This is done with the help of the grid concept shown in Figure 6.4 that shows four reference points surrounding the desired point.

The technique relies on defining new coordinates such that

$$t = \frac{x_1 - x_{1a}[j]}{x_{1a}[j+1] - x_{1a}[j]} \tag{6.27}$$

$$u = \frac{x_2 - x_{2a}[k]}{x_{2a}[k+1] - x_{2a}[k]} \tag{6.28}$$

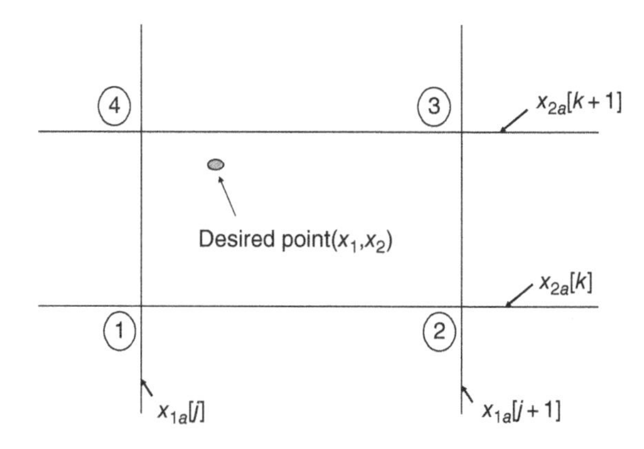

Figure 6.4 Grid concept for spline interpolation.

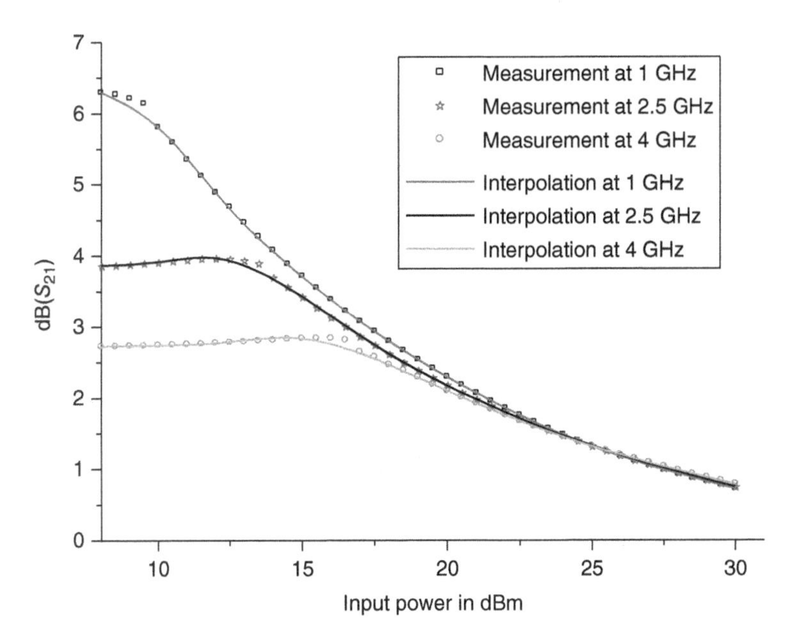

Figure 6.5 Two-dimensional interpolation of S_{21} using only nine reference points.

The cubic spline interpolant function can then be written as

$$y(x_1,x_2) = \sum_{i=1}^{4}\sum_{j=1}^{4} c_{ij} t^{i-1} u^{j-1} \tag{6.29}$$

where the c_{ij} coefficients are determined using the continuity property of the spline functions and their derivatives. This requires the knowledge for each point on the grid of

$$y_a[j,k] = y(x_{1a}[j], x_{2a}[k]) \tag{6.30}$$

and

$$\frac{\partial y}{\partial x_1}, \frac{\partial y}{\partial x_2} \text{ and } \frac{\partial^2 y}{\partial x_1 \partial x_2}$$

The two-dimensional interpolation algorithm based on the aforementioned procedures has been independently tested prior to being used in the power amplifier design program.

Figure 6.5 shows S_{21} interpolation results for a CLY5 power transistor from Infineon that used only nine reference points. It was found that when using nine reference points the interpolation results had about a 4% error with actual measurements and provided an acceptable accuracy for the technique.

6.4 Multistage Transducer Gain

The design procedure requires to first design the power transistor stage that is located next to the load. This means that the classic RFT technique of Chapter 2 needs to be modified to start from the load rather than form the source. This section provides the new formulas for the multistage transducer power gain when starting from the load.

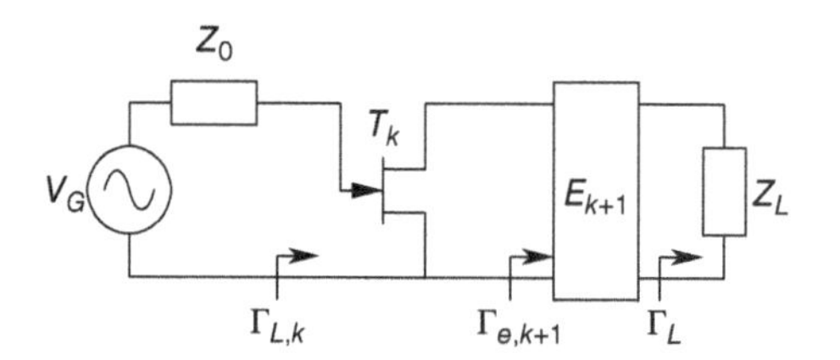

Figure 6.6 Last transistor stage connected to load Z_L.

We start with the last transistor stage where the source load Z_G is taken as Z_0 as shown in Figure 6.6.

The transducer gain for this last stage is given by

$$G_{T,k+1} = \frac{|e_{21,k+1}|^2 |S_{21,k}|^2}{|1 - S_{22,k}\Gamma_{e,k+1}|^2 |1 - \Gamma_L e_{22,k+1}|^2} \left(1 - |\Gamma_L|^2\right) \tag{6.31}$$

with

$$\Gamma_{e,k+1} = e_{11,k+1} + \frac{e_{21,k+1}^2 \Gamma_L}{1 - e_{22,k+1}\Gamma_L} \tag{6.32}$$

$$\Gamma_L = \frac{Z_L - Z_0}{Z_L + Z_0} \tag{6.33}$$

where

$\Gamma_{e,k+1}$ is the input reflection coefficient of equalizer $k+1$ when loaded on output by Z_L
Γ_L is the reflection coefficient of the load impedance Z_L
$e_{ij,k+1}$ are the scattering parameters of equalizer $k+1$

The process is repeated until we reach the source using

$$\Gamma_{L,k} = S_{11,k} + \frac{S_{12,k}S_{21,k}\Gamma_{e,k+1}}{1 - S_{22,k}\Gamma_{e,k+1}} \tag{6.34}$$

where

$\Gamma_{L,k}$ is the input reflection coefficient of transistor k when loaded on output by $\Gamma_{e,k+1}$
$S_{ij,k}$ are the scattering parameters of transistor k

When reaching the source, we first consider the situation where the source load Z_G is still taken as the characteristic impedance Z_0 as shown in Figure 6.7

The transducer gain at this point is given by

$$G_{T,2} = G_{T,3} \frac{|e_{21,2}|^2 |S_{21,1}|^2}{|1 - S_{22,1}\Gamma_{e,2}|^2 |1 - \Gamma_{L,2} e_{22,2}|^2} \tag{6.35}$$

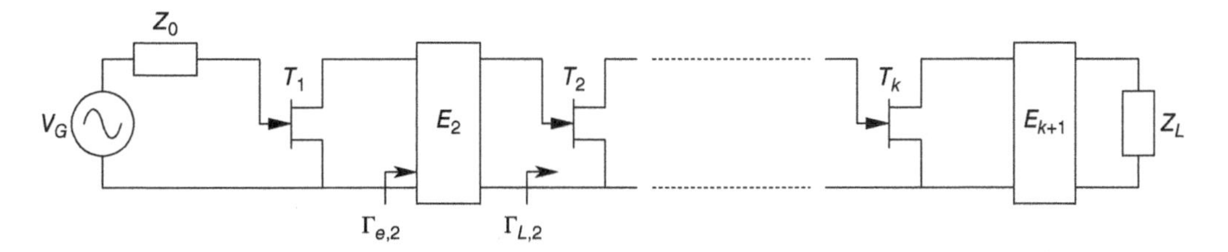

Figure 6.7 Situation at the second equalizer.

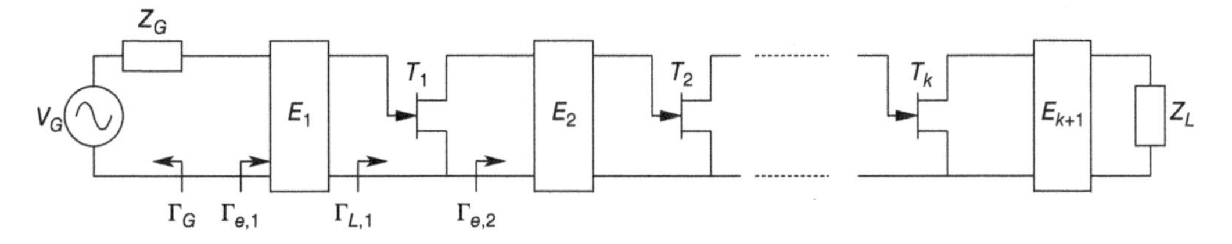

Figure 6.8 Matching the source impedance.

with

$$\Gamma_{e,2} = e_{11,2} + \frac{e_{21,2}^2 \Gamma_{L,2}}{1 - e_{22,2}\Gamma_{L,2}} \tag{6.36}$$

$$\Gamma_{L,2} = S_{11,2} + \frac{S_{12,2}S_{21,2}\Gamma_{e,3}}{1 - S_{22,2}\Gamma_{e,3}} \tag{6.37}$$

Finally, the first equalizer is added to match the arbitrary source impedance Z_G as shown in Figure 6.8.

The overall transducer gain is then given by

$$G_T = G_{T,2} \frac{|e_{21,1}|^2}{|1 - \Gamma_G \Gamma_{e,1}|^2 |1 - \Gamma_{L,1} e_{22,1}|^2} \left(1 - |\Gamma_G|^2\right) \tag{6.38}$$

with

$$\Gamma_{e,1} = e_{11,1} + \frac{e_{21,1}^2 \Gamma_{L,1}}{1 - e_{22,1}\Gamma_{L,1}} \tag{6.39}$$

$$\Gamma_{L,1} = S_{11,1} + \frac{S_{12,1}S_{21,1}\Gamma_{e,2}}{1 - S_{22,1}\Gamma_{e,2}} \tag{6.40}$$

$$\Gamma_G = \frac{Z_G - Z_0}{Z_G + Z_0} \tag{6.41}$$

where

$\Gamma_{e,1}$ is the input reflection coefficient of equalizer 1 when loaded on output by $\Gamma_{L,1}$
$\Gamma_{L,1}$ is the input reflection coefficient of transistor 1 when loaded on output by $\Gamma_{e,2}$
Γ_G is the reflection coefficient of the source impedance Z_G

6.5 Multistage VSWR

In the classic RFT technique of Chapter 2, the input and output VSWR were computed using a stage-by-stage approach that started from the source. This section provides the new formulas for the multistage input and output VSWR when starting from the load.

Figure 6.9 is used to compute the VSWR.

The input VSWR is given by

$$\mathrm{VSWR}_{IN} = \frac{1+|\Gamma_{e,1}|}{1-|\Gamma_{e,1}|} \tag{6.42}$$

where $\Gamma_{e,1}$ is the reflection coefficient at the input of the first equalizer and is given by (6.39).

The output VSWR is given as

$$\mathrm{VSWR}_{OUT} = \left.\frac{1+|\Gamma_{s,k+1}|}{1-|\Gamma_{s,k+1}|}\right|_{k+1\ stages} \tag{6.43}$$

and is computed using the recursive formula

$$\begin{aligned}
\Gamma_{g,2} &= S_{22,1} \\
\Gamma_{s,2} &= e_{22,2} + \frac{e_{21,2}^2\Gamma_{g,2}}{1-e_{11,2}\Gamma_{g,2}} \\
\Gamma_{g,3} &= S_{22,2} + \frac{S_{12,2}S_{21,2}\Gamma_{s,2}}{1-S_{11,2}\Gamma_{s,2}} \\
&\vdots \\
\Gamma_{s,k} &= e_{22,k} + \frac{e_{21,k}^2\Gamma_{g,k}}{1-e_{11,k}\Gamma_{g,k}} \\
\Gamma_{g,k+1} &= S_{22,k} + \frac{S_{12,k}S_{21,k}\Gamma_{s,k}}{1-S_{11,k}\Gamma_{s,k}}
\end{aligned} \tag{6.44}$$

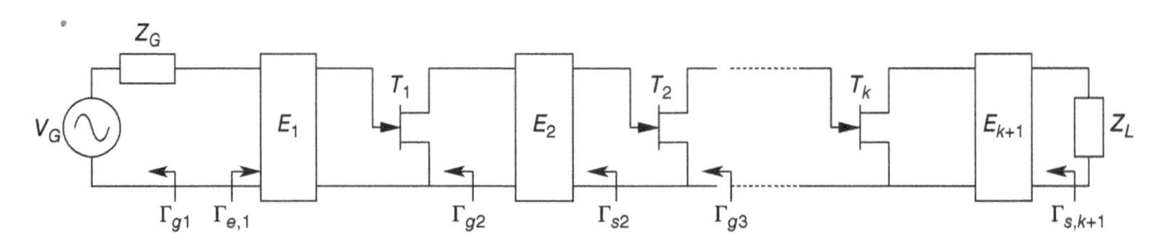

Figure 6.9 Defining the input and output VSWR.

where

$\Gamma_{g,k}$ is the output reflection coefficient of transistor $k-1$ when loaded on input by $\Gamma_{s,k-1}$
$\Gamma_{s,k}$ is the output reflection coefficient of equalizer k when loaded on input by $\Gamma_{g,k}$

6.6 Optimization Process

The optimization algorithm is used to define polynomial coefficients $h_0, h_1, \ldots, h_n$ of $h(t)$ of an equalizer that minimize a least mean square error function that includes desired target functions with different weight factors.

For the power amplifiers of this chapter, the error function for all but the last equalizer includes transducer power gain and VSWR:

$$E=\sum_{i=1}^{m}\left[W_1\left(\frac{G_{T,k}(\Omega_i,\boldsymbol{H})}{G_{T0,k}}-1\right)^2+W_2\left(\frac{\text{VSWR}_k(\Omega_i,\boldsymbol{H})}{\text{VSWR}_{0,k}}-1\right)^2\right] \tag{6.45}$$

where the functions are given in terms of Richard's frequency $\Omega_i=\tan\left(\frac{\pi}{2}\frac{\omega_i}{\omega_0}\right)$ and where ω_i are the actual frequencies and ω_0 is the normalization frequency used in the Richard's transform.

$G_{T,k}(\Omega_i, \boldsymbol{H})$ is the transducer power gain for k equalizer–transistor stages given by the equations of Section 6.4 and used to optimize the kth equalizer. The transducer power gain depends on frequency through the scattering parameters $e_{ij}(j\Omega)$ of the equalizer and the scattering parameters $S_{ij}(j\Omega)$ of the transistor. The target transducer power gain $G_{T0,k}$ for optimizing the kth equalizer is taken as a constant in the frequency band being optimized. The expressions for the target transducer gain are provided in Section 6.8.
$\text{VSWR}_{IN,k}(\Omega_i, \boldsymbol{H})$ is the input VSWR at stage k given by the equations in Section 6.5. The target $\text{VSWR}_{IN0,k}$ is a constant over the frequency band being optimized. The value is often between 1 and 2. Note that the input VSWR is generally optimized using the first equalizer.
$\text{VSWR}_{OUT,k}(\Omega_i, \boldsymbol{H})$ is the output VSWR at stage k given by the equations in Section 6.5. The target $\text{VSWR}_{OUT0,k}$ is a constant over the frequency band being optimized. The value is often between 1 and 2. Note that the output VSWR is generally optimized using the last equalizer.

The vector $\boldsymbol{H}$ includes the polynomial coefficients $h_0, h_1, \ldots, h_n$ and is used as the $\boldsymbol{X}$ vector in the Levenberg–More optimization algorithms of Chapter 2 and Appendix B.

The main computation in the optimization routine is to define the step $\Delta\boldsymbol{H}$ to add to the elements of vector $\boldsymbol{H}$ at the next iteration given by

$$\Delta\boldsymbol{H}=-\left[\boldsymbol{J}^T\boldsymbol{J}+\lambda\boldsymbol{D}^T\boldsymbol{D}\right]^{-1}\boldsymbol{J}^T e(\boldsymbol{H}) \tag{6.46}$$

where

$e(\boldsymbol{H})$ is the error vector
J is the Jacobian matrix of the error with elements $\partial e_j/\partial h_i$
D is a diagonal matrix
λ is the Levenberg–Marquardt–More parameter described in Appendix B

An important element in the optimization routine is the Jacobian matrix. It is based on derivatives of the error functions. For the transducer power gain analytical expressions for the derivatives that can be determined are rather tedious and lengthy to express and are written into a program. In the case of the output VSWR, the analytical expressions become very cumbersome and it is preferred to use a numerical derivative such that

$$\frac{\partial \mathrm{VSWR}_{OUT}(\Omega_j, \boldsymbol{H})}{\partial h_i} = \frac{\mathrm{VSWR}_{OUT}(\Omega_j, \boldsymbol{H} + \Delta h_i) - \mathrm{VSWR}_{OUT}(\Omega_j, \boldsymbol{H})}{\Delta h_i} \tag{6.47}$$

where a step of $\Delta h_i = 10^{-5}$ was found satisfactory when starting with $h(t)$ polynomial coefficients initialized at $h_i = \pm 1$.

The last equalizer is optimized to match the output load that provides the maximum added power. It uses the procedures given in Section 6.3.1. It is the only transistor that requires two-dimensional interpolation for large-signal S data. But once the optimum input power level has been defined, then only one-dimensional large-signal S parameter data are needed for this last transistor. It is given for the optimum input power level and the frequency points used in the optimization of the other transistor stages. Since the optimization starts from the load and goes toward the source, the last transistor S parameter data are critical for computing the transducer gain formulas.

6.7 Design Procedures

The design procedure starts with the design of the last equalizer and the power transistor stage. Then the transducer gain is defined from the load toward the source. The transducer gain formulas at a given stage k starting from the load are given in Section 6.4. In this section, we provide the target transducer gain values to be used when starting from the load toward the source.

Starting with the last transistor stage as shown in Figure 6.10.

The last equalizer is used to adapt the output load that maximizes the added power. We match the output of transistor k to the input of equalizer $k+1$ loaded on Z_L:

$$\Gamma_{e,k+1} = S_{22,k}^* \tag{6.48}$$

Then assuming as shown in Figure 6.6 that the source impedance is taken as the characteristic impedance Z_0 (most often 50 Ω), then the maximum transducer gain for transistor T_k is given by

$$G_{T,k+1\ MAX} = \frac{|S_{21,k}|^2}{|1 - S_{22,k}|^2} \tag{6.49}$$

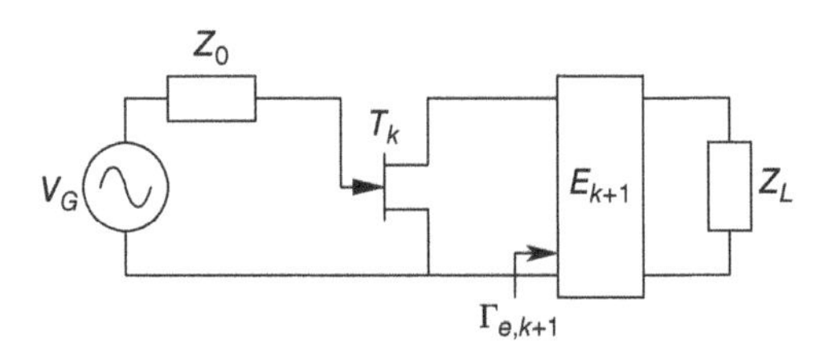

Figure 6.10 Last transistor stage loaded by Z_L.

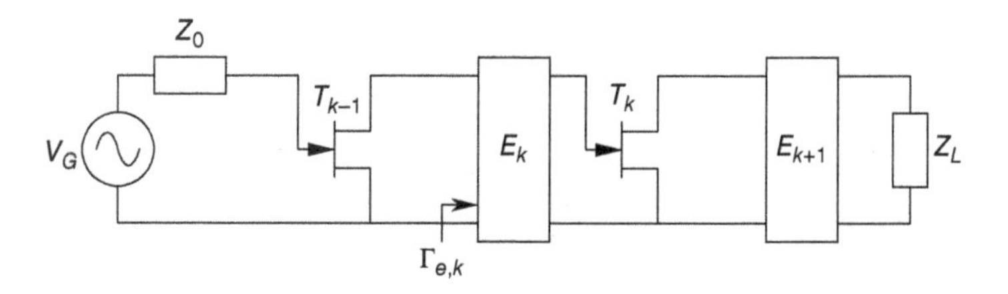

Figure 6.11 Optimizing equalizer E_k.

Note that since the last equalizer is used to adapt the load according to (6.48), it is not used to optimize the transducer power gain of (6.31). The maximum gain expression in (6.49) is used to define the recurrence formulas for the target transducer gain for the other equalizers.

The next step is to optimize equalizer E_k as shown in Figure 6.11.

In this case, the maximum transducer gain is given by

$$G_{T,k\ MAX} = \frac{|S_{21,k-1}|^2}{|1 - S_{22,k-1}|^2} G_{d\max,k} \tag{6.50}$$

where

$$G_{d\max,k} = \frac{|S_{21,k}|^2}{\left(1 - |S_{11,k}|^2\right)\left(1 - |S_{22,k}|^2\right)} \frac{1}{1 - \dfrac{S_{11,k}^* S_{22,k}^* S_{12,k} S_{21,k}}{\left(1 - |S_{11,k}|^2\right)\left(1 - |S_{22,k}|^2\right)}} \tag{6.51}$$

One should note that $G_{T,k\ MAX}$ changes over the frequency band and we use the minimum value of $G_{T,k\ MAX}$ over the optimization frequency band a target constant value to be used in (6.45). Under these conditions, the target value for transducer gain at the kth equalizer is given by

$$G_{T0,k} = \min\{G_{T,k\ MAX}(\omega_i)\} \tag{6.52}$$

where the $\omega_i' s$ are the frequency points used in the optimization.

The process is then repeated with the other stages and for equalizer E_2 the maximum transducer gain is given by

$$G_{T,2\ MAX} = \frac{|S_{21,1}|^2}{|1 - S_{22,1}|^2} G_{d\max,2} G_{d\max,3} \cdots G_{d\max,k} \tag{6.53}$$

and the target value for the transducer gain for optimizing the equalizer E_2 is

$$G_{T0,2} = \min\{G_{T,2\ MAX}(\omega_i)\} \tag{6.54}$$

Finally, after adding equalizer E_1 the overall maximum transducer gain is given by

$$G_{T,1\ MAX} = G_{d\max,1} G_{d\max,2} G_{d\max,3} \cdots G_{d\max,k} \tag{6.55}$$

and the target value for the transducer gain for optimizing equalizer E_1 is given by

$$G_{T0,1} = \min\{G_{T,1\ MAX}(\omega_i)\} \tag{6.56}$$

Note that the input VSWR is most often optimized using the equalizer close to the source. The target value for VSWR is a constant value chosen between 1 and 2.

6.8 Realizations

This new approach was implemented in a program called RFCAD-POWER. It was used for S-band power amplifier design. Even if the method has the intrinsic limitation of not controlling harmonics, the objective was to determine the level of performance that can be obtained using this procedure.

6.8.1 Realization of a One-Stage Power Amplifier

In order to validate the added-power optimization, a one-stage power amplifier operating in the S-band at 2.245 GHz was realized on 254 μm Teflon/glass substrate [6] and is shown in Figure 6.12.

The power transistor was a CLY5 solid-state discrete device from Infineon. The transistor is biased according to a class AB. Using a SOT223 package model for the CLY5 device, bias and stabilization circuits have been designed to ensure unconditional stability while minimizing gain degradation. For bias concerns, high characteristic impedance quarter-wavelength stubs with shunt capacitors were used as gate and drain bias circuits. To prevent undesirable in-band and out-band oscillations, capacitors and resistors have been placed in the gate bias circuits or directly at the transistor gate.

This first realization was mainly to validate the procedure, so the expected performances were loosely defined as 1 W output power with 30% PAE and an input return loss better than 20 dB.

The equalizer circuits consisted of cascaded quarter-wavelength transmission lines whose characteristic impedances and resonance frequencies could be modified. The output equalizer was

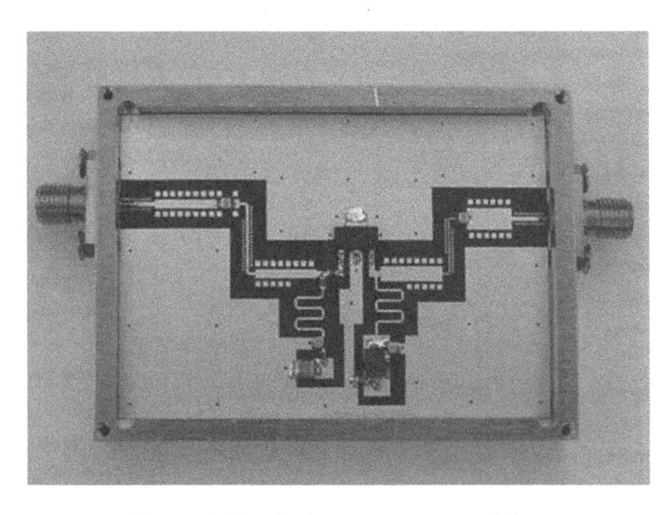

Figure 6.12 Single-stage power amplifier.

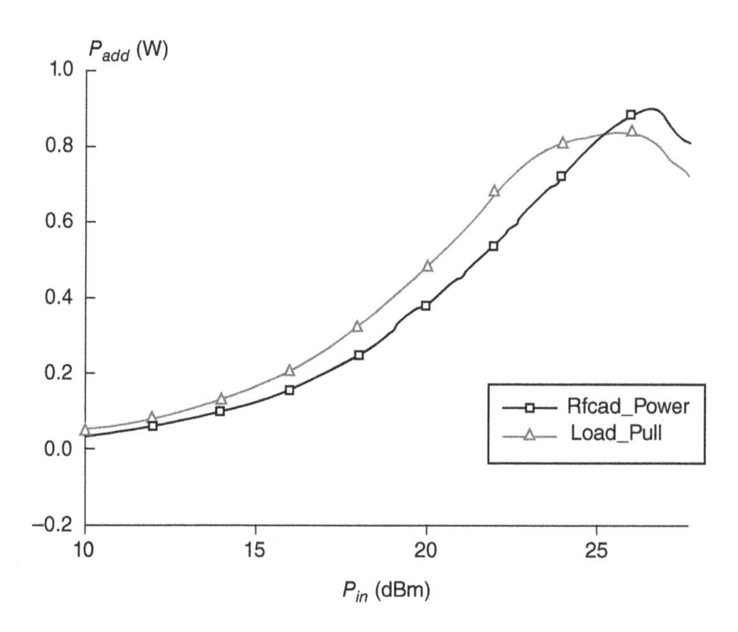

Figure 6.13 Added power simulations using RFCAD-POWER and load-pull simulations using HPADS.

designed to transform 50 Ω to the optimum load impedance for optimum added power. The input equalizer was designed to achieve maximum linear gain and reduce input return losses.

The first step was to check whether the results from RFCAD-POWER and from load-pull simulations using HPADS software were in agreement. Figure 6.13 shows the results. According to RFCAD-POWER simulations, the added power reaches a maximum of 0.9 W at 26.5 dBm input power. This was the input power value used in the optimization of the power amplifier. Overall, the agreement with load-pull simulations is acceptable.

Figures 6.14, 6.15, and 6.16 show the simulated results using RFCAD-POWER and the measured results for the single-stage amplifier. In the RFCAD-POWER simulations, the features of the Teflon/glass substrate have been integrated into the transmission lines behavior. The frequency response of the Murata capacitor has been also taken into account. The quarter-wavelength transmission lines used in the equalizers and bias circuits have been scaled to reduce overall circuit dimension.

Figure 6.14 shows that the measured added power reaches a maximum of 0.62 W at the 25.7 dBm input power level, thus leading to 30 dBm of output power. At this operating input power, the amplifier reaches about 1.7 dB compression and the PAE is higher than 36%.

Figure 6.15 shows that the measured PAE is at a maximum of 47.7% at 26 dBm input power. The measured small-signal gain of the single-stage amplifier is about 6 dB as seen in Figure 6.16. Figure 6.17 shows that the measured input return loss is better than 20 dB from 2.2 to 2.9 GHz. The measured output return loss was found to be about 6 dB due to large-signal output matching.

6.8.2 Realization of a Three-Stages Power Amplifier

A three-stage power amplifier operating in the S-band has been designed using the RFCAD-POWER program based on the modified RFT procedures described in this chapter [7, 8]. The power transistor used in the last stage is the CLY5 transistor from Infineon. In the other two stages,

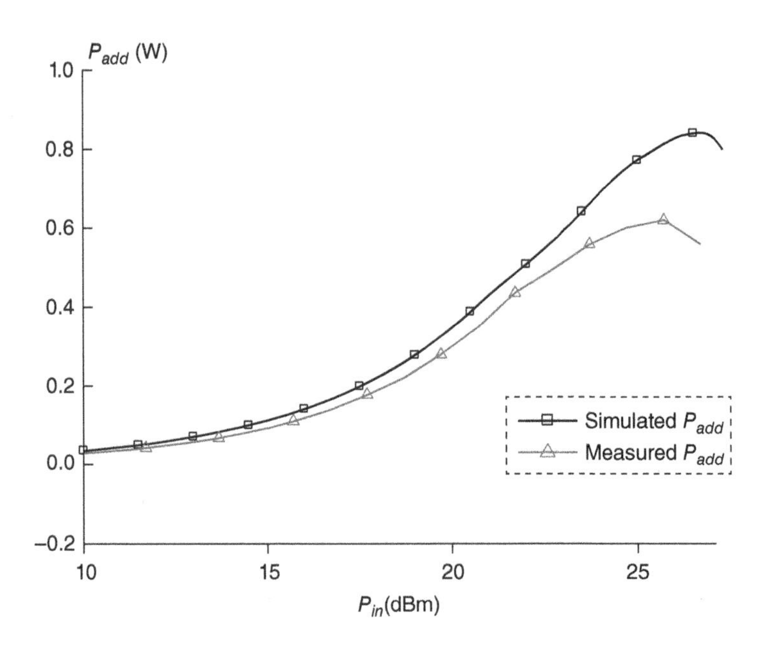

Figure 6.14 Measured and simulated added power.

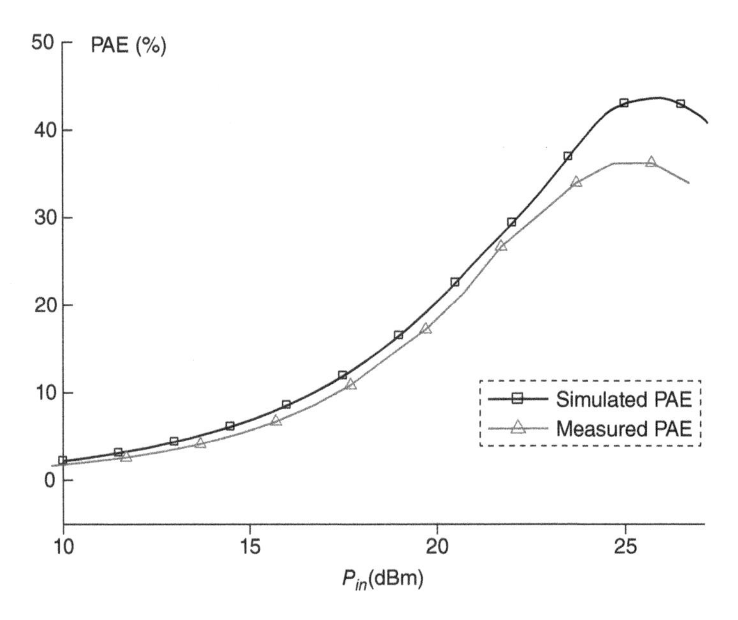

Figure 6.15 Measured and simulated PAE.

CLY5 transistors from Infineon have been selected for their high linear gain. The objectives for this three-stage amplifier were 1 W output power with 30% PAE, a linear gain of 30 dB, and an input return loss better than 20 dB.

The 2.245 GHz three-stage power amplifier was realized on 254 μm Teflon/glass substrate and is shown Figure 6.18.

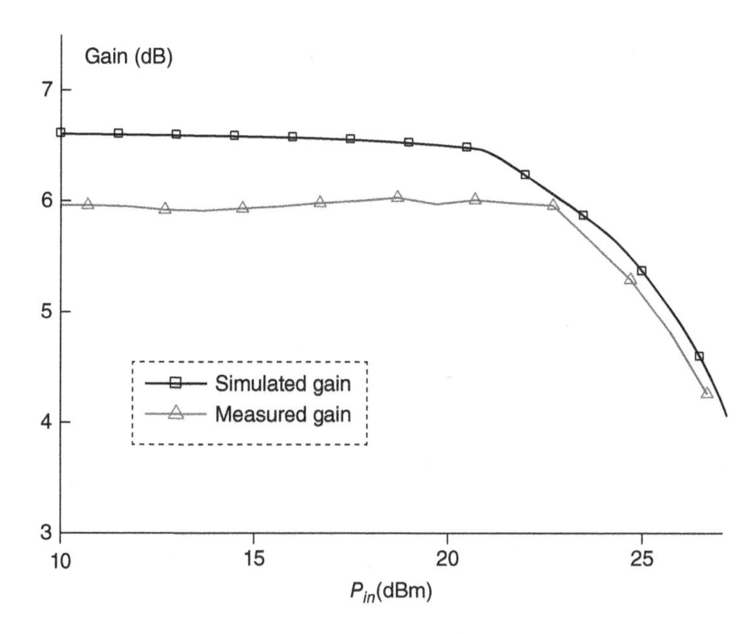

Figure 6.16 Measured and simulated gain.

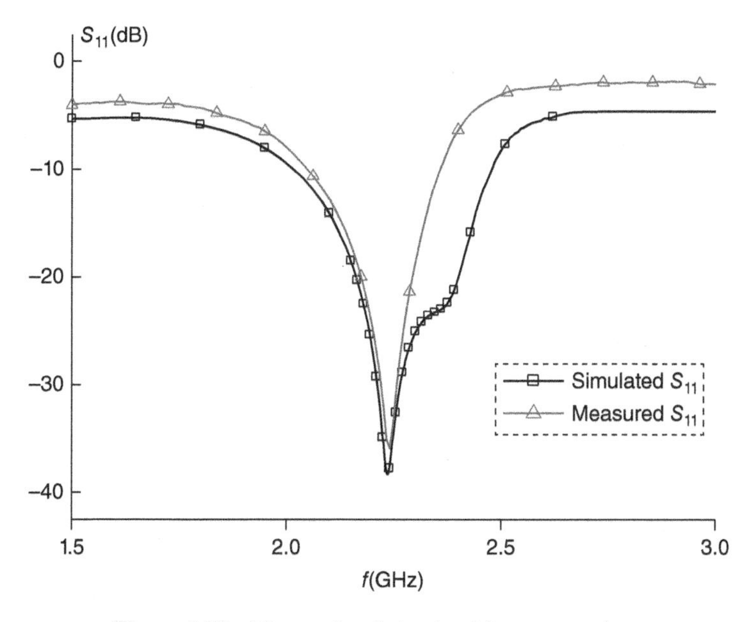

Figure 6.17 Measured and simulated input return loss.

To ensure stability and efficient bias, high characteristic impedance quarter-wavelength stubs, capacitors, and resistors have been added to the circuits. The equalizers were designed using the modified RFT described in this chapter by first designing the output equalizer circuit, then the inter-stage equalizers, and finally, the input equalizer. The equalizers were realized using cascaded quarter-wavelength transmission lines. The output equalizer was designed to optimize the added

Figure 6.18 Three-stage power amplifier.

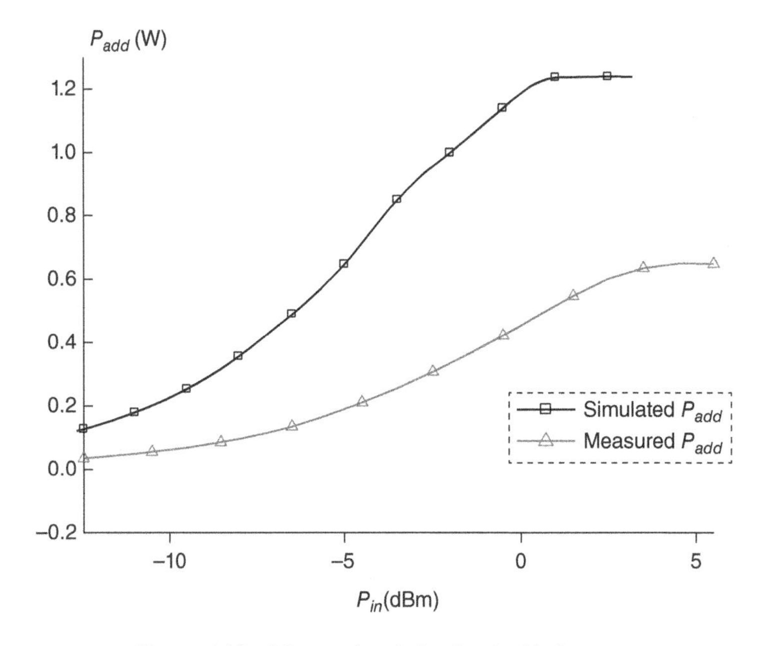

Figure 6.19 Measured and simulated added power.

power. Next, the first interstage equalizer is synthesized to provide the optimal input power level to the final stage without consuming too much energy to not degrade the PAE. Finally, the input equalizer is used for impedance transformation to 50 Ω while preserving a high gain and minimizing the input return loss.

The measured and simulated responses of the three-stage power amplifier are shown in Figures 6.19, 6.20, and 6.21.

From Figures 6.19 and 6.20, at 4.5 dBm input power (3.4 dB gain compression with respect to the small-signal gain), the measured output power reaches 28.1 dBm, leading to a maximum added power of 0.65 W. The PAE reaches 30% at this input power level.

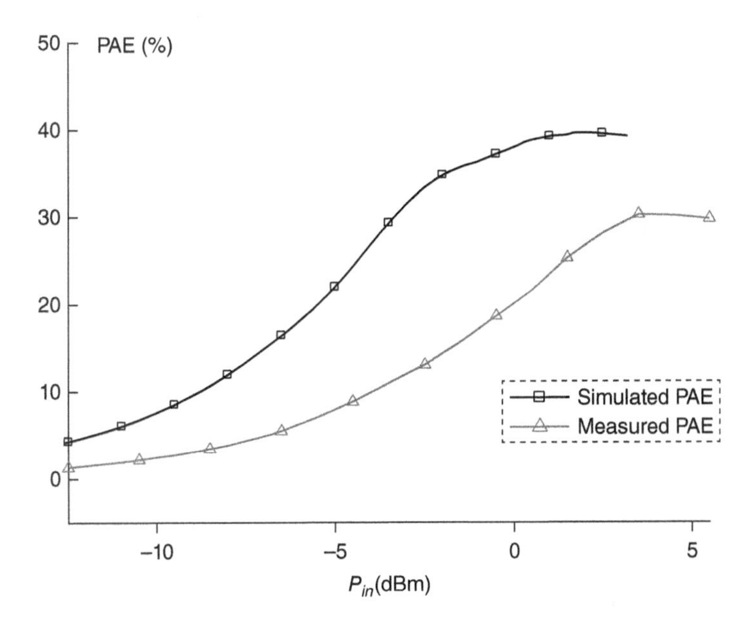

Figure 6.20 Measured and simulated PAE.

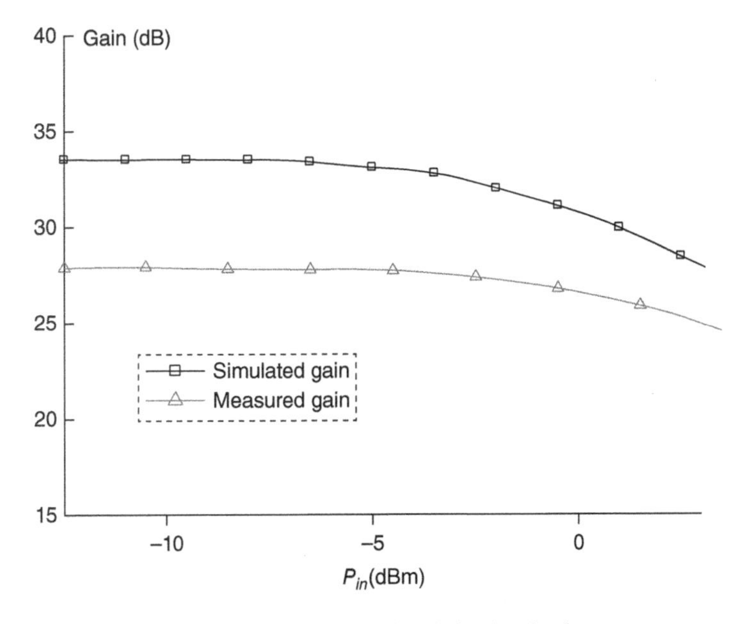

Figure 6.21 Measured and simulated gain.

Figure 6.21 shows that the measured gain of the three-stage amplifier is about 27.9 dB. The measured input return loss was found to be better than 11 dB.

Note that the differences between simulated and measured responses were due to the interstage transistor no longer operating in its linear region when providing the optimum input power level for the power transistor. The RFCAD-POWER program at the time did not include two-dimensional large signal interpolation for transistors other than that of the last stage.

6.9 Linear Power Amplifiers

6.9.1 Theory

The conception of linear power amplifiers with good VSWR is possible by using several identical modules placed in parallel as shown in Figure 6.22 [9, 10].

The n amplifier modules are modeled as $M_1, M_2, \ldots, M_n$ and are characterized using the same scattering matrix S.

The overall scattering matrix Σ of this system is given by

$$\Sigma = [S-\alpha][I-\alpha S]^{-1} \tag{6.57}$$

where

$$\alpha = \frac{n-1}{n+1} \tag{6.58}$$

and

$$I = \begin{pmatrix} 1 & 0 \\ 0 & 1 \end{pmatrix}$$

The overall scattering matrix Σ can then be expressed as

$$\Sigma = \frac{1}{1-\alpha(S_{11}+S_{22})+\alpha^2\Delta} \begin{pmatrix} S_{11}-\alpha(1+\Delta)+\alpha^2 S_{22} & (1-\alpha^2)S_{12} \\ (1-\alpha^2)S_{21} & S_{22}-\alpha(1+\Delta)+\alpha^2 S_{11} \end{pmatrix} \tag{6.59}$$

with $\Delta = S_{11}S_{22} - S_{12}S_{21}$

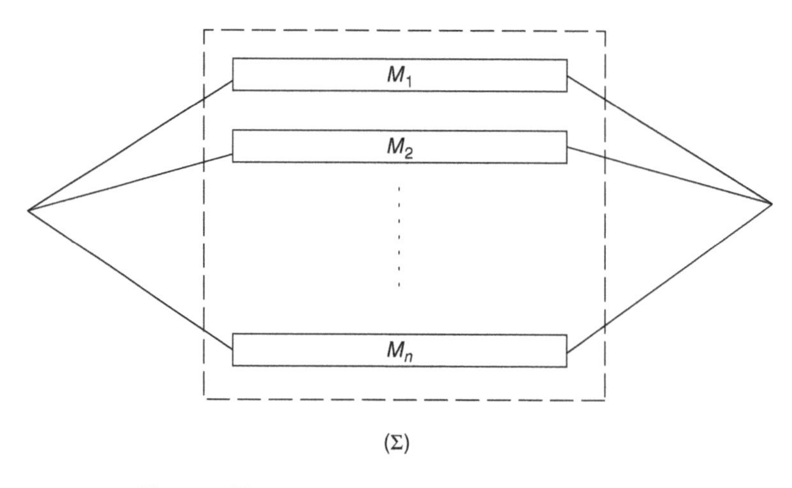

Figure 6.22 n identical amplifier modules in parallel.

In the unilateral case ($S_{12}=0$), we have

$$\Sigma=\frac{1}{(1-\alpha S_{11})(1-\alpha S_{22})}\begin{pmatrix}(S_{11}-\alpha)(1-\alpha S_{22}) & 0\\ (1-\alpha^2)S_{21} & (S_{22}-\alpha)(1-\alpha S_{11})\end{pmatrix} \tag{6.60}$$

Figure 6.23 shows a single module adapted for maximum unilateral transducer gain.

For perfect matching of the load, one takes $\rho_L=S_{22}^*$, and for perfect matching of the source, one takes $\rho_S=S_{11}^*$. This leads to a maximum unilateral transducer gain for a single module that is perfectly symmetric and given by

$$g_{Tu,\max}=\frac{1}{1-|S_{11}|^2}|S_{21}|^2\frac{1}{1-|S_{22}|^2} \tag{6.61}$$

Figure 6.24 shows n identical modules in parallel adapted for maximum unilateral transducer gain.

For perfect matching of the load, one takes $\Gamma_L=\Sigma_{22}^*$, and for perfect matching of the source, one takes $\Gamma_S=\Sigma_{11}^*$. This leads to a maximum unilateral transducer gain for n identical modules in parallel that is perfectly symmetric and is given by

$$G_{Tu,\max}=\frac{1}{1-|\Sigma_{11}|^2}|\Sigma_{21}|^2\frac{1}{1-|\Sigma_{22}|^2} \tag{6.62}$$

The matching conditions for the n modules of Figure 6.24 can be written as

$$\Gamma_S=\frac{S_{11}^*-\alpha}{1-\alpha S_{11}^*}=\frac{\rho_S-\alpha}{1-\alpha\rho_S} \tag{6.63}$$

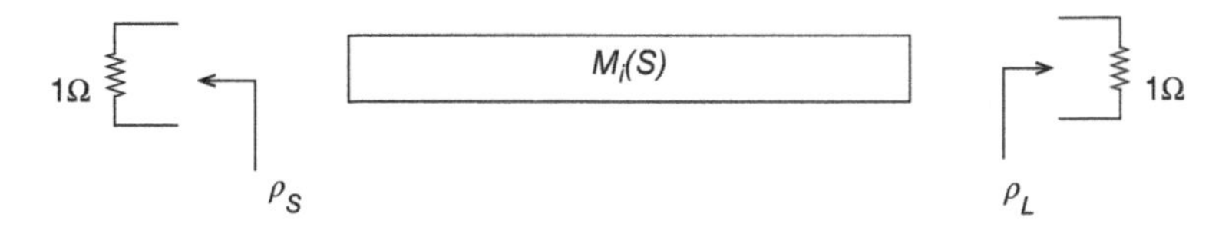

Figure 6.23 Maximum unilateral transducer gain for one module $M_i(S)$.

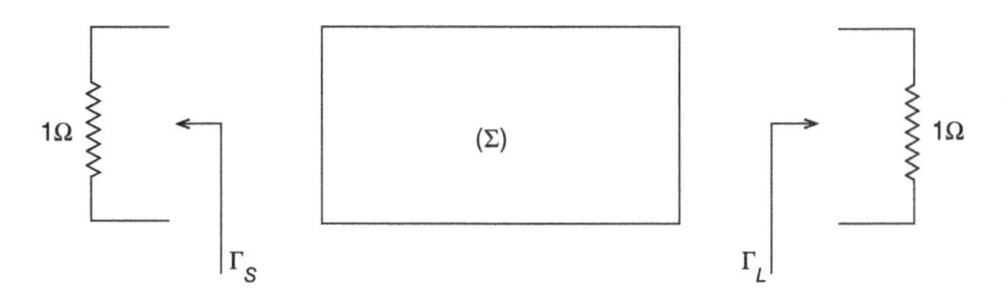

Figure 6.24 n identical modules in parallel characterized by scattering matrix Σ.

$$\Gamma_L = \frac{S_{22}^* - \alpha}{1 - \alpha S_{22}^*} = \frac{\rho_L - \alpha}{1 - \alpha \rho_L} \tag{6.64}$$

The input VSWR for the case of a single module is given by

$$\text{VSWR}_{IN} = \frac{1 + |\rho_S|}{1 - |\rho_S|} \tag{6.65}$$

and the input VSWR for the case of n identical modules in parallel is given by

$$\text{VSWR}_{n\ IN} = \frac{1 + |\Gamma_S|}{1 - |\Gamma_S|} \tag{6.66}$$

The input VSWR of n identical modules in parallel will be minimum and equal to 1 when $\Gamma_S = 0$ or when

$$\Gamma_S = \frac{S_{11}^* - \alpha}{1 - \alpha S_{11}^*} = \frac{\rho_S - \alpha}{1 - \alpha \rho_S} = 0$$

This occurs when

$$\rho_S = \alpha \tag{6.67}$$

or when the input VSWR of a single module is such that

$$\text{VSWR}_{IN} = \frac{1 + |\rho_S|}{1 - |\rho_S|} = \frac{1 + \alpha}{1 - \alpha} = n \tag{6.68}$$

The output VSWR for the case of a single module is given by

$$\text{VSWR}_{OUT} = \frac{1 + |\rho_L|}{1 - |\rho_L|} \tag{6.69}$$

and the output VSWR for the case of n identical modules in parallel is given by

$$\text{VSWR}_{n\ OUT} = \frac{1 + |\Gamma_L|}{1 - |\Gamma_L|} \tag{6.70}$$

The output VSWR of n identical modules in parallel will be minimum and equal to 1 when $\Gamma_L = 0$ or when

$$\Gamma_L = \frac{S_{22}^* - \alpha}{1 - \alpha S_{22}^*} = \frac{\rho_L - \alpha}{1 - \alpha \rho_L} = 0$$

This occurs when

$$\rho_L = \alpha \tag{6.71}$$

or when the output VSWR of a single module is such that

$$\mathrm{VSWR}_{OUT} = \frac{1+|\rho_S|}{1-|\rho_S|} = \frac{1+\alpha}{1-\alpha} = n \tag{6.72}$$

In conclusion, the VSWR of the system of n identical modules will be minimum and equal to 1 when the VSWR of a single module is equal to n. This property is used in the next section.

6.9.2 Arborescent Structures

The arborescent structure is shown in Figure 6.25. It is made of four identical blocks made of a transistor T cascaded with an equalizer $E2$ and two input equalizers $E1$. Two identical blocks are placed in parallel and then cascaded with an input equalizer $E1$.

Since the arborescent power amplifier in Figure 6.25 has one principal symmetrical axis AA' and two secondary symmetrical axis aa' and bb', the equivalent configuration of Figure 6.26 can be used where $E1'$ represents two equalizers $E1$ in parallel, $E'2$ represents four equalizers $E2$ in parallel, and T' represents four transistors T in parallel.

Note that the transistor T can represent a transistor with feedback and biasing circuitry.

The situation in Figure 6.26 corresponds to the case of a single-stage amplifier that can be optimized for gain and VSWR using the classic RFT of Chapter 2. The RFT will provide the optimal

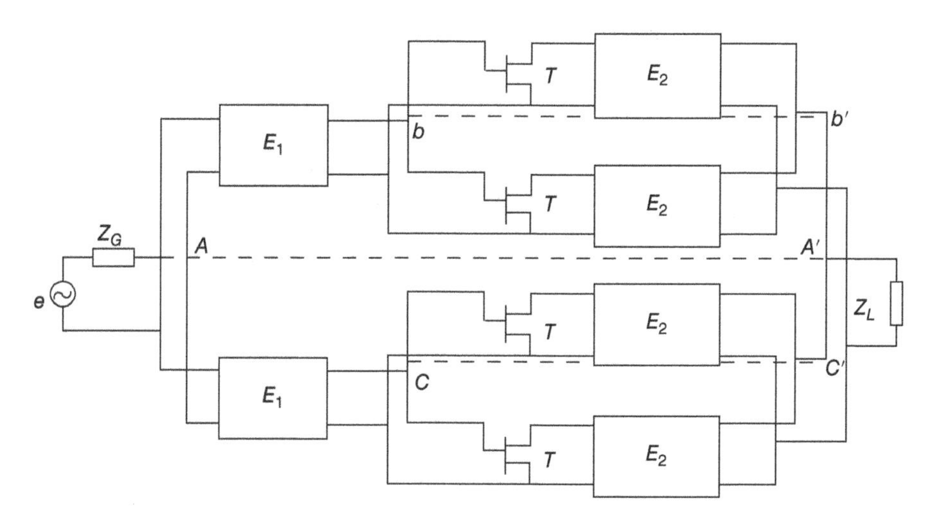

Figure 6.25 Arborescent power amplifier using four identical amplifier modules.

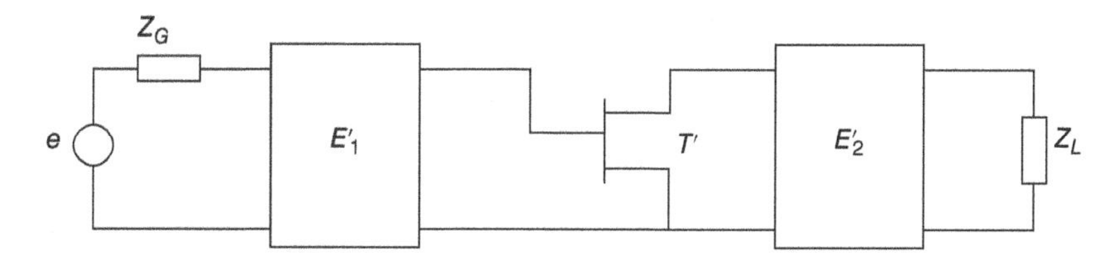

Figure 6.26 Equivalent configuration.

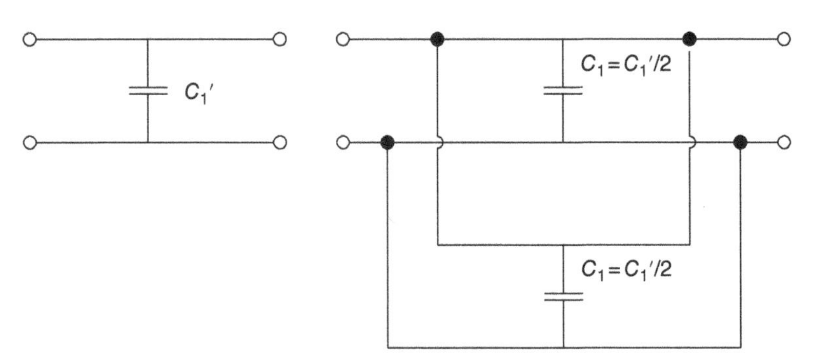

Figure 6.27 Capacitors for equalizers $E1$.

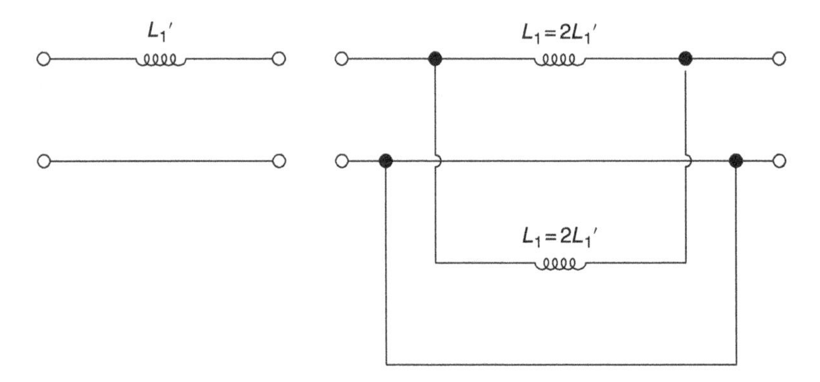

Figure 6.28 Inductors for equalizers $E1$.

$E1'$ and $E2'$ equalizers. From there, one needs to find a procedure to define the equalizer $E1$ from $E1'$ and define the equalizer $E2$ from $E2'$. This procedure is described later.

In the case of lumped equalizers $E1'$ and $E2'$ made of inductors L' and capacitors C', lumped equalizers $E1$ and $E2$ made of inductors L and capacitors C can be found using the equalities given later.

For the equalizer $E1'$ that represents two equalizers $E1$ in parallel, a capacitor C_1' will represent two capacitors in parallel such that $C_1 = C_1'/2$ as shown in Figure 6.27.

And an inductor L_1' will represent two inductors in parallel such that $L_1 = 2L_1'$ as shown in Figure 6.28.

For the equalizer $E2'$ that represents four equalizers $E2$ in parallel, a capacitor C_2' will represent four capacitors in parallel such that $C_2 = C_2'/4$ and an inductor L_2' will represent four inductors in parallel such that $L_2 = 4L_2'$.

6.9.3 Example of an Arborescent Linear Power Amplifier

This theoretical example corresponds to a 1 W amplifier in the 5.9–6.4 GHz band using an arborescence of 2 and 4 paths with 3 cascaded stages as shown in Figure 6.29.

The transistors are characterized by the measured S parameter data from THOMSON given in Table 6.1. The second stage uses a transistor with feedback impedance Z_c formed by a 350 Ω resistor and a 1 pF capacitor.

After optimization for gain and VSWR, the arborescent power amplifier has the following features in the passband:

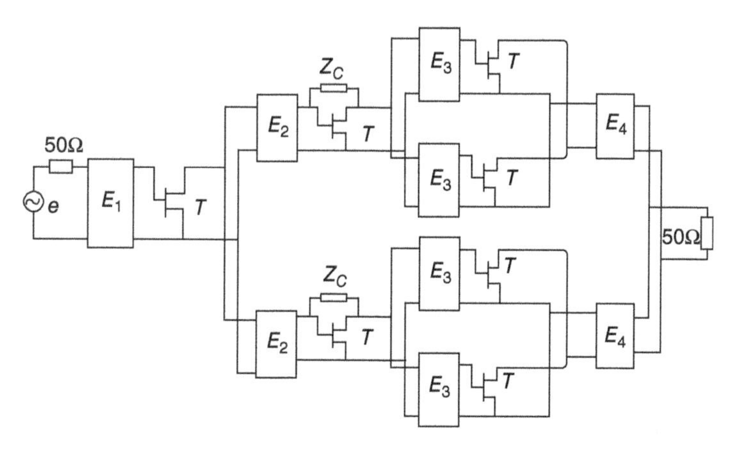

Figure 6.29 Arborescence of 2 and 4 paths with 3 cascaded stages.

Table 6.1 *S* parameters for the transistors

	S_{11}		S_{12}		S_{21}		S_{22}	
FREQ (GHz)	MOD	Phase	MOD	Phase	MOD	Phase	MOD	Phase
5.9	0.783	−141	0.118	15	1.81	79	0.343	−96
6.0	0.783	−142	0.118	14	1.79	78	0.344	−97
6.1	0.782	−143	0.118	14	1.76	77	0.345	−97
6.2	0.782	−143	0.118	14	1.74	76	0.346	−98
6.3	0.781	−144	0.118	14	1.71	75	0.347	−98
6.4	0.781	−145	0.118	13	1.69	75	0.349	−99

- Gain $\geq 28\text{dB}$
- Input VSWR ≤ 2
- Output VSWR ≤ 2

The optimized reflexion coefficients $e_{11,j}(s)$ of equalizers E_1, E_2, E_3, E_4 are given below along with the values used for weighting factor W_1 representing gain and weighting factor W_2 representing input VSWR used during optimization:

$$W_1 \quad W_2$$

$$e_{111}(s) = \frac{0 + 1.817s + 1.114s^2 - 1.395s^3 - 0.966s^4}{1 + 3.192s + 3.444s^2 + 1.884s^3 + 0.966s^4} \qquad 1 \quad 2$$

$$e_{112}(s) = \frac{0 - 1.908s + 1.363s^2 - 0.903s^3 - 0.438s^4}{1 + 3.227s + 3.387s^2 + 1.089s^3 + 0.438s^4} \qquad 1 \quad 2$$

$$e_{113}(s) = \frac{0 - 3.069s + 1.489s^2 - 0.579s^3 - 0.159s^4}{1 + 4000s + 3.289s^2 + 0.625s^3 + 0.159s^4} \qquad 1 \quad 0$$

$$e_{114}(s) = \frac{0 - 14.64s + 2.531s^2 - 16.38s^3 + 4.309s^4}{1 + 15.08s + 6.622s^2 + 17.42s^3 + 4.309s^4} \qquad 1 \quad 0$$

Figure 6.30 shows the gain increase after a new transistor stage is added.

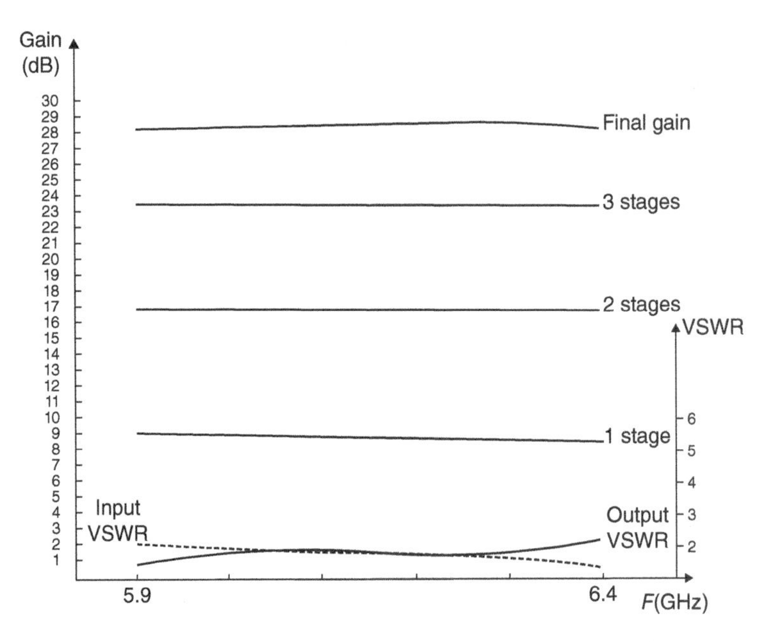

Figure 6.30 Optimized responses of the linear power amplifier.

Table 6.2 Inductor (*nH*) and capacitor (*pF*) values for the equalizers

E_1:	$C_1 = 0.525$	$L_1 = 0.977$	$C_2 = 1.966$	$L_2 = 0.733$
E_2:	$C_1 = 0.120$	$L_1 = 2.198$	$C_2 = 1.156$	$L_2 = 1.082$
E_3:	$C_1 = 0.033$	$L_1 = 3.312$	$C_2 = 0.846$	$L_2 = 1.316$
E_4:	$L_1 = 0.634$	$C_1 = 5.337$	$L_2 = 0.474$	$C_2 = 2.055$

The equalizers are realized in the form of *LC* ladders. The *L* and *C* values for each equalizer are given in Table 6.2.

Note that the three first equalizers begin with a capacitor.

The RFT gave rapidly results that were in good agreement with results obtained using commercial software.

In conclusion, the chapter provides a contribution to the study of multistage microwave power amplifiers. We can see that the RFT can be extended to power amplifiers when introducing the large-signal scattering parameters. Software, including a two-dimensional algorithm, based on the spline functions of the large-signal scattering parameters, has been written in C language to implement the new iterative technique.

Due to the importance of load terminating conditions, the method goes from the load to the source. Consequently, the computer program enables the optimization of added power, linear gain, and input VSWR. The validity and advantages of this new formalism are demonstrated by two examples of power amplifier realization. The power amplifiers were made for CNES with commercial transistors from Infineon. Another method based on arborescent structures was also presented. It provided some key properties when placing amplifiers in parallel.

References

[1] M. Hazouard, Conception and Realization of Power Microwave Amplifiers Using the Real Frequency Method, Ph.D. dissertation, University Bordeaux 1, January 7, 2002.

[2] W. H. Leighton, R. J. Chaffin, J. G. Webb, "RF amplifier design with signal S-parameters," IEEE Trans. Microw. Theory Tech., vol. MTT 21, pp. 809–814, 1973.

[3] R. F. Gilmore, F. Rosenbaum, "An analytic approach to optimum oscillator design using S-parameters," IEEE Trans. Microw. Theory Tech., vol. MTT 31, pp. 633–639, 1983.

[4] V. Rizzoli, A. Lipparini, F. Mastri, "Computation of large-signal S-parameters by harmonic-balance technique," Electron. Lett., vol. 24, pp. 329–330, 1988.

[5] M. Hazouard, E. Kerherve, P. Jarry, "Multistage solid-state power amplifier design by a new alternative synthesis technique," Int. J. Microw. Millimeter-Wave Comput. Aided Eng., vol. 14, no. 2, pp. 87–98, March 2004.

[6] M. Hazouard, E. Kerherve, P. Jarry, "New solid-state power amplifier design using large signal S parameters," 31st European Microwave Conference 2001, London, pp. 381–383, September 2001.

[7] E. Kerherve, M. Hazouard, L. Courcelle, P. Jarry, "Large signal S-parameters CAD technique applied to power amplifier design," IEEE Conference ISCAS, May 2003, Bangkok, pp. 441–444.

[8] M. Hazouard, E. Kerherve, L. Courcelle, P. Jarry, "CAD technique applied based on the large signal S-parameters applied to power amplifier design," IEEE Workshop on Power Amplifiers for Wireless Communications, September 9 and 10, 2002, San Diego, CA.

[9] A. Perennec, Synthesis and Realizations of Microwave Amplifiers by the Real Frequency Method, Ph.D. thesis, Ph.D. dissertation, Brest University, July 1988.

[10] P. Jarry, A. Perennec, "Optimisation of gain and VSWR in multistage microwave amplifiers using the real frequency method," European Conference on Circuit Theory and Design, ECCTD'87, pp. 203–208, September 1987.

7

Multistage Active Microwave Filters

7.1 Introduction

Microwave filters are commonly located in the front end of the telecommunication system [1]. In the reception path they are immediately followed by a low-noise amplifier. In order to reduce the size and development cost of these two high-frequency circuits, a combined design approach under the form of an active filter has been proposed [2, 3]. However, improving both functions simultaneously in an active filter has proven to be time-consuming, tedious, and without any guarantee of optimization convergence. In the case of multistage microwave active circuit design, one must solve the classical matching problem: construct a matching network between a generator and a load such that the transfer of power is maximized over a prescribed frequency band. In designing an active filter, if the optimum topology is unknown, the task of designing the matching networks becomes difficult with commercially available computer programs.

Microwave Amplifier and Active Circuit Design Using the Real Frequency Technique, First Edition.
Pierre Jarry and Jacques N. Beneat.

The real frequency method (RFT) technique presented in this book has shown to be very suitable for other problems than the classical multistage amplifier design of Chapter 2 [4, 5]. In this chapter, we show that the RFT can be modified to successfully design active filters.

The technique has been implemented in a program called FREELCD to synthesize equalizers optimized for transducer gain in the passband and that include transmission zeros to shape the stopbands of the active filters.

Simulation results for two broadband microwave active filters are presented: a low-pass case from 0 to 5 GHz with two transmission zeros and a bandpass case from 3 to 7 GHz with four transmission zeros.

A bandpass active filter from 7.7 to 8.1 GHz with two transmission zeros was realized in monolithic microwave integrated circuit (MMIC) technology and is also given.

7.2 Multistage Active Filter Representation

The multistage active filter is represented in Figure 7.1 [4]. It assumes that there are k transistors. An equalizer is associated with each transistor. There is one final equalizer placed after the kth transistor. The source and load impedances are assumed arbitrary, complex, and frequency dependent. The transistor can represent a transistor alone or a transistor with added but known circuitry as long as the S parameters can be provided.

The equalizer of the kth stage is characterized by the scattering matrix

$$S_{E_k} = \begin{pmatrix} e_{11,k}(s) & e_{12,k}(s) \\ e_{21,k}(s) & e_{22,k}(s) \end{pmatrix} \tag{7.1}$$

The transistor circuit of the kth stage is characterized by the scattering matrix

$$S_{T_k} = \begin{pmatrix} S_{11,k}(s) & S_{12,k}(s) \\ S_{21,k}(s) & S_{22,k}(s) \end{pmatrix} \tag{7.2}$$

In this chapter the equalizer is assumed to be a lossless and reciprocal lumped two-port so that its scattering matrix can be expressed as

$$S_E = \begin{pmatrix} e_{11}(s) = \dfrac{h(s)}{g(s)} & e_{12}(s) = \dfrac{f(s)}{g(s)} \\ e_{21}(s) = \dfrac{f(s)}{g(s)} & e_{22}(s) = \dfrac{-\varepsilon h(-s)}{g(s)} \end{pmatrix} \tag{7.3}$$

where $f(s)$ is an even or odd function. If $f(s)$ is even, then $\varepsilon = 1$; and if $f(s)$ is odd, then $\varepsilon = -1$. In addition, the lossless condition $|e_{11}|^2 + |e_{12}|^2 = 1$ leads to

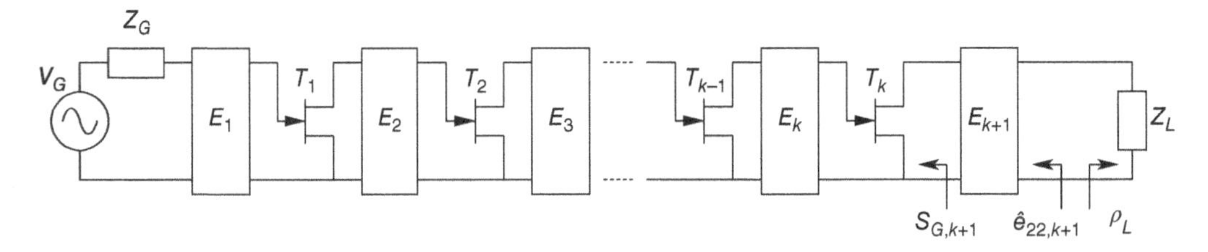

Figure 7.1 Multistage active filter representation.

$$g(s)g(-s) = h(s)h(-s) + f(s)f(-s) \tag{7.4}$$

$h(s)$ and $g(s)$ are polynomials of order n with real-valued coefficients:

$$h(s) = h_0 + h_1 s + \cdots + h_{n-1}s^{n-1} + h_n s^n$$

$$g(s) = g_0 + g_1 s + \cdots + g_{n-1}s^{n-1} + g_n s^n$$

and in this chapter, $f(s)$ is taken as

$$f(s) = \pm s^k \prod_{j=1}^{m} \left(s^2 + \omega_j^2\right) \tag{7.5}$$

$$\text{and} \quad \varepsilon = (-1)^k$$

where k represents the number of transmission zeros at $\omega = 0$ and where there are m number of finite transmission zeros $\omega_j' s$ and with $k + 2m \leq n$ leading to

$$g(s)g(-s) = h(s)h(-s) + (-1)^k s^{2k} \prod_{j=1}^{m} \left(s^2 + \omega_j^2\right)^2 \tag{7.6}$$

For a given equalizer, the problem consists of defining the unknown polynomial coefficients h_0, $h_1, h_2, \ldots, h_n$ that optimize the response of the system. Once $h(s)$ is known, $g(s)g(-s)$ can be computed using

$$G(s^2) = g(s)g(-s) = G_0 + G_1 s^2 + \cdots + G_n s^{2n}$$

For example, when we have three transmission zeros ω_1, ω_2, and ω_3 so that $m = 3$, $G(s^2)$ is given by

$$\begin{aligned}
&G_0 = h_0^2 \\
&G_1 = -h_1^2 + 2h_2h_0 \\
&\vdots \\
&G_i = (-1)^i h_i^2 + 2h_{2i}h_0 + 2\sum_{j=2}^{i} (-1)^{j-1} h_{j-1}h_{2j-i+1} \\
&\vdots \\
&G_k = G_i|_{i=k} + (-1)^k S_3^2 \\
&G_{k+1} = G_i|_{i=k+1} + 2(-1)^k S_2 S_3 \\
&G_{k+2} = G_i|_{i=k+2} + (-1)^k \left(2S_1S_3 + S_2^2\right) \\
&G_{k+3} = G_i|_{i=k+3} + 2(-1)^k (S_1S_2 + S_3) \\
&G_{k+4} = G_i|_{i=k+4} + (-1)^k \left(S_1^2 + S_2\right) \\
&G_{k+5} = G_i|_{i=k+5} + 2(-1)^k S_1 \\
&G_{k+6} = G_i|_{i=k+6} + (-1)^k \\
&\vdots \\
&G_n = (-1)^n h_n^2
\end{aligned} \tag{7.7}$$

where

$$S_1 = \omega_1^2 + \omega_2^2 + \omega_3^2$$
$$S_2 = \omega_1^2\omega_2^2 + \omega_2^2\omega_3^3 + \omega_3^2\omega_1^2$$
$$S_3 = (\omega_1\omega_2\omega_3)^2$$

The left-half roots of $g(s)g(-s)$ are retained to form a Hurwitz polynomial $g(s)$. At this point, the equalizer scattering parameters $e_{11}(s)$, $e_{12}(s)$, $e_{21}(s)$, and $e_{22}(s)$ can be computed. During optimization, they will be needed for computing the error function to be optimized, and after optimization, for example, $e_{11}(s)$ can be used to realize the equalizer using the synthesis procedures of Section 7.7.

7.3 Multistage Transducer Gain

The overall transducer gain T_{GT} of the multistage active filter when terminated on arbitrary terminations Z_G and Z_L as shown in Figure 7.1 is given by

$$T_{GT} = T_k \frac{|e_{21,k+1}|^2\left(1-|\rho_L|^2\right)}{|1-e_{11,k+1}S_{G,k+1}|^2|1-\hat{e}_{22,k+1}\rho_L|^2} \tag{7.8}$$

where

$$\hat{e}_{22,k+1} = e_{22,k+1} + \frac{e_{21,k+1}^2 S_{G,k+1}}{1-e_{11,k+1}S_{G,k+1}} \tag{7.9}$$

$$S_{G,k+1} = S_{22,k} + \frac{S_{12,k}S_{21,k}\hat{e}_{22,k}}{1-S_{11,k}\hat{e}_{22,k}} \tag{7.10}$$

$$\rho_L = \frac{Z_L - Z_0}{Z_L + Z_0} \tag{7.11}$$

where

$S_{G,k+1}$ is the output reflection coefficient of transistor k when loaded on input
$\hat{e}_{22,k+1}$ is the output reflection of equalizer $k+1$ when loaded on input
ρ_L is the reflection coefficient of the load impedance Z_L

and where T_k is the transducer gain of the first k equalizer–transistor stages terminated on Z_0 as shown in Figure 7.2. Z_0 is the reference used for the scattering parameters. For example, $Z_0 = 50\Omega$ is often used to provide the measured scattering parameters of a transistor.

The transducer gain T_k of the first k equalizer–transistor stages is computed recursively by

$$T_k = T_{k-1} \frac{|e_{21,k}|^2|S_{21,k}|^2}{|1-e_{11,k}S_{G,k}|^2|1-\hat{e}_{22,k}S_{11,k}|^2} \quad \text{for } k>1 \tag{7.12}$$

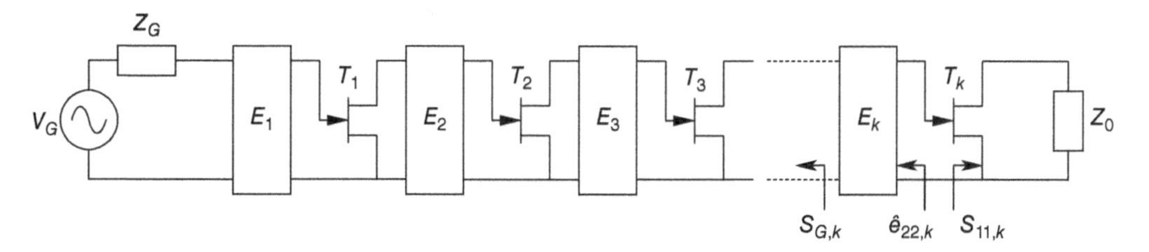

Figure 7.2 Transducer gain for k equalizer–transistor stages terminated on Z_0.

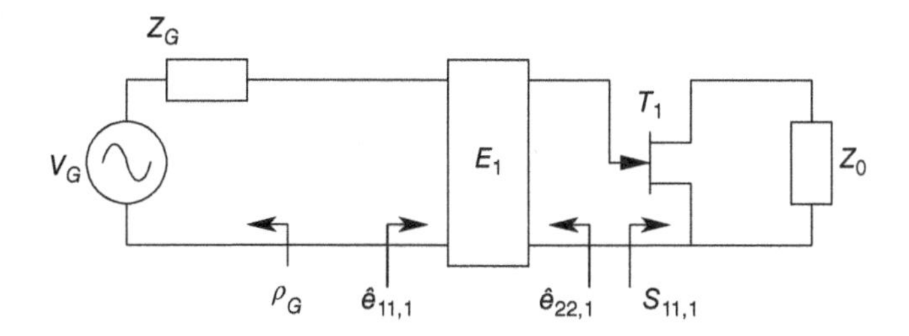

Figure 7.3 Transducer gain for the first stage terminated on Z_0.

with

$$\hat{e}_{22,k} = e_{22,k} + \frac{e_{21,k}^2 S_{G,k}}{1 - e_{11,k} S_{G,k}} \tag{7.13}$$

$$S_{G,k} = S_{22,k-1} + \frac{S_{12,k-1} S_{21,k-1} \hat{e}_{22,k-1}}{1 - S_{11,k-1} \hat{e}_{22,k-1}} \tag{7.14}$$

where

T_{k-1} is the transducer gain of the first $k-1$ equalizer–transistor stages
$e_{ij,k}$ are the scattering parameters of equalizer k
$S_{ij,k}$ are the scattering parameters of transistor k
$\hat{e}_{22,k}$ is the output reflection coefficient of equalizer k when loaded on input
$S_{G,k}$ is the output reflection coefficient of transistor $k-1$ when loaded on input

The recursion starts with the first stage ($k=1$) as shown in Figure 7.3.

The transducer gain of the first stage is given by

$$T_1 = \left(1 - |\rho_G|^2\right) \frac{|e_{21,1}|^2 |S_{21,1}|^2}{|1 - e_{11,1} S_{G,1}|^2 |1 - \hat{e}_{22,1} S_{11,1}|^2} \tag{7.15}$$

with

$$\hat{e}_{22,1} = e_{22,1} + \frac{e_{21,1}^2 S_{G,1}}{1 - e_{11,1} S_{G,1}} \tag{7.16}$$

$$S_{G,1} = \rho_G \tag{7.17}$$

$$\rho_G = \frac{Z_G - Z_0}{Z_G + Z_0} \tag{7.18}$$

where

$S_{G,1}$ would be the output reflection coefficient of a transistor 0, but since there is no transistor 0, it is equal to the reflection coefficient of the source ρ_G
$\hat{e}_{22,1}$ is the output reflection of equalizer 1 when loaded on input
ρ_G is the reflection coefficient of the source impedance Z_G

The transducer gain equations are demonstrated in Appendix A.

7.4 Multistage VSWR

The input voltage standing wave ratio (VSWR) corresponding to the k equalizer–transistor stages terminated on Z_0 as shown in Figure 7.3 is computed using

$$\text{VSWR}_{IN} = \frac{1 + |\hat{e}_{11,1}|}{1 - |\hat{e}_{11,1}|} \tag{7.19}$$

As transistor–equalizers stages are added to form the situation of Figure 7.2, the input VSWR is calculated recursively using

$$\begin{aligned}
S_{L,k} &= S_{11,k} \\
\hat{e}_{11,k} &= e_{11,k} + \frac{e_{21,k}^2 S_{L,k}}{1 - e_{22,k} S_{L,k}} \\
S_{L,k-1} &= S_{11,k-1} + \frac{S_{12,k-1} S_{21,k-1} \hat{e}_{11,k}}{1 - S_{22,k-1} \hat{e}_{11,k}} \\
\hat{e}_{11,k-1} &= e_{11,k-1} + \frac{e_{21,k-1}^2 S_{L,k-1}}{1 - e_{22,k-1} S_{L,k-1}} \\
&\vdots \\
\hat{e}_{11,1} &= e_{11,1} + \frac{e_{21,1}^2 S_{L,1}}{1 - e_{22,1} S_{L,1}}
\end{aligned} \tag{7.20}$$

where

$\hat{e}_{11,k}$ is the input reflection coefficient of equalizer k when loaded on output
$S_{L,k}$ is the input reflection coefficient of transistor k when loaded on output

After the last $(k+1)$th equalizer has been added for compensating arbitrary load impedance Z_L as shown in Figure 7.1, the input VSWR is computed using the recursive formula starting with

$$\begin{aligned}
\hat{e}_{11,k+1} &= e_{11,k+1} + \frac{e_{21,k+1}^2 \rho_L}{1 - e_{22,k+1} \rho_L} \\
S_{L,k} &= S_{11,k} + \frac{S_{12,k} S_{21,k} \hat{e}_{11,k+1}}{1 - S_{22,k} \hat{e}_{11,k+1}} \\
\rho_L &= \frac{Z_L - Z_0}{Z_L + Z_0}
\end{aligned} \tag{7.21}$$

The output VSWR including the $(k+1)$th equalizer and terminated on Z_L as shown in Figure 7.1 is given by

$$\mathrm{VSWR}_{OUT} = \frac{1+|\hat{e}_{22,k+1}|}{1-|\hat{e}_{22,k+1}|} \tag{7.22}$$

with

$$\begin{aligned}\hat{e}_{22,k+1} &= e_{22,k+1} + \frac{e_{21,k+1}^2 S_{G,k+1}}{1-e_{11,k+1}S_{G,k+1}}\\ S_{G,k+1} &= S_{22,k} + \frac{S_{12,k}S_{21,k}\hat{e}_{22,k}}{1-S_{11,k}\hat{e}_{22,k}}\end{aligned} \tag{7.23}$$

where

$\hat{e}_{22,k+1}$ is the output reflection coefficient of equalizer $k+1$ when loaded on input
$S_{G,k+1}$ is the output reflection coefficient of transistor k when loaded on input

and one uses (2.13) and (2.14) for recursively computing the output VSWR.

7.5 Multistage Phase and Group Delay

For some applications such as impulse radar, it is important to optimize the phase or the group. In these cases, a linear phase or a constant group delay is needed in a frequency band of interest [4, 6].

From Appendix A, the overall forward transmission coefficient $S_{T21,k}(j\omega)$ at stage k of the multistage active filter is found to be

$$S_{T21,k}(j\omega) = S_{T21,k-1}(j\omega)\frac{e_{21,k}(j\omega)S_{21,k}(j\omega)}{(1-e_{11,k}(j\omega)S_{G,k}(j\omega))(1-\hat{e}_{22,k}(j\omega)S_{11,k}(j\omega))} \tag{7.24}$$

From (7.24) one can see that the phase of the forward transmission coefficient is the sum and difference of the various phases:

$$\varphi_{ST21,k} = \varphi_{ST21,k-1} + \varphi_{e21,k} + \varphi_{S21,k} - (\varphi_1 + \varphi_2) \tag{7.25}$$

with

$$\begin{aligned}\varphi_1 &= -\left(\varphi_{e11,k} + \varphi_{SG,k}\right)\\ \varphi_2 &= -\left(\varphi_{\hat{e}22,k} + \varphi_{S11,k}\right)\end{aligned} \tag{7.26}$$

where

$\varphi_{e11,k}$ is the phase of $e_{11,k}(j\omega)$ of the kth equalizer
$\varphi_{e21,k}$ is the phase of $e_{21,k}(j\omega)$ of the kth equalizer
$\varphi_{S11,k}$ is the phase of $S_{11,k}(j\omega)$ of the kth transistor (or transistor block)
$\varphi_{S21,k}$ is the phase of $S_{21,k}(j\omega)$ of the kth transistor (or transistor block)
$\varphi_{\hat{e}22,k}$ is the phase of $\hat{e}_{22,k}(j\omega)$ reflection coefficient at the output of equalizer k

$\varphi_{SG,k}$ is the phase of $S_{G,k}(j\omega)$ reflection coefficient at the output of transistor $k-1$
$\varphi_{ST21,k-1}$ is the phase of $S_{T21,k-1}(j\omega)$ the overall forward transmission coefficient at stage $k-1$
$\varphi_{ST21,k}$ is the phase of $S_{T21,k}(j\omega)$ the overall forward transmission coefficient at stage k

It is seen from (7.25) that the phase of the forward transmission coefficient at stage k follows a simple recurrence relation in terms of the overall phase at stage $(k-1)$ and the phases of the kth equalizer and the kth transistor.

The group delay is defined as the derivative of the phase such that

$$\tau(\omega) = -\frac{d\varphi(\omega)}{d\omega} \tag{7.27}$$

The group delay characterizes the time the energy at frequency ω propagates through the system. In practice, it is desired to have a constant group delay in the frequency band of operation. To have a constant group delay means to have a linear phase in frequency band of operation.

The computation of the group delay is numerical since only measured phase characteristics $\varphi_{S21,k}(j\omega_i)$ and $\varphi_{S11,k}(j\omega_i)$ at the ω_i discrete frequency points are available for transistor k. The group delay is computed using a numerical derivative at the ω_i frequencies in the passband of the active filter such that

$$\tau(\omega_i) = -\frac{\varphi(\omega_i+\Delta\omega_i)-\varphi(\omega_i)}{\Delta\omega_i} \tag{7.28}$$

where $\Delta\omega_i$ is the frequency interval between two consecutive frequency points used in the optimization process.

7.6 Optimization Process

The optimization algorithm is used to define polynomial coefficients $h_0, h_1, \ldots, h_n$ of $h(t)$ for an equalizer that minimize a least mean square error function that includes desired target functions with different weight factors.

For the active filters of this chapter, the error function consists of the transducer gain, the VSWR, and the group delay.

$$E = \sum_{i=1}^{m}\left[W_1\left(\frac{G_{T,k}(\omega_i,\boldsymbol{H})}{G_{T0,k}}-1\right)^2 + W_2\left(\frac{\mathrm{VSWR}_k(\omega_i,\boldsymbol{H})}{\mathrm{VSWR}_{0,k}}-1\right)^2 + W_3\left(\frac{\tau_k(\omega_i,\boldsymbol{H})}{\tau_{0,k}}-1\right)^2\right] \tag{7.29}$$

where ω_i are the frequencies used for optimization.

$G_{T,k}(\omega_i, \boldsymbol{H})$ is the transducer power gain for k equalizer–transistor stages given by the equations in Section 7.3 and used to optimize the kth equalizer. The transducer power gain depends on frequency through the scattering parameters $e_{ij}(j\omega)$ of the equalizer and the scattering parameters $S_{ij}(j\omega)$ of the transistor. The target transducer power gain $G_{T0,k}$ for optimizing the kth equalizer is taken as a constant in the frequency band being optimized. The expressions for the target transducer gain are provided in Section 7.8.

$\mathrm{VSWR}_{IN,k}(\omega_i, \boldsymbol{H})$ is the input VSWR at stage k given by the equations in Section 7.4. The target $\mathrm{VSWR}_{IN0,k}$ is a constant over the frequency band being optimized. The value is often a number between 1 and 2. Note that the input VSWR is generally optimized using the first equalizer.

$\text{VSWR}_{OUT,k}(\omega_i, \boldsymbol{H})$ is the output VSWR at stage k given by the equations in Section 7.4. The target $\text{VSWR}_{OUT0,k}$ is a constant over the frequency band being optimized. The value is often a number between 1 and 2. Note that the output VSWR is generally optimized using the last equalizer.
$\tau_k(\omega_i, \boldsymbol{H})$ is the group delay for k equalizer–transistor stages given by the equations in Section 7.5 and is used to optimize the kth equalizer. The group delay $\tau_{0,k}$ for optimizing the kth equalizer is taken as a constant in the frequency band being optimized.

The vector $\boldsymbol{H}$ includes the polynomial coefficients $h_0, h_1, \ldots, h_n$ and is used as the $\boldsymbol{X}$ vector in the Levenberg–More optimization algorithms of Chapter 2 and Appendix B.

The main computation in the optimization routine is to define the step $\Delta\boldsymbol{H}$ to add to the elements of vector $\boldsymbol{H}$ at the next iteration given by

$$\Delta\boldsymbol{H} = -\left[\boldsymbol{J}^T\boldsymbol{J} + \lambda\boldsymbol{D}^T\boldsymbol{D}\right]^{-1}\boldsymbol{J}^T e(\boldsymbol{H}) \tag{7.30}$$

where

$e(\boldsymbol{H})$ is the error vector
J is the Jacobian matrix of the error with elements $\partial e_j/\partial h_i$
D is a diagonal matrix
λ is the Levenberg–Marquardt–More parameter described in Appendix B

An important element is the Jacobian matrix. It is based on derivatives of the error functions. For the transducer power gain, the analytical expressions for the derivatives that can be determined are rather tedious and lengthy to express and are written into a program. In the case of the input VSWR, the analytical expressions become very cumbersome and it is preferred to use a numerical derivative such that

$$\frac{\partial \text{VSWR}_{IN}(\omega_j, \boldsymbol{H})}{\partial h_i} = \frac{\text{VSWR}_{IN}(\omega_j, \boldsymbol{H} + \Delta h_i) - \text{VSWR}_{IN}(\omega_j, \boldsymbol{H})}{\Delta h_i} \tag{7.31}$$

where a step of $\Delta h_i = 10^{-5}$ was found satisfactory when starting with $h(t)$ polynomial coefficients initialized at $h_i = \pm 1$.

The same is the case for the group delay, where numerical derivatives are also preferred:

$$\frac{\partial \tau_k(\omega_j, \boldsymbol{H})}{\partial h_i} = \frac{\tau_k(\omega_j, \boldsymbol{H} + \Delta h_i) - \tau_k(\omega_j, \boldsymbol{H})}{\Delta h_i} \tag{7.32}$$

7.7 Synthesis Procedures

The synthesis procedure described in this section is based on consecutive extractions of an inductor or a capacitor or a resonating LC cell from the transfer matrix of the equalizer. The transfer matrix itself is obtained from converting the scattering matrix of the equalizer defined in Section 7.2, and it is completely defined by the three polynomials $f(s)$, $g(s)$, and $h(s)$. Complete definitions and explanations used for this synthesis method are given in Appendix D and are summarized here.

Starting with a transfer matrix such that

$$f(s) = \pm s^k \prod_{j=1}^{m} \left(s^2 + \omega_j^2\right)$$

$$g(s) = g_n s^n + g_{n-1} s^{n-1} + \cdots + g_1 s + g_0$$

$$h(s) = h_n s^n + h_{n-1} s^{n-1} + \cdots + h_1 s + h_0$$

If $h_n = g_n$, then one extracts a series inductor such that

$$L = \frac{2g_n}{g_{n-1} - h_{n-1}} \tag{7.33}$$

and the remaining transfer matrix is given by

$$\begin{aligned} f_r(s) &= f(s) \\ g_r(s) &= g(s) + \frac{Ls}{2}(h(s) - g(s)) \\ h_r(s) &= h(s) + \frac{Ls}{2}(h(s) - g(s)) \end{aligned} \tag{7.34}$$

If $h_n = -g_n$, then one extracts a parallel capacitor such that

$$C = \frac{2g_n}{g_{n-1} + h_{n-1}} \tag{7.35}$$

and the remaining transfer matrix is given by

$$\begin{aligned} f_r(s) &= f(s) \\ g_r(s) &= g(s) - \frac{Cs}{2}(h(s) + g(s)) \\ h_r(s) &= h(s) + \frac{Cs}{2}(h(s) + g(s)) \end{aligned} \tag{7.36}$$

If $h_0 = -g_0$, then one extracts a parallel inductor such that

$$L = \frac{g_1 + h_1}{2g_0} \tag{7.37}$$

and the remaining transfer matrix is given by

$$\begin{aligned} f_r(s) &= \frac{f(s)}{s} \\ g_r(s) &= \frac{sg(s) - \frac{1}{2L}(h(s) + g(s))}{s^2} \\ h_r(s) &= \frac{sh(s) + \frac{1}{2L}(h(s) + g(s))}{s^2} \end{aligned} \tag{7.38}$$

If $h_0 = g_0$, then one extracts a series capacitor such that

$$C = \frac{g_1 - h_1}{2g_0} \tag{7.39}$$

and the remaining transfer matrix is given by

$$\begin{aligned} f_r(s) &= \frac{f(s)}{s} \\ g_r(s) &= \frac{sg(s) + \frac{1}{2C}(h(s) - g(s))}{s^2} \\ h_r(s) &= \frac{sh(s) + \frac{1}{2C}(h(s) - g(s))}{s^2} \end{aligned} \tag{7.40}$$

If $h(s) - g(s)$ is divisible by the factor $s^2 + \omega_0^2$, then one extracts a series LC cell such that

$$\begin{aligned} C &= \left[\frac{s}{s^2 + {\omega_0}^2} \frac{g(s) - h(s)}{2g(s)} \right]_{s = j\omega_0} \\ L &= \frac{1}{C{\omega_0}^2} \end{aligned} \tag{7.41}$$

and the remaining transfer matrix is given by

$$\begin{aligned} f_r(s) &= \frac{f(s)}{s^2 + \omega_0^2} \\ g_r(s) &= \frac{(s^2 + \omega_0^2)g(s) + \frac{s}{2C}(h(s) - g(s))}{(s^2 + \omega_0^2)^2} \\ h_r(s) &= \frac{(s^2 + \omega_0^2)h(s) + \frac{s}{2C}(h(s) - g(s))}{(s^2 + \omega_0^2)^2} \end{aligned} \tag{7.42}$$

If $h(s) + g(s)$ is divisible by the factor $s^2 + \omega_0^2$, then one extracts a parallel LC cell such that

$$\begin{aligned} L &= \left[\frac{s}{s^2 + \omega_0^2} \frac{g(s) + h(s)}{2g(s)} \right]_{s = j\omega_0} \\ C &= \frac{1}{L\omega_0^2} \end{aligned} \tag{7.43}$$

and the remaining transfer matrix is given by

$$\begin{aligned} f_r(s) &= \frac{f(s)}{s^2+\omega_0^2} \\ g_r(s) &= \frac{\left(s^2+\omega_0^2\right)g(s) - \frac{s}{2L}(h(s)+g(s))}{\left(s^2+\omega_0^2\right)^2} \\ h_r(s) &= \frac{\left(s^2+\omega_0^2\right)h(s) + \frac{s}{2L}(h(s)+g(s))}{\left(s^2+\omega_0^2\right)^2} \end{aligned} \tag{7.44}$$

What follows is an example to illustrate the technique. Let us assume that the optimization produced the following $e_{11}(s)$:

$$e_{11}(s) = \frac{12s^5 - s^4 + 52s^3 - 4.5s^2 + 10.25s}{12s^5 + 7s^4 + 54s^3 + 29.5s^2 + 18.25s + 4}$$

from which

$$g(s) = 12s^5 + 7s^4 + 54s^3 + 29.5s^2 + 18.25s + 4$$

$$h(s) = 12s^5 - s^4 + 52s^3 - 4.5s^2 + 10.25s$$

and for this equalizer, we had specified

$$f(s) = \left(s^2+4\right)$$

Since $g_n = h_n$, we first extract a series inductor

$$L = \frac{2g_n}{g_{n-1} - h_{n-1}} = \frac{2\times 12}{7-(-1)} = 3$$

and the remaining transfer matrix is such that

$$\begin{aligned} f_r(s) &= f(s) = \left(s^2+4\right) \\ g_r(s) &= g(s) + \frac{Ls}{2}(h(s)-g(s)) = 4s^4 + 3s^3 + 17.5s^2 + 12.25s + 4 \\ h_r(s) &= h(s) + \frac{Ls}{2}(h(s)-g(s)) = -4s^4 + s^3 - 16.5s^2 + 4.25s \end{aligned}$$

This can be done using the following MATLAB code:

```
clc;
clear;
close all;
g = [12 7 54 29.5 18.25 4];
```

```
h = [12 -1 52 -4.5 10.25 0];
L = (2*g(1))/(g(2)-h(2));
gr = [0g] + (1/2)*[(h-g) 0]*L;
hr = [0h] + (1/2)*[(h-g) 0]*L;
clear g h
g = gr(3:length(gr));
h = hr(3:length(hr));
```

Now $g_n = -h_n$, so we extract a parallel capacitor

$$C = \frac{2g_n}{g_{n-1} + h_{n-1}} = \frac{2\times 4}{3+1} = 2$$

and the remaining transfer matrix is such that

$$f_r(s) = f(s) = \left(s^2 + 4\right)$$

$$g_r(s) = g(s) - \frac{Cs}{2}(h(s) + g(s)) = 2s^3 + s^2 + 8.25s + 4$$

$$h_r(s) = h(s) + \frac{Cs}{2}(h(s) + g(s)) = 2s^3 + 8.25s$$

This can be done using the following MATLAB code:

```
clc;
clear;
close all;
g = [4 3 17.5 12.25 4];
h = [-4 1 -16.5 4.25 0];
C = (2*g(1))/(g(2)+h(2));
gr = [0g] - (1/2)*[(h+g) 0]*C;
hr = [0h] + (1/2)*[(h+g) 0]*C;
g = gr(3:length(gr));
h = hr(3:length(hr));
```

Now we have $g(s) - h(s) = s^2 + 4$ that is divisible by the factor $s^2 + \omega_0^2$ with $\omega_0^2 = 4$, so we can extract a series LC cell

$$C = \left[\frac{s}{s^2 + \omega_0^2}\frac{g(s) - h(s)}{2g(s)}\right]_{s = j\omega_0} = \left[\frac{s}{s^2 + 4}\frac{s^2 + 4}{2(2s^3 + s^2 + 8.25s + 4)}\right]_{s = j2} = \frac{2j}{2(-16j - 4 + 16.5j + 4)} = 2$$

$$L = \frac{1}{C\omega_0^2} = \frac{1}{2\times 4} = 0.125$$

and the remaining transfer matrix is such that

$$f_r(s) = \frac{f(s)}{s^2 + \omega_0^2} = \frac{(s^2+4)}{(s^2+4)} = 1$$

$$g_r(s) = \frac{\left(s^2+\omega_0^2\right)g(s) + \dfrac{s}{2C}(h(s)-g(s))}{\left(s^2+\omega_0^2\right)^2} = 2s+1$$

$$h_r(s) = \frac{\left(s^2+\omega_0^2\right)h(s) + \dfrac{s}{2C}(h(s)-g(s))}{\left(s^2+\omega_0^2\right)^2} = 2s$$

This can be done using the following MATLAB code:

```
clc;
clear;
close all;
syms s
gs = 2*s^3 + s^2 + (33*s)/4 + 4;
hs = 2*s^3 + (33*s)/4;
wo = 2;
num = simplify((gs-hs)/(s^2+wo^2));
den = gs + hs;
s = j*wo;
C = eval(s*num/den);
L = 1/(C*wo^2);
syms s
grs0 = simplify(gs + (1/(2*C))*s*(hs-gs)/(s^2+wo^2));
grs = simplify(grs0/(s^2+wo^2));
hrs0 = simplify(hs + (1/(2*C))*s*(hs-gs)/(s^2+wo^2));
hrs = simplify(hrs0/(s^2+wo^2));
fprintf('C = %6.4f \n', C);
fprintf('L = %6.4f \n', L);
disp(' ');
fprintf('g(s) = ');
disp(grs)
fprintf('h(s) = ');
disp(hrs)
```

Now $g_n = h_n$, so we can extract a series inductor

$$L = \frac{2g_n}{g_{n-1} - h_{n-1}} = \frac{2\times 2}{1-0} = 4$$

and the remaining transfer matrix is such that

$$f_r(s) = f(s) = 1$$

$$g_r(s) = g(s) + \frac{Ls}{2}(h(s)-g(s)) = 1$$

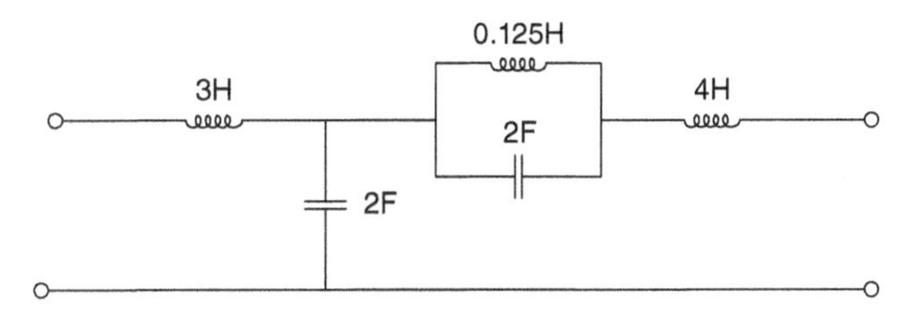

Figure 7.4 The normalized to 1 Ω equalizer. The unit of an inductance is the Henry (H) and the unit of a capacitance is the Farad (F).

$$h_r(s) = h(s) + \frac{Ls}{2}(h(s) - g(s)) = 0$$

This can be done using the following MATLAB code:

```
clc;
clear;
close all;
g = [2 1];
h = [2 0];
L = (2*g(1))/(g(2)-h(2));
gr = [0g] + (1/2)*[(h-g) 0]*L;
hr = [0h] + (1/2)*[(h-g) 0]*L;
clear g h
g = gr(3:length(gr));
h = hr(3:length(hr));
```

At this stage, the transfer matrix is equal to the identity matrix and according to (7.3) the remaining equalizer scattering matrix is

$$S_E = \begin{pmatrix} \frac{h(s)}{g(s)} & \frac{f(s)}{g(s)} \\ \frac{f(s)}{g(s)} & \frac{-\varepsilon h(-s)}{g(s)} \end{pmatrix} = \begin{pmatrix} 0 & 1 \\ 1 & 0 \end{pmatrix}$$

In summary, the synthesized equalizer is given in Figure 7.4.

Note that these are normalized to 1 Ω and often the frequencies have been scaled to the maximum frequency in the passband to allow for initial guess for the $h(s)$ polynomial to be taken as $h_i = \pm 1$. To denormalize the values, one uses the technique in Section 2.8.

7.8 Design Procedures

The design of the multistage active filter starts with the first equalizer–transistor stage terminated on reference impedance Z_0 (e.g., 50 Ω) as shown in Figure 7.5.

The transducer gain for the first stage is given by

$$T_1 = \left(1 - |\rho_G|^2\right) \frac{|e_{21,1}|^2 |S_{21,1}|^2}{|1 - e_{11,1}\rho_G|^2 |1 - \hat{e}_{22,1} S_{11,1}|^2} \tag{7.45}$$

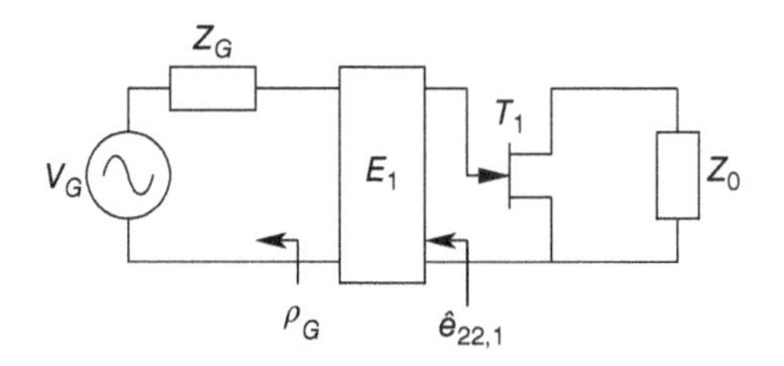

Figure 7.5 Optimizing the first equalizer.

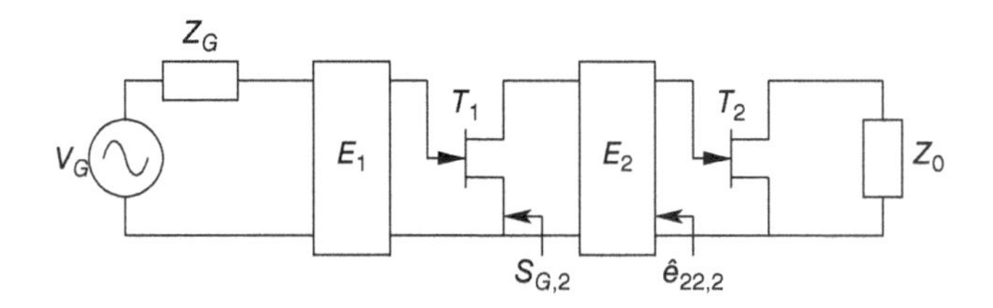

Figure 7.6 Optimizing the second equalizer.

with

$$\hat{e}_{22,1} = e_{22,1} + \frac{e_{21,1}^2 \rho_G}{1 - e_{11,1}\rho_G} \tag{7.46}$$

$$\rho_G = \frac{Z_G - Z_0}{Z_G + Z_0} \tag{7.47}$$

and the target transducer gain for the first stage is given by

$$T_{0,1} = \min\left\{ \frac{|S_{21,1}(j\omega)|^2}{\left|1 - |S_{11,1}(j\omega)|^2\right|} \right\} \tag{7.48}$$

Once the first equalizer has been optimized, the second equalizer–transistor stage is placed between the first transistor and the load Z_0 as shown in Figure 7.6. At this time, the first equalizer is fixed and only $e_{11,2}(t)$, $e_{22,2}(t)$, and $e_{12,2}(t) = e_{21,2}(t)$ of the second equalizer vary between iterations of the optimization.

The transducer gain T_2 for the two first equalizer–transistor stages is

$$T_2 = T_1 \frac{|e_{21,2}|^2 |S_{21,2}|^2}{|1 - e_{11,2}S_{G,2}|^2 |1 - \hat{e}_{22,2}S_{11,2}|^2} \tag{7.49}$$

with

$$\hat{e}_{22,2} = e_{22,2} + \frac{e_{21,2}^2 S_{G,2}}{1 - e_{11,2}S_{G,2}} \tag{7.50}$$

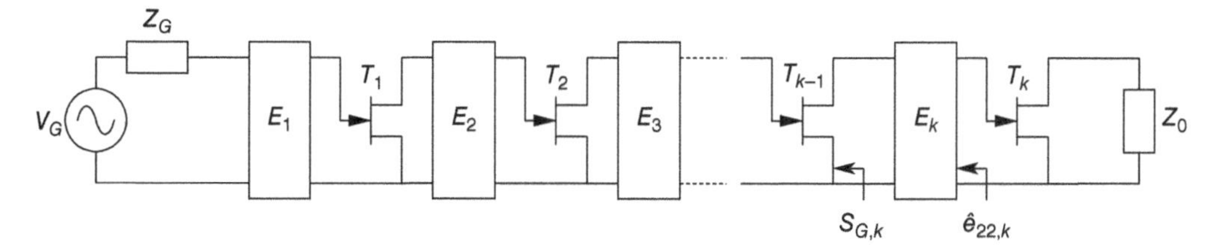

Figure 7.7 Optimization the kth equalizer.

$$S_{G,2} = S_{22,1} + \frac{S_{12,1}S_{21,1}\hat{e}_{22,1}}{1 - S_{11,1}\hat{e}_{22,1}} \tag{7.51}$$

and the target transducer gain for the second stage is given by

$$T_{0,2} = \min\left\{ T_{MAX,1}\frac{|S_{21,2}|^2}{\left|1 - |S_{11,2}|^2\right|} \right\} \tag{7.52}$$

where

$$T_{MAX,1} = \frac{|S_{21,1}|^2}{\left|1 - |S_{11,1}|^2\right|\left|1 - |S_{22,1}|^2\right|} \frac{1}{\left|1 - \frac{S_{12,1}S_{21,1}S_{11,1}^*S_{22,1}^*}{\left(1 - |S_{11,1}|^2\right)\left(1 - |S_{22,1}|^2\right)}\right|^2} \tag{7.53}$$

The procedure is repeated, and Figure 7.7 shows the situation for kth equalizer–transistor terminated on Z_0.

The transducer gain T_k for the first k equalizer–transistor stages is

$$T_k = T_{k-1}\frac{|e_{21,k}|^2|S_{21,k}|^2}{|1 - e_{11,k}S_{G,k}|^2|1 - \hat{e}_{22,k}S_{11,k}|^2} \tag{7.54}$$

with

$$\hat{e}_{22,k} = e_{22,k} + \frac{e_{21,k}^2 S_{G,k}}{1 - e_{11,k}S_{G,k}} \tag{7.55}$$

$$S_{G,k} = S_{22,k-1} + \frac{S_{12,k-1}S_{21,k-1}\hat{e}_{22,k-1}}{1 - S_{11,k-1}\hat{e}_{22,k-1}} \tag{7.56}$$

and the target transducer gain for the kth stage is given by

$$T_{0,k} = \min\left\{ T_{MAX,1} \times T_{MAX,2} \times \cdots \times T_{MAX,k-1}\frac{|S_{21,k}|^2}{\left|1 - |S_{11,k}|^2\right|} \right\} \tag{7.57}$$

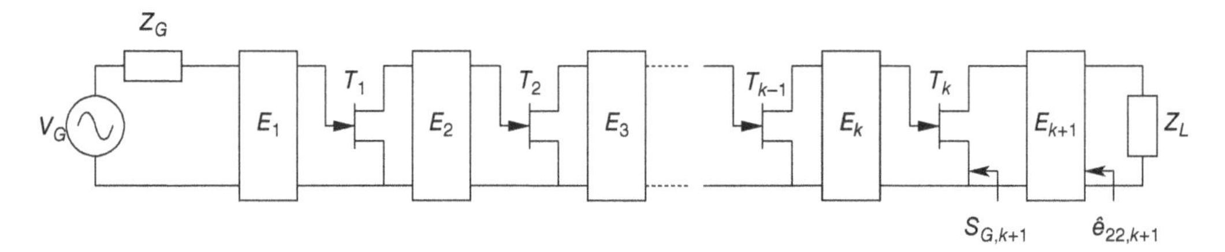

Figure 7.8 Optimizing the $(k+1)$th equalizer.

where

$$T_{MAX,i} = \frac{|S_{21,i}|^2}{\left|1-|S_{11,i}|^2\right|\left|1-|S_{22,i}|^2\right|} \frac{1}{\left|1-\frac{S_{12,i}S_{21,i}S_{11,i}^*S_{22,i}^*}{\left(1-|S_{11,i}|^2\right)\left(1-|S_{22,i}|^2\right)}\right|^2} \tag{7.58}$$

Once the kth equalizer has been optimized, a last equalizer E_{k+1} can be added to compensate for an arbitrary load Z_L different than Z_0 as shown in Figure 7.8.

The overall transducer gain T_{GT} in the case of Figure 7.8 is given by

$$T_{GT} = T_k \frac{|e_{21,k+1}|^2\left(1-|\rho_L|^2\right)}{|1-e_{11,k+1}S_{G,k+1}|^2|1-\hat{e}_{22,k+1}\rho_L|^2} \tag{7.59}$$

with

$$\hat{e}_{22,k+1} = e_{22,k+1} + \frac{e_{21,k+1}^2 S_{G,k+1}}{1-e_{11,k+1}S_{G,k+1}} \tag{7.60}$$

$$S_{G,k+1} = S_{22,k} + \frac{S_{12,k}S_{21,k}\hat{e}_{22,k}}{1-S_{11,k}\hat{e}_{22,k}} \tag{7.61}$$

$$\rho_L = \frac{Z_L - Z_0}{Z_L + Z_0} \tag{7.62}$$

and the target transducer gain for the $(k+1)$th equalizer is given by

$$T_{0,k+1} = \min\{T_{MAX,1} \times T_{MAX,2} \times \cdots \times T_{MAX,k-1} \times T_{MAX,k}\} \tag{7.63}$$

7.9 Simulations and Realizations

7.9.1 Two-Stage Low-Pass Active Filter

The procedures described in this chapter were implemented in a program called FREELCD. The first active filter design example is a low-pass active filter with two stages [4–7]. The transistor was an NEC NE20283A FET biased at $V_{DS} = 2\text{V}$ and $I_{DS} = 10.0\text{mA}$. Table 7.1 provides the measured small-signal scattering parameters of the transistor for a few frequency points.

Table 7.1 Measured S parameters of the NEC NE20283A FET transistor

Freq	S_{11}		S_{12}		S_{21}		S_{22}	
(GHz)	MOD	Phase	MOD	Phase	MOD	Phase	MOD	Phase
0.5	0.993	−9.1	0.001	78.6	4.308	170.9	0.697	−7.1
1.5	0.974	−27.0	0.030	67.8	4.223	153.3	0.686	−21.1
3.0	0.918	−53.0	0.058	49.7	3.993	128	0.653	−41.1
4.0	0.869	−70.1	0.069	38	3.807	111.6	0.625	−53.8
5.0	0.818	−87.0	0.080	28.6	3.596	95.9	0.595	−66.1
6.0	0.770	−103.4	0.087	16.0	3.368	80.7	0.568	−77.8
7.0	0.731	−118.6	0.092	6.3	3.134	66.4	0.547	−88.9
8.0	0.698	−132.4	0.094	−2.1	2.921	53.0	0.533	−98.4

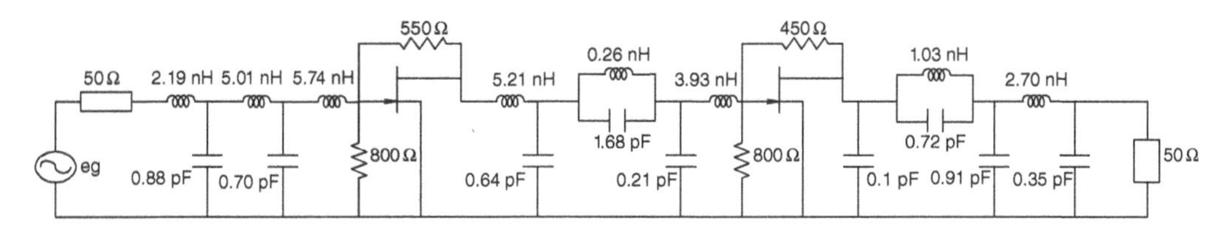

Figure 7.9 Synthesized two-stage low-pass active filter.

The first equalizer E_1 had an order $n=5$, with $k=0$ and no transmission zeros so that $f(s)=1$.
The second equalizer E_2 had an order $n=5$, with $k=0$ and one transmission zero at $f_{01}=7.5\,\text{GHz}$ so that $f(s)=s^2+\omega_{01}^2$.
The third equalizer E_3 had an order $n=5$, with $k=0$ and one transmission zero at $f_{02}=5.8\,\text{GHz}$ so that $f(s)=s^2+\omega_{02}^2$.

The active filter was optimized for gain, VSWR, and group delay in a 1.5–4.5 GHz passband. After optimization, the following equalizer input reflection coefficients were obtained:

$$e_{11,1}(s)=\frac{1.46s-0.86s^2+1.84s^3-0.39s^4+0.45s^5}{1+2.76s+2.74s^2+2.52s^3+0.88s^4+0.45s^5}$$

$$e_{11,2}(s)=\frac{12.8s-0.15s^2+5.85s^3+0.0097s^4+0.39s^5}{8.33+15.75s+6.05s^2+6.14s^3+0.47s^4+0.39s^5}$$

$$e_{11,3}(s)=\frac{0.27s-0.39s^2-0.15s^3-0.12s^4-0.11s^5}{4.98+5.82s+4.39s^2+2.14s^3+0.69s^4+0.11s^5}$$

Using the synthesis procedures of Section 7.8, the equalizers were synthesized from $e_{11}(s)$ and denormalized for 50 Ω and scaling frequency of 14 GHz. Figure 7.9 shows the final synthesized active filter.

Figure 7.10 shows that in the 1.5–4.5 GHz optimized passband the simulated gain is about $20\pm0.5\,\text{dB}$. It also shows the transmission zeros at 5.8 and 7.5 GHz. The simulated input and output VSWR are less than 1.88 (equivalent to a −10.3 dB return loss).

Figure 7.11 shows that the group delay is about 0.5 ns in the passband.

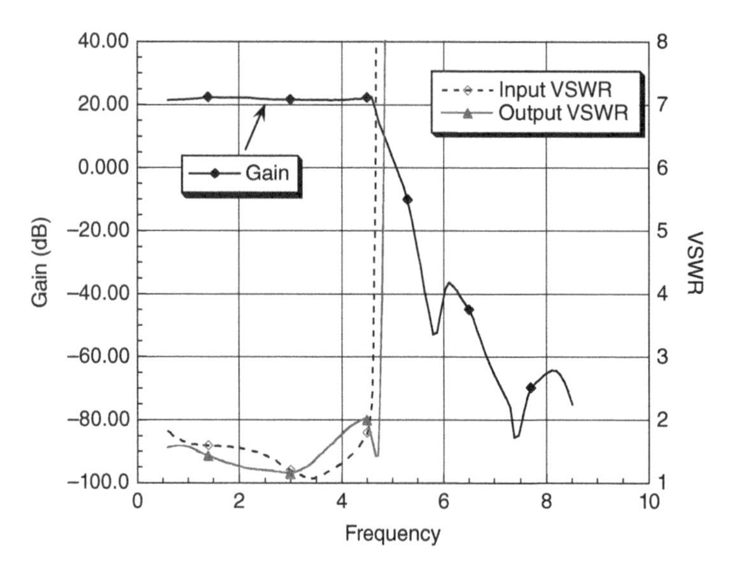

Figure 7.10 Simulated gain and VSWR of the low-pass active filter.

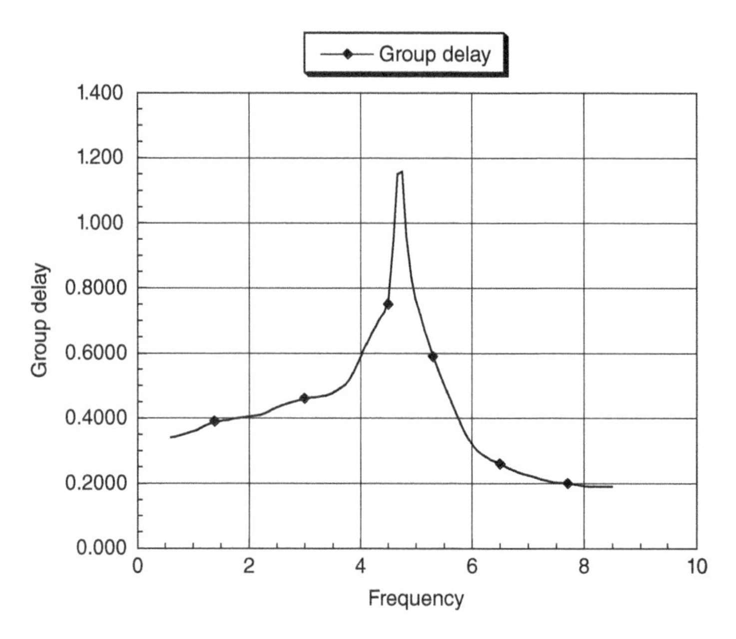

Figure 7.11 Simulated group delay (ns) of the low-pass active filter.

7.9.2 Single-Stage Bandpass Active Filter

The second active filter design example is a single-stage bandpass active filter [4–7]. It uses a Fujitsu FHX04X MESFET transistor. Table 7.2 provides the measured small-signal scattering parameters of the transistor for a few frequency points.

The first equalizer E_1 had an order $n = 9$, with $k = 2$ and a transmission zero at $f_{01} = 2.3\text{GHz}$ before the passband and a transmission zero at $f_{02} = 8.5\text{GHz}$ after the passband so that

Table 7.2 Measured S parameters of the Fujitsu FHX04X MESFET

Freq (GHz)	S_{11} MOD	S_{11} Phase	S_{12} MOD	S_{12} Phase	S_{21} MOD	S_{21} Phase	S_{22} MOD	S_{22} Phase
0.5	0.999	−4.99	0.004	86.93	2.612	176.00	0.854	−1.67
1.0	0.998	−9.97	0.009	83.87	2.602	172.0	0.853	−3.33
3.0	0.986	−29.29	0.027	72.01	2.511	156.41	0.843	−9.79
5.0	0.967	−46.98	0.043	61.20	2.354	141.87	0.828	−15.76
7.0	0.945	−62.51	0.055	51.74	2.165	128.70	0.811	−21.13
9.0	0.924	−75.75	0.065	43.65	1.973	116.91	0.795	−25.96
11.0	0.906	−86.92	0.072	36.77	1.792	106.37	0.783	−30.39
13.0	0.891	−96.27	0.077	30.92	1.630	96.88	0.775	−34.52

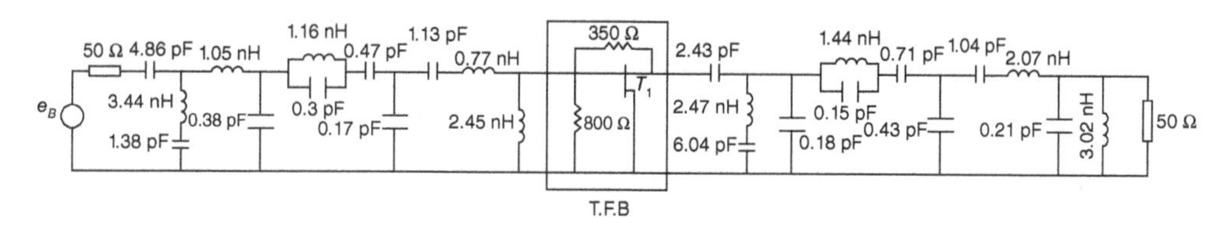

Figure 7.12 Synthesized single-stage bandpass active filter.

$$f(s) = s^2\left(s^2 + \omega_{01}^2\right)\left(s^2 + \omega_{02}^2\right)$$

The first equalizer E_1 had an order $n=9$, with $k=2$ and a transmission zero at $f_{03} = 1.3\,\text{GHz}$ before the passband, and a transmission zero at $f_{04} = 10.5\,\text{GHz}$ after the passband so that

$$f(s) = s^2\left(s^2 + \omega_{03}^2\right)\left(s^2 + \omega_{04}^2\right)$$

The active filter was optimized for gain, VSWR, and group delay in a 3–7 GHz passband. After optimization, the following equalizer input reflection coefficients were obtained:

$$e_{11,1}(s) = \frac{0.97 - 0.68s + 2.37s^2 - 0.56s^3 + 0.38s^4 + 1.5s^5 - 0.06s^6 + 1.04s^7 + 0.13s^8 + 0.18s^9}{0.97 + 3.48s + 8.34s^2 + 1.35s^3 + 16s^4 + 12.6s^5 + 7.59s^6 + 3.31s^7 + 0.91s^8 + 0.18s^9}$$

$$e_{11,2}(s) = \frac{0.89 - 0.7s + 2.9s^2 - 1.67s^3 + 2.88s^4 - 0.45s^5 + 0.39s^6 - 0.66s^7 - 0.2s^8 - 0.32s^9}{0.89 + 3.88s + 11s^2 + 20.4s^3 + 26.7s^4 + 23.2s^5 + 14.5s^6 + 6.52s^7 + 1.95s^8 + 0.32s^9}$$

Figure 7.12 shows the final synthesized single-stage bandpass active filter.

Figure 7.13 shows the simulated gain and return losses. The transducer power gain is about $10.2 \pm 0.2\,\text{dB}$ and the input and output VSWRs are less than 1.9. The four transmission zeros at 1.3, 2.3, 8.5, and 10.5 GHz can also be seen in Figure 7.13. Figure 7.14 shows the simulated group delay is fairly flat in the 3–7 GHz passband.

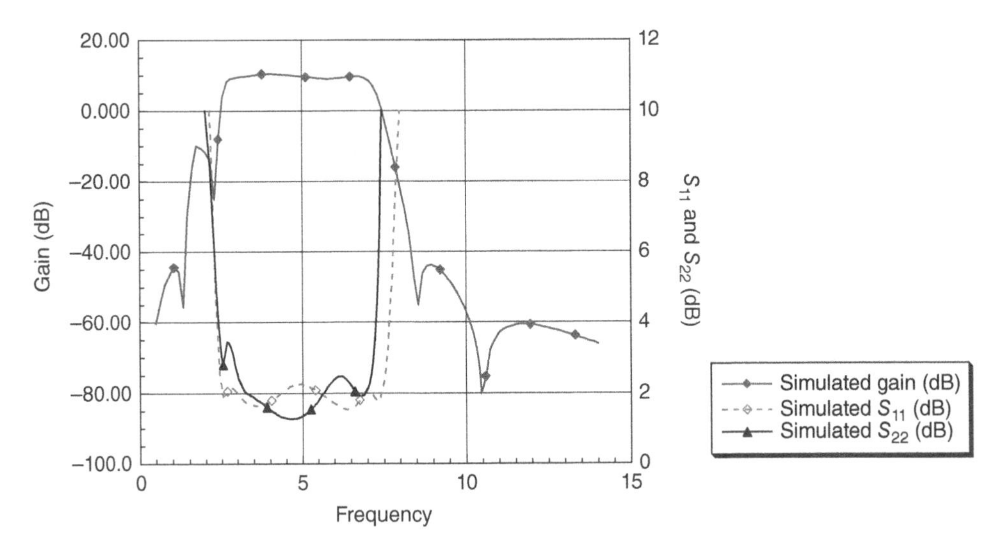

Figure 7.13 Simulated gain and VSWR of the single-stage bandpass active filter.

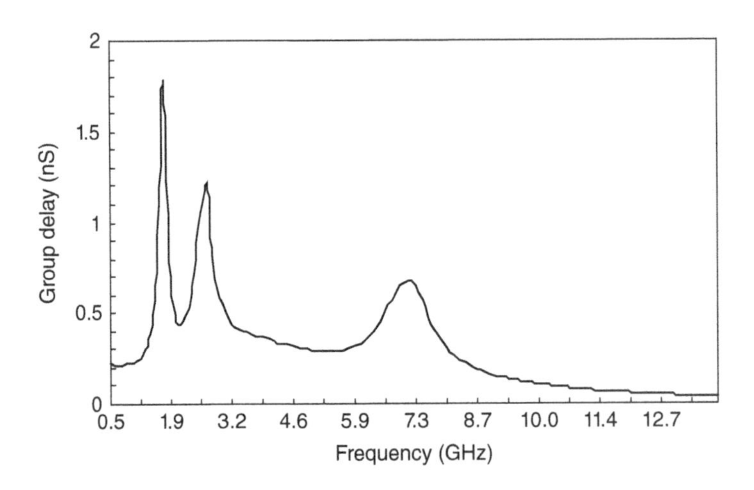

Figure 7.14 Simulated group delay of the single-stage bandpass active filter.

7.9.3 Single-Stage Bandpass Active Filter MMIC Realization

In this example, a microwave bandpass active filter using a single CEPD223 transistor and lumped passive components was designed using the FREELCD program. It was optimized for gain and VSWR in a 7.7–8.1 GHz passband [8–10]. It was designed to have two finite transmission zeros: one before the passband at 6.8 GHz and one after the passband at 9.2 GHz. Figure 7.15 shows the equalizer element values obtained after optimization.

Figures 7.16, 7.17, and 7.18 show the simulated results using the FREELCD optimization program that does not take into account losses and using HP-MDS software, where GaAs substrate characteristics are introduced using the HB200 UMS-THOMSON design kit in anticipation of an MMIC implementation. It is seen that the losses reduce the gain and gain flatness in the passband. There is, however, good agreement on the location of the transmission zeros at 6.8 and 9.2 GHz, and the return losses in the passband.

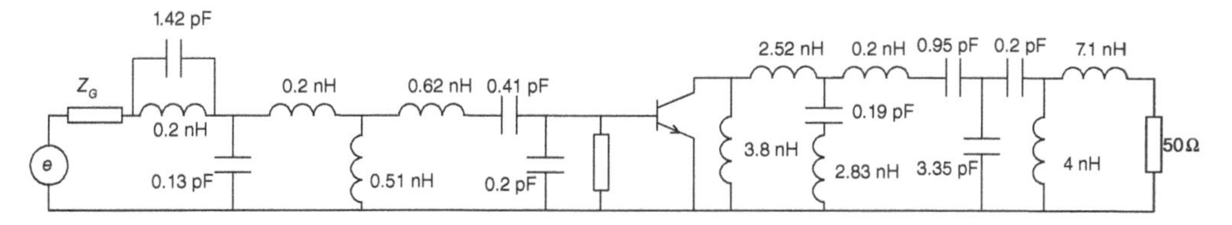

Figure 7.15 Single-stage bandpass active filter.

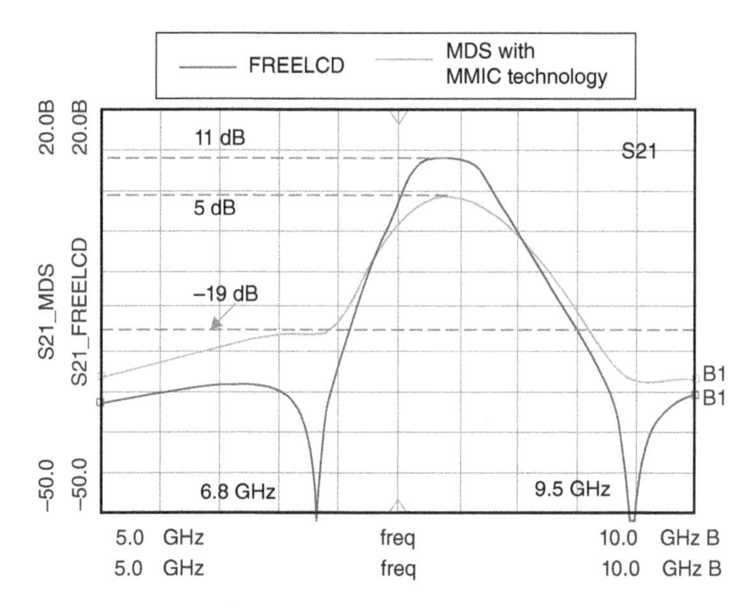

Figure 7.16 Simulated gain of the active filter.

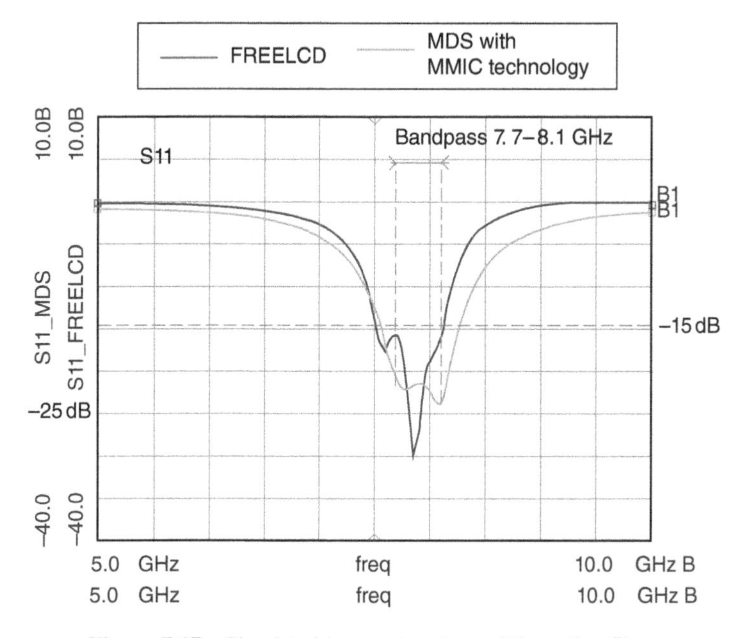

Figure 7.17 Simulated input return loss of the active filter.

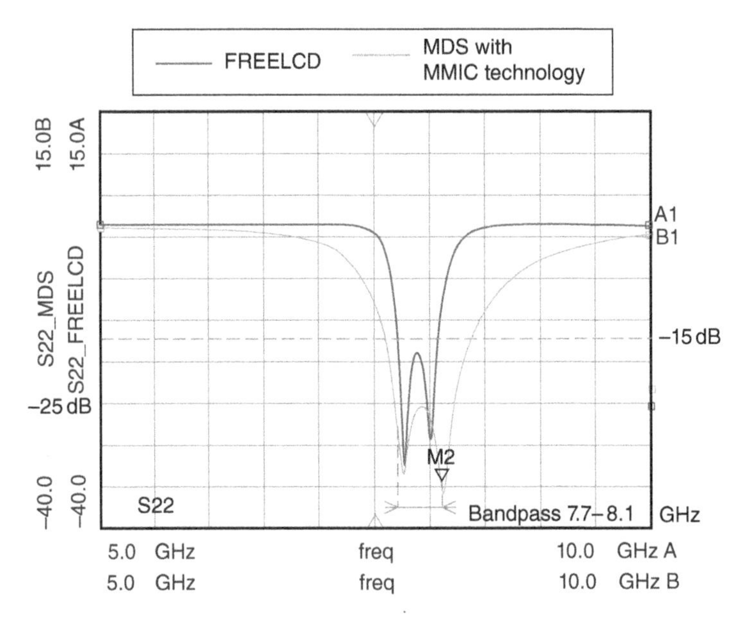

Figure 7.18 Simulated output return loss of the active filter.

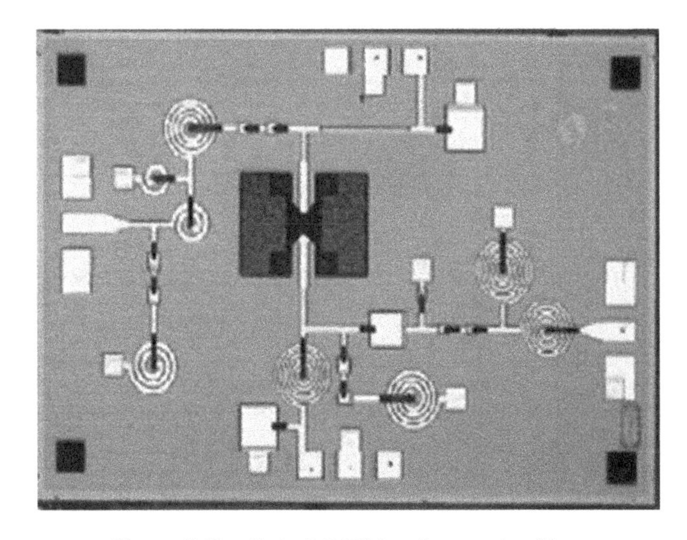

Figure 7.19 GaAs MMIC bandpass active filter.

The active filter was then realized as an MMIC on a 4 mm^2 GaAs substrate using a CEPD223 transistor and lumped passive components from the UMS THOMSON foundry. A picture of the MMIC active filter is shown in Figure 7.19.

Figures 7.20, 7.21, and 7.22 show the measured and simulated responses of the MMIC bandpass active filter.

There is overall a frequency shift between measured and simulated gains. The measured output VSWR does show a noticeable degradation compared to simulations. Note that these were university research results using experimental technology at the time with higher losses than expected. But we can note that the performances remain in concordance with the specifications.

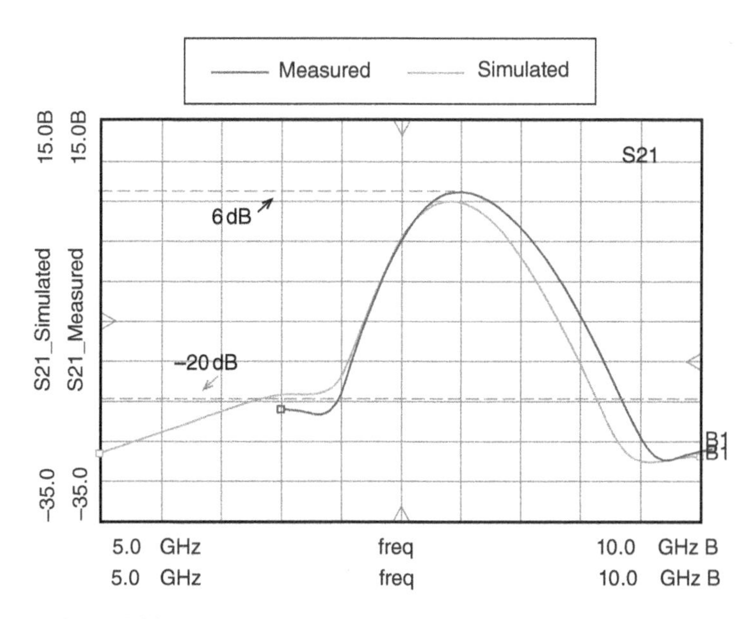

Figure 7.20 Measured and simulated gain of the MMIC active filter.

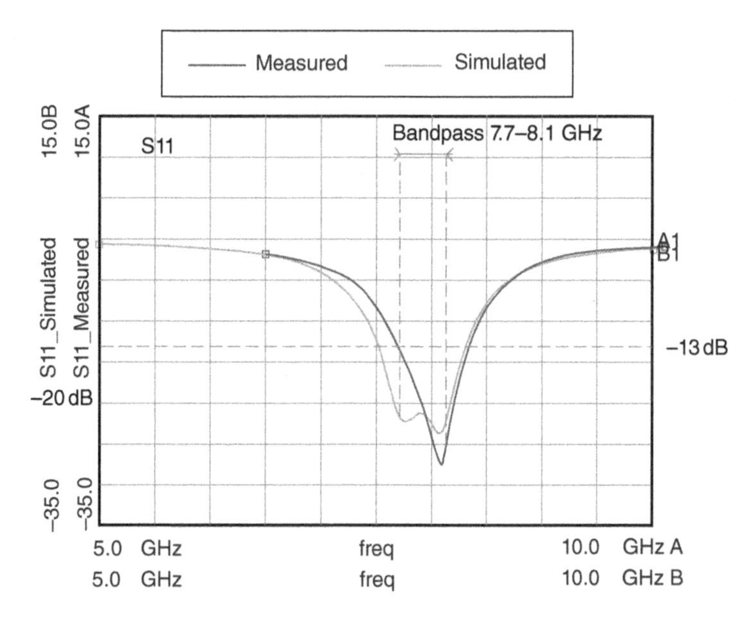

Figure 7.21 Measured and simulated input return loss of the MMIC active filter.

In conclusion, this work has provided a contribution to the study of integrated active microwave filters. The technique based on the RFT has the added advantages of generality, being applicable to all matching problems, and universality, as it involves neither equalizer values nor a predefined equalizer topology. The formalism of this technique has been first used in the case of amplifiers. It allows without complex calculations to optimize power gain, VSWR, and VSWR performance

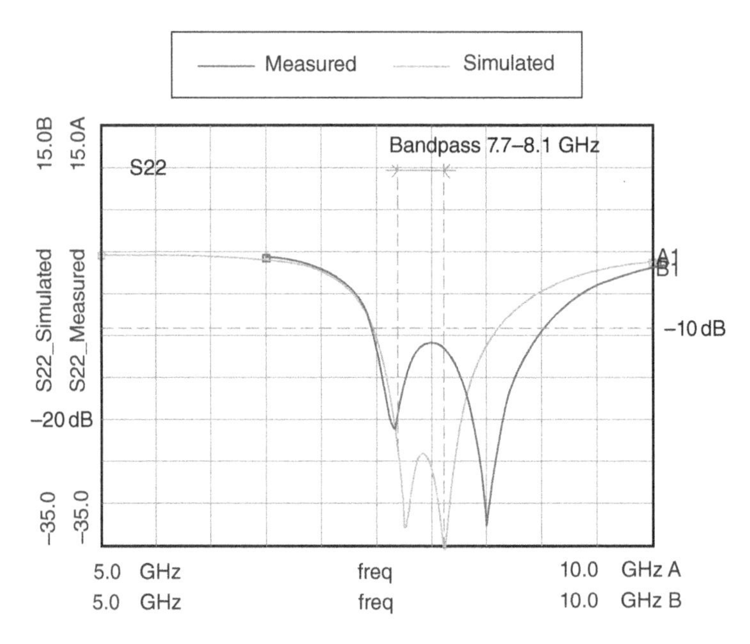

Figure 7.22 Measured and simulated output return loss of the MMIC active filter.

parameters of a microwave active filter by introducing a function $f(s)$ in the Belevitch's representation of the matching network. This formalism allows obtaining transmission zero at finite frequencies. The validity of the proposed method has been demonstrated by a narrow band MMIC active filter realization. The experimental results are in good agreement with the theoretical calculations. The results obtained confirm the synthesis of microwave active filters and a computer program FREELCD was created. The filters were simulated for CNES-France and ALCATEL SPACE, and we also thank the UMS THOMSON foundry for the circuit realization.

References

[1] P. Jarry, J. Beneat, Advanced Design Techniques and Realizations of Microwave and RF Filters, Wiley-IEEE Press, New York, 2008.

[2] S.E. Sussman-Fort, "Design concepts for microwave GaAs FET active filters," IEEE Trans. Microw. Theory Tech., vol. 37, pp. 1418–1424, 1989.

[3] G.F. Zhang, M.L. Villegas, C.S. Ripoll, "Microwave active filter using GaAs monolithic floating active inductor," Microw. Opt. Technol. Lett., vol. 5, pp. 381–388, 1992.

[4] E. EL Hendaoui, The Simplified Real-Frequency Method Applied to the Active Filters Synthesis, Ph.D. dissertation, University of Bordeaux, October 1992.

[5] E. EL Hendaoui, A. Perennec, P. Jarry, "The simplified real frequency method applied to active filters synthesis," IEEE MTT Symposium, Boston, MA, pp. 455–457, 1991.

[6] E. EL Hendaoui, P. Jarry, "The design of microwave integrated multistage active filters," PIERS Pasadena, California, USA, 12–16 July 1993.

[7] P. Jarry, E. Kerherve, E. EL Hendaoui, H. Tertuliano, "CAD of microwave integrated active filters by the real frequency—Computer program FREELCD," Ann. Telecommun., vol. 51, no. 5–6, pp. 191–205, 1996.

[8] E. Kerherve, P. Jarry, "Efficient numerical method to design of microwave active circuits," IEEE MTT Symposium, Denver, USA, pp. 1611–1614, 8–13 June 1997.

[9] P. Jarry, E. Kerherve, M. Hazouard, J. M. Pham, "CAD of microwave integrated multistage active filters by the real frequency method," IEEE RAWCOM (Radio and Wireless Conference), Colorado Spring, USA, pp. 339–342, August 1998.

[10] E. Kerherve, P. Jarry, "The real frequency technique applied to the synthesis of MMIC active filters with attenuation poles at finite frequencies," Analog Integr. Circ. Sig. Process. Int. J, vol. 36, no. 1 & 2, pp. 145–152, July/August 2003.

8

Passive Microwave Equalizers for Radar Receiver Design

8.1 Introduction

Receiver design for digital radar and optical communication involves optimization and realization of an equalizing filter to provide a suitable signal at the input of the decision circuit. This is a frequently encountered problem in filter design, where one needs to realize a network capable of having an arbitrary response in a given frequency band [1, 2]. In this chapter, our approach is to approximate the target frequency response for a single equalizer using the real frequency technique (RFT) [3, 4]. Since the RFT optimizes equalizer functions that are realizable, it can also directly synthesize the elements of the equalizing filter. The solution uses a Levenberg–Marquardt–More optimization technique that was proven to be successful in amplifier design. The optimization produces realizable functions without an a priori choice of equalizer topology and does not require an analytic form of the system transfer function to be assumed. This allows for an equalizer to be realized using many forms (localized, distributed, etc.).

The procedures described in this chapter were implemented in a program called SRC.

The method is illustrated by the design of a raised cosine correction filter for a radar communication receiver given by the CNES (French National Center of Space Studies). Examples of microwave equalizers for radar applications are given in the 8 GHz range.

Microwave Amplifier and Active Circuit Design Using the Real Frequency Technique, First Edition.
Pierre Jarry and Jacques N. Beneat.

8.2 Equalizer Needs for Radar Application

The RFT has been used successfully for optimizing the transducer power gain, the input and output VSWR, the noise figure, the transimpedance gain and group delay, but in each case the target function for each of these quantities was assumed to be a constant in a band of frequencies.

In the case of the radar application, the objective is to use the equalizer to provide an output with a raised cosine spectrum to minimize intersymbol interference. Compared to the traditional RFT, this means that the target function is no longer a constant but it is more generally a curve provided over a given frequency band.

In our case, the target curve is given for the forward transmission coefficient $|S_{21}(j\omega)|_{\mathrm{dB}}$ of the equalizer and is shown in Figure 8.1. It peaks at two frequencies F_1 and F_2 symmetrical around the center frequency and dips at the center frequency. The equalizer is used to compensate for a function taken as $\omega/\sin\omega\cdot\sqrt{\cos\omega}$ to provide the output spectrum as shown in Figure 8.1.

Besides the need of a target function that is no longer a constant over the band of frequencies, we are no longer interested in an active network. However, rather than creating an entirely new program to design an equalizer for the radar application, the transistor can be virtually eliminated from the current RFT by using appropriate values for its scattering parameters.

Figure 8.2 shows the first-stage configuration used in the classic RFT (see Chapter 2).

The transducer gain of the first stage is given by

$$T_1 = \left(1-|\rho_G|^2\right)\frac{|e_{21,1}|^2|S_{21,1}|^2}{|1-e_{11,1}S_{G,1}|^2|1-\hat{e}_{22,1}S_{11,1}|^2} \tag{8.1}$$

$$\hat{e}_{22,1} = e_{22,1} + \frac{e_{21,1}^2 S_{G,1}}{1-e_{11,1}S_{G,1}} \tag{8.2}$$

$$S_{G,1} = \rho_G \tag{8.3}$$

$$\rho_G = \frac{Z_G - Z_0}{Z_G + Z_0} \tag{8.4}$$

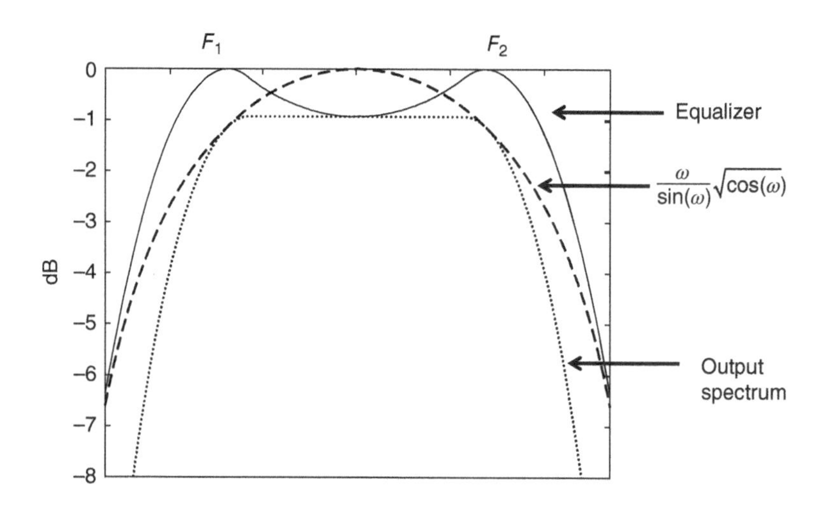

Figure 8.1 Target $|S_{21}(j\omega)|_{\mathrm{dB}}$ response for the equalizer.

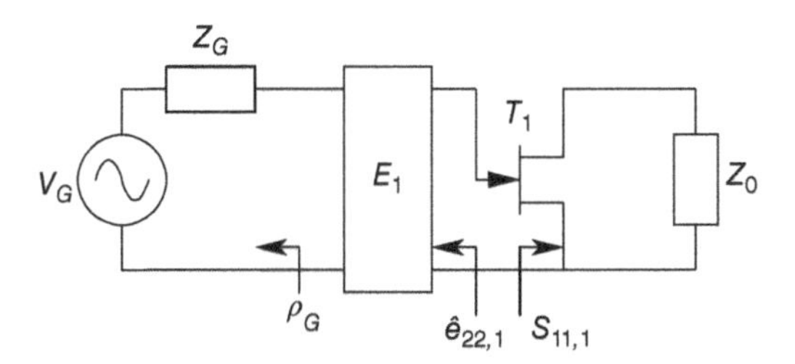

Figure 8.2 Transducer gain for the first stage terminated on Z_0.

If we choose to have a "dummy" transistor device such that [5]

$$S_{21,1}(j\omega) = S_{12,1}(j\omega) = 1 \tag{8.5}$$

$$S_{11,1}(j\omega) = S_{22,1}(j\omega) = 0 \tag{8.6}$$

then

$$T_1 = \left(1 - |\rho_G|^2\right) \frac{|e_{21,1}|^2}{|1 - e_{11,1}\rho_G|^2} \tag{8.7}$$

This is the same as the transducer gain of a single two-port when the output is adapted. In addition, if the input is also perfectly matched, then $Z_G = Z_0$, $\rho_G = 0$, and the transducer gain is directly equal to magnitude squared of the forward transmission coefficient of the equalizer:

$$T_1 = |e_{21,1}|^2 \tag{8.8}$$

Another way at looking at the issue is that the RFT is based on the cascade of scattering matrices. The cascade of scattering matrices does not follow the properties of matrix multiplication, but the cascade of their corresponding transfer matrices does. The transfer matrix of the dummy transistor is given by

$$T = \frac{1}{S_{21}} \begin{pmatrix} S_{12}S_{21} - S_{11}S_{22} & S_{11} \\ -S_{22} & 1 \end{pmatrix} = \begin{pmatrix} 1 & 0 \\ 0 & 1 \end{pmatrix}$$

And we can check that this is the identity matrix. Therefore, the effects of the dummy transistor on the response of the cascaded system are nonexistent.

8.3 Passive Equalizer Representation

The passive equalizer structure is represented as shown in Figure 8.3 [5, 6]. It only includes one equalizer–transistor stage. The transistor is of no consequence since it will be specified with scattering parameters that do not contribute to the response of the overall system. One also assumes the output to be connected to the characteristic impedance used for the scattering parameters, most generally taken as 50 Ω.

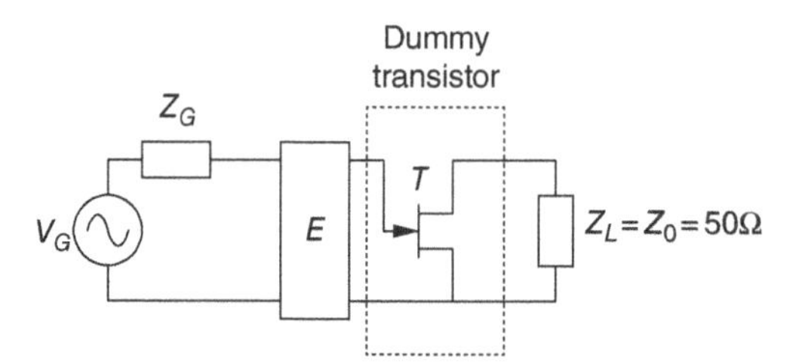

Figure 8.3 Passive equalizer structure.

The equalizer is characterized by the scattering matrix

$$S_E = \begin{pmatrix} e_{11}(s) & e_{12}(s) \\ e_{21}(s) & e_{22}(s) \end{pmatrix} \tag{8.9}$$

The transistor is characterized by the scattering matrix

$$S_T = \begin{pmatrix} 0 & 1 \\ 1 & 0 \end{pmatrix} \quad \text{for all frequencies} \tag{8.10}$$

In this chapter, the equalizer is assumed to be a lossless and reciprocal lumped two-port so that its scattering matrix can be expressed as [7]

$$S_E = \begin{pmatrix} e_{11}(s) = \dfrac{h(s)}{g(s)} & e_{12}(s) = \dfrac{f(s)}{g(s)} \\ e_{21}(s) = \dfrac{f(s)}{g(s)} & e_{22}(s) = \dfrac{-\varepsilon h(-s)}{g(s)} \end{pmatrix} \tag{8.11}$$

where $f(s)$ is an even or odd function. If $f(s)$ is even, then $\varepsilon = 1$; and if $f(s)$ is odd, then $\varepsilon = -1$. In addition, the lossless condition $|e_{11}|^2 + |e_{12}|^2 = 1$ leads to

$$g(s)g(-s) = h(s)h(-s) + f(s)f(-s) \tag{8.12}$$

$h(s)$ and $g(s)$ are polynomials of order n with real-valued coefficients:

$$h(s) = h_0 + h_1 s + \cdots + h_{n-1}s^{n-1} + h_n s^n$$

$$g(s) = g_0 + g_1 s + \cdots + g_{n-1}s^{n-1} + g_n s^n$$

and in this chapter, $f(s)$ is taken as

$$f(s) = \pm s^k \prod_{j=1}^{m} \left(s^2 + \omega_j^2\right) \tag{8.13}$$

$$\text{and} \quad \varepsilon = (-1)^k$$

where k represents the number of transmission zeros at $\omega = 0$ and where there are m number of finite transmission zeros $\omega_j' s$ and with $k + 2m \le n$ leading to

$$g(s)g(-s) = h(s)h(-s) + (-1)^k s^{2k} \prod_{j=1}^{m} \left(s^2 + \omega_j^2\right)^2 \tag{8.14}$$

For a given equalizer, the problem consists of defining the unknown polynomial coefficients h_0, $h_1, h_2, \ldots, h_n$ that optimize the response of the system. Once $h(s)$ is known, $g(s)g(-s)$ can be computed using

$$G\left(s^2\right) = g(s)g(-s) = G_0 + G_1 s^2 + \cdots + G_n s^{2n}$$

For example, when we have three transmission zeros ω_1, ω_2, and ω_3 so that $m = 3$, $G(s^2)$ is given by

$$\begin{aligned}
&G_0 = h_0^2 \\
&G_1 = -h_1^2 + 2h_2 h_0 \\
&\vdots \\
&G_i = (-1)^i h_i^2 + 2h_{2i} h_0 + 2\sum_{j=2}^{i} (-1)^{j-1} h_{j-1} h_{2j-i+1} \\
&\vdots \\
&G_k = G_i\big|_{i=k} + (-1)^k S_3^2 \\
&G_{k+1} = G_i\big|_{i=k+1} + 2(-1)^k S_2 S_3 \\
&G_{k+2} = G_i\big|_{i=k+2} + (-1)^k \left(2S_1 S_3 + S_2^2\right) \\
&G_{k+3} = G_i\big|_{i=k+3} + 2(-1)^k (S_1 S_2 + S_3) \\
&G_{k+4} = G_i\big|_{i=k+4} + (-1)^k \left(S_1^2 + S_2\right) \\
&G_{k+5} = G_i\big|_{i=k+5} + 2(-1)^k S_1 \\
&G_{k+6} = G_i\big|_{i=k+6} + (-1)^k \\
&\vdots \\
&G_n = (-1)^n h_n^2
\end{aligned} \tag{8.15}$$

where

$$\begin{aligned}
&S_1 = \omega_1^2 + \omega_2^2 + \omega_3^2 \\
&S_2 = \omega_1^2 \omega_2^2 + \omega_2^2 \omega_3^3 + \omega_3^2 \omega_1^2 \\
&S_3 = (\omega_1 \omega_2 \omega_3)^2
\end{aligned}$$

The left-half roots of $g(s)g(-s)$ are retained to form a Hurwitz polynomial $g(s)$. At this point, the equalizer scattering parameters $e_{11}(s)$, $e_{12}(s)$, $e_{21}(s)$, and $e_{22}(s)$ can be computed. During optimization, they will be needed for computing the error function to be optimized, and after optimization, for example, $e_{11}(s)$ can be used to realize the equalizer.

8.4 Optimization Process

The optimization algorithm is used to define polynomial coefficients $h_0, h_1, \ldots, h_n$ of $h(t)$ for the equalizer that minimize a least mean square error function that includes the desired target functions with different weight factors.

For the passive equalizer of this chapter, the error function consists of the transducer gain, the VSWR, and the group delay:

$$E=\sum_{i=1}^{m}\left[W_1\left(\frac{T_1(\omega_i,\boldsymbol{H})}{T_{1,0}(\omega_i)}-1\right)^2+W_2\left(\frac{\text{VSWR}(\omega_i,\boldsymbol{H})}{\text{VSWR}_0}-1\right)^2+W_3\left(\frac{\tau(\omega_i,\boldsymbol{H})}{\tau_0}-1\right)^2\right] \tag{8.16}$$

where

ω_i are the frequencies used for optimization.

$T_1(\omega_i, \boldsymbol{H})$ is the transducer power gain of the equalizer such that $T_1 = |e_{21}|^2$. The target transducer power gain $T_{1,0}(\omega_i)$ is assumed to be available as a data file (see Fig. 8.4).

$\text{VSWR}_{IN}(\omega_i, \boldsymbol{H})$ is the input VSWR. The target VSWR_{IN0} is a constant over the frequency band being optimized. The value is often between 1 and 2.

$\text{VSWR}_{OUT}(\omega_i, \boldsymbol{H})$ is the output VSWR. The target VSWR_{OUT0} is a constant over the frequency band being optimized. The value is often between 1 and 2.

$\tau(\omega_i, \boldsymbol{H})$ is the group delay of the system and was defined in Section 7.5. The target group delay τ_0 is taken as a constant in the frequency band being optimized.

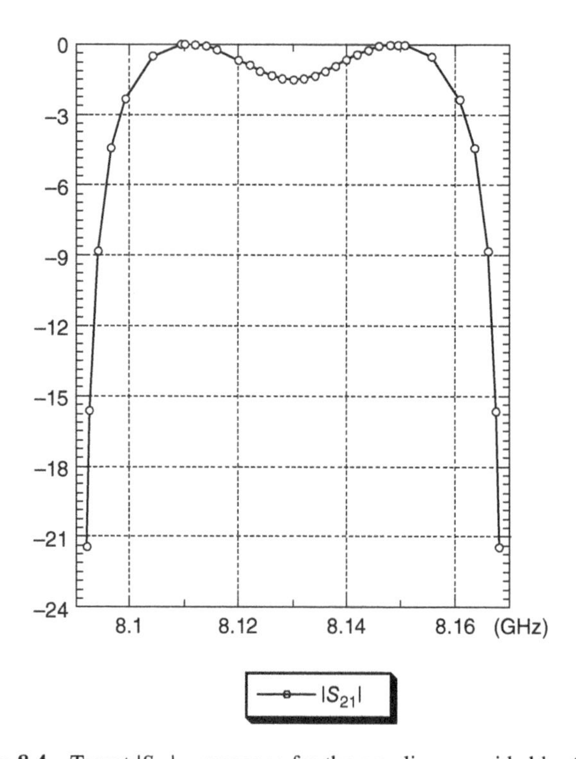

Figure 8.4 Target $|S_{21}|_{\text{dB}}$ response for the equalizer provided by CNES.

The vector $\boldsymbol{H}$ includes the polynomial coefficients $h_0, h_1, \ldots, h_n$ and is used as the $\boldsymbol{X}$ vector in the Levenberg–More optimization algorithms of Chapter 2 and Appendix B.

The main computation in the optimization routine is to define the step $\Delta\boldsymbol{H}$ to add to the elements of vector $\boldsymbol{H}$ at the next iteration given by

$$\Delta\boldsymbol{H} = -\left[\boldsymbol{J}^T\boldsymbol{J} + \lambda\boldsymbol{D}^T\boldsymbol{D}\right]^{-1}\boldsymbol{J}^T e(\boldsymbol{H}) \tag{8.17}$$

where

$e(\boldsymbol{H})$ is the error vector
J is the Jacobian matrix of the error with elements $\partial e_j/\partial h_i$
D is a diagonal matrix
λ is the Levenberg–Marquardt–More parameter described in Appendix B

An important element is the Jacobian matrix. It is based on derivatives of the error functions. For the transducer power gain due to the numerical nature of the target gain curve, it is difficult to define simple analytical expressions and one needs to use numerical derivatives. For example,

$$\frac{T(\omega_j,\boldsymbol{H})}{\partial h_i} = \frac{T(\omega_j,\boldsymbol{H}+\Delta h_i) - T(\omega_j,\boldsymbol{H})}{\Delta h_i} \tag{8.18}$$

with

$$T(\omega_i,\boldsymbol{H}) = \frac{T_1(\omega_i,\boldsymbol{H})}{T_{1,0}(\omega_i)} - 1 \tag{8.19}$$

where a step of $\Delta h_i = 10^{-5}$ was found satisfactory when starting with $h(t)$ polynomial coefficients initialized at $h_i = \pm 1$. The same approach can be used for the VSWR and the group delay.

8.5 Examples of Microwave Equalizers for Radar Receivers

The procedures described in this chapter were implemented in a program called SRC [6, 7]. The goal is to approximate the target $|S_{21}(j\omega)|_{\text{dB}}$ response provided by the CNES and is shown in Figure 8.4. The pulse at the equalizer filter output is required to have a raised cosine spectrum to minimize intersymbol interference. The passband extends from 8.096 to 8.163 GHz with a center frequency of 8.133 GHz. The attenuation at the center frequency should be 1.45 dB. The equalizer is also required to ensure linear phase. However, due to the smoothness of the amplitude phase, distortion of the equalizer was tolerable.

8.5.1 Sixth-Order Equalizer with No Transmission Zeros

In the first example, the target response of Figure 8.4 was approximated using a sixth-order equalizer with no transmission zeros ($f(s) = 1$). The program used 50 points equally distributed in the band 8.0966–8.1634 GHz. The attenuation of the target response was about −4.45 dB at the limits of the frequency interval.

Table 8.1 Polynomials coefficients for the sixth order, no transmission zeros equalizer

$f(s) = (f_0, f_1, f_2, \ldots, f_n)$
$f[0] = 1.0000$
$f[1,2,3,4,5,6] = 0.0000$
$g(s) = (g_0, g_1, g_2, \ldots, g_n)$
$g[0] = 1.0000$
$g[1] = 34442.0309$
$g[2] = 482.1493$
$g[3] = 69499.6502$
$g[4] = 779.2629$
$g[5] = 35059.1080$
$g[6] = 296.4540$
$h(s) = (h_0, h_1, h_2, \ldots, h_n)$
$h[0] = 0.0000$
$h[1] = 34442.0169$
$h[2] = 287.3138$
$h[3] = 69497.4794$
$h[4] = 583.7161$
$h[5] = 35057.4544$
$h[6] = 296.4540$

The results are given for normalized frequencies, and the initial data required for the program include band edges, number of elements, and response of the radar. After optimization, the SRC program can also provide the amplitude, VSWR, and group delay responses beyond the optimization points.

After optimization placing most of the weight on the gain, the optimized equalizer was such that

$$S_E = \begin{pmatrix} e_{11}(s) = \dfrac{h(s)}{g(s)} & e_{12}(s) = \dfrac{1}{g(s)} \\ e_{21}(s) = \dfrac{1}{g(s)} & e_{22}(s) = \dfrac{h(-s)}{g(s)} \end{pmatrix}$$

where the coefficient values for polynomials $f(s)$, $g(s)$, and $h(s)$ are provided in Table 8.1.

Figure 8.5 shows the equalizer and target responses. The equalizer is a close fit to the target in the passband and has 1.5 dB attenuation at the center frequency. The difference error in the optimization band 8.0966–8.1634 GHz is provided in Figure 8.6. It is about 0.079 dB in the passband. The main issue is the very steep slope of the target response in the stopband. Next, transmission zeros are added to the equalizer to improve the attenuation beyond the passband.

8.5.2 *Sixth-Order Equalizer with Two Transmission Zeros*

In the second example, the target response of Figure 8.4 was approximated using a sixth-order equalizer with two transmission zeros, one on the left of the passband at 7.998 GHz and the other on the right of the passband at 8.269 GHz to improve the attenuation in the stopband. The transmission zeros are symmetrical about the center frequency. The program used 50 points equally

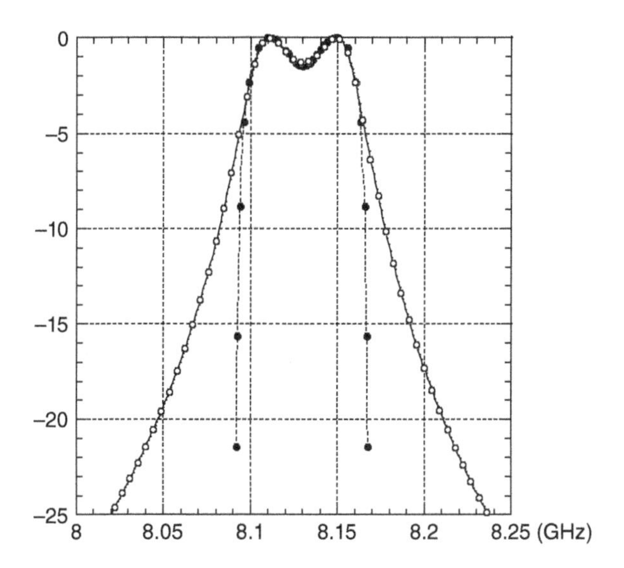

Figure 8.5 $|S_{21}|_{dB}$ responses: target (–∘–) and equalizer (–•–).

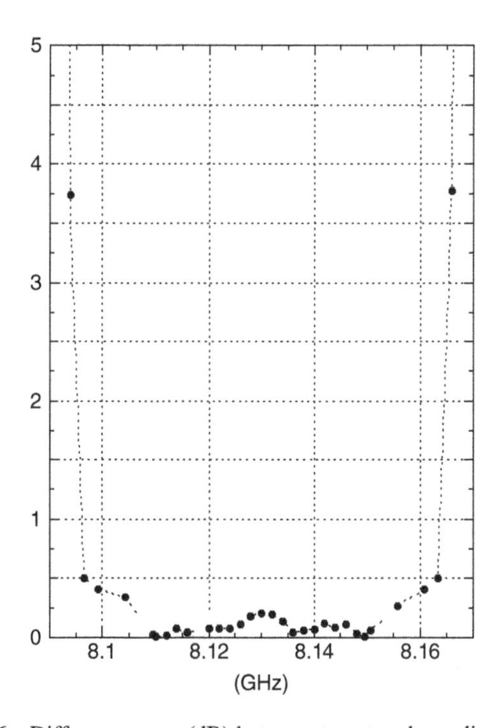

Figure 8.6 Difference error (dB) between target and equalizer response.

Table 8.2 Polynomials coefficients for the sixth order, one symmetrical transmission zero equalizer

$f(s) = (f_0, f_1, f_2, \ldots, f_n)$
$f[0] = 1.0000$
$f[1] = 162.8130$
$f[2] = 6627.0221$
$f[3,4,5,6] = 0.0000$
$g(s) = (g_0, g_1, g_2, \ldots, g_n)$
$g[0] = 1.0000$
$g[1] = 17360.5087$
$g[2] = 14053.7656$
$g[3] = 35116.7490$
$g[4] = 28254.1216$
$g[5] = 17757.6962$
$g[6] = 14201.0119$
$h(s) = (h_0, h_1, h_2, \ldots, h_n)$
$h[0] = 0.0000$
$h[1] = 24155.4124$
$h[2] = -19393.1069$
$h[3] = 48732.9596$
$h[4] = -39152.0371$
$h[5] = 24578.8490$
$h[6] = -19760.2099$

distributed in the band 8.0925–8.1675 GHz. The attenuation of the target response is about −14.5 dB at the limits of the frequency interval.

After optimization placing most of the weight on the gain, the optimized equalizer is such that

$$S_E = \begin{pmatrix} e_{11}(s) = \dfrac{h(s)}{g(s)} & e_{12}(s) = \dfrac{f(s)}{g(s)} \\ e_{21}(s) = \dfrac{f(s)}{g(s)} & e_{22}(s) = \dfrac{h(-s)}{g(s)} \end{pmatrix}$$

The coefficient values for polynomials $f(s)$, $g(s)$, and $h(s)$ are provided in Table 8.2.

Figure 8.7 shows the response with the two transmission zeros, the response from the first example without transmission zeros and the target response. It is seen that the transmission zeros bring an improvement to the stop band without deteriorating the response in the passband.

Figure 8.8 shows that placing the transmission zeros closer to the center frequency does improve the attenuation in stopband and a better match to the target response.

In conclusion, the RFT based on the Levenberg–Marquardt–More optimization algorithm is extremely useful and practical to design equalizers in the case of efficient telecommunications receivers (RADAR for microwaves and also LIDAR for optical waves).

The SRC program is a practical tool that is easy to use by mainly supplying the scattering parameters of the target response. The formalism is quite simple and gives simultaneous optimization on several useful parameters (gain, VSWR, group delay). This work was done with CNES (French National Center of Space Studies).

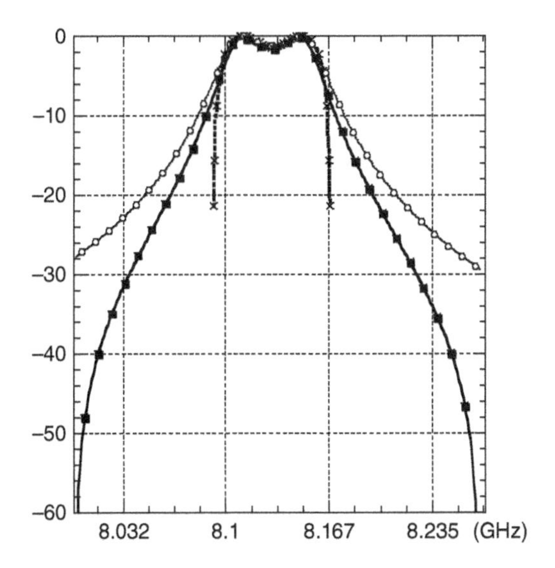

Figure 8.7 $|S_{21}|_{dB}$ responses: target (– × –), equalizer with no transmission zeros (–∘–), equalizer with two transmission zeros (–■–).

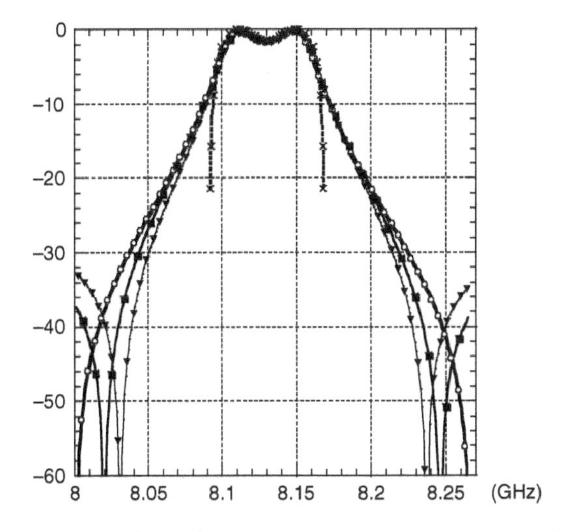

Figure 8.8 $|S_{21}|_{dB}$ responses: target (– × –), equalizer with transmission zero at 8.0 GHz (–∘–), equalizer with transmission zero at 8.02 GHz (–■–), equalizer with transmission zeros at 8.03 GHz (–▲–).

References

[1] H. J. Carlin, J. J. Komiak, "A new method of broadband equalization applied to microwave amplifiers," IEEE Trans. Microw. Theory Tech., Vol. MTT 27, pp.93–99, February 1979.

[2] A. Perennec, R. Soares, P. Jarry, P. Legaud, M. Goloubkoff, "Computer aided design of hybrid and monolithic broadband amplifiers for optoelectronic receivers," IEEE Trans. Microw. Theory Tech., Vol. MTT 37, pp. 1475–1478, September 1989.

[3] E. Kerherve, P. Jarry, P. M. Martin, "Design of broad-band matching network with lossy junctions using the real frequency technique," IEEE Trans. Microw. Theory Tech., Vol. MTT 46, pp. 242–249, March 1998.
[4] P. Jarry, E. Kerherve, H. EL Hendaoui, H. Tertuliano, "CAD of microwave integrated active filters by the real frequency-computer program FREELCD," Ann. Telecommun., Vol. 51, no. 5–6, pp. 191–205, 1996.
[5] P. Jarry, E. Kerherve, M. Lecouve, J. M. Pham, Ch. Zanchi, L. Lapierre, J. Sombrin, "Computer aided design of microwave equalizers using the real frequency method. Application to the design of high flow radar communication receivers," IEEE IMOC, Iguaçu Brazil, pp. 957–960, September 19–23, 2003.
[6] M. Lecouve, Amplitude Equalization using the Levenberg-Marquardt's Algorithm, Ph.D. dissertation, University Bordeaux I, October 18, 2000.
[7] V. Belevitch, "Elementary application of the scattering formalism to network design," IRE Trans. Circuit Theory, vol. CT 3, pp. 97–104, June 1956.

9

Synthesis of Microwave Antennas

9.1 Introduction

During radar cross-sectional measurements, when using a single antenna for transmission and reception of the radar pulse there are two undesired energy returns. The first comes from the reflection on the circulator, and the second comes from the reflection on the antenna. While the circulator reflection might be acceptable, it is not generally the case for the antenna. Therefore, there is a need for designing an equalizer that solves this antenna-matching problem.

The real frequency technique (RFT) was adapted to optimize an antenna return loss $S_{11}(j\omega)$ over a specified frequency band using measured characteristics of the antenna. This modification was implemented in a program called Antenna Computer-Aided Design (ACAD).

The technique was applied to the case of a 250–350 MHz star antenna band, and in the case of a 100–1200 MHz wideband horn antenna. In this second example, a multiplexing technique is proposed in addition to using the RFT to improve the return losses of wideband antennas.

9.2 Antenna Needs

The electromagnetic energy received by radar depends on radar–target distance, gain of the antenna, polarization of the incident wave, and the value of the radar cross section (RCS). The latter parameter is dependent of the target object (area, shape, orientation, directional reflectivity).

Microwave Amplifier and Active Circuit Design Using the Real Frequency Technique, First Edition.
Pierre Jarry and Jacques N. Beneat.

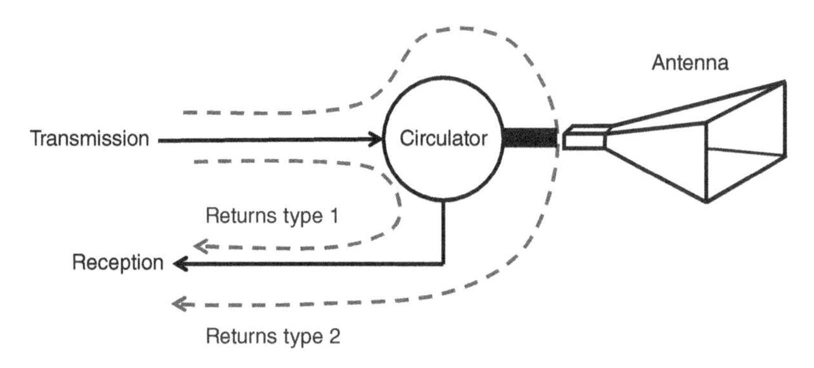

Figure 9.1 Undesirable returns during transmission.

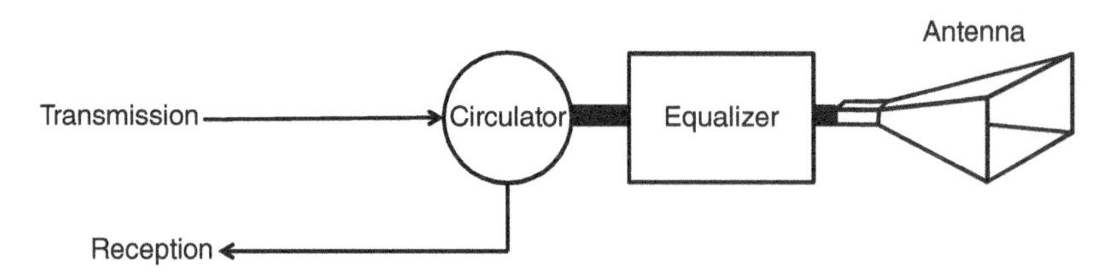

Figure 9.2 Optimizing an equalizer between circulator and antenna.

Therefore, with everything else equal, it is more difficult to detect a return from an object with a small RCS value [1]. For these small returns, it is important to minimize the effects of transmission so that do not produce false alarms. When using the same antenna for transmission and reception, there are two main sources of undesirable returns during transmission. These are shown in Figure 9.1.

The first source is due to direct reflection off the circulator, and the second source is due to reflection off the antenna.

In the VHF-UHF frequency ranges, a high-performance circulator gives in general a return loss of 40 dB, whereas antennas can be much worse. Figure 9.6 shows the example of antenna return loss for a broadband horn antenna. The return loss is not constant and is about 10 dB in most of the frequency bands.

In this chapter, we propose to add an equalizer between antenna and circulator as shown in Figure 9.2 so that the RFT could be used. We assume that the circulator provides a perfectly adapted input to the equalizer while the antenna forms an arbitrary impedance load.

An RFT approach for solving the antenna-matching problem was first introduced by Ramahi and Mittra [2] to overcome the limitations of analytical methods. Unlike analytical and conventional CAD optimization techniques, the RFT uses measured data from the physical devices to optimize the overall system [3]. Building from the optimization of passive equalizers of Chapter 8, the modifications to the RFT program are rather modest. The main objective is to improve the $S_{11}(j\omega)$ parameter of the equalizer for a load impedance representing the measured characteristics of the antenna over a given frequency band.

9.3 Antenna Equalizer Representation

The antenna-matching equalizer structure is represented in Figure 9.3. It only includes one equalizer that is connected to source characteristic impedance Z_0 generally taken as 50 Ω and the arbitrary load Z_L defined by the measured characteristics of the antenna.

The equalizer is characterized by the scattering matrix

$$S_E = \begin{pmatrix} e_{11}(s) & e_{12}(s) \\ e_{21}(s) & e_{22}(s) \end{pmatrix} \tag{9.1}$$

In this chapter, the equalizer is assumed to be a lossless and reciprocal lumped two-port so that its scattering matrix can be expressed as [4]

$$S_E = \begin{pmatrix} e_{11}(s) = \dfrac{h(s)}{g(s)} & e_{12}(s) = \dfrac{f(s)}{g(s)} \\ e_{21}(s) = \dfrac{f(s)}{g(s)} & e_{22}(s) = \dfrac{-\varepsilon h(-s)}{g(s)} \end{pmatrix} \tag{9.2}$$

where $f(s)$ is an even or odd function. If $f(s)$ is even, then $\varepsilon = 1$; and if $f(s)$ is odd, then $\varepsilon = -1$. In addition, the lossless condition $|e_{11}|^2 + |e_{12}|^2 = 1$ leads to

$$g(s)g(-s) = h(s)h(-s) + f(s)f(-s) \tag{9.3}$$

$h(s)$ and $g(s)$ are polynomials of order n with real-valued coefficients:

$$h(s) = h_0 + h_1 s + \cdots + h_{n-1}s^{n-1} + h_n s^n$$
$$g(s) = g_0 + g_1 s + \cdots + g_{n-1}s^{n-1} + g_n s^n$$

and in this chapter, $f(s)$ is taken as

$$f(s) = \pm s^k \quad \text{and} \quad \varepsilon = (-1)^k \tag{9.4}$$

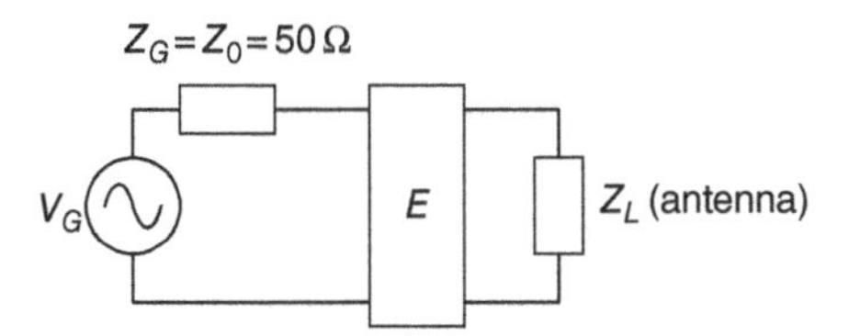

Figure 9.3 Antenna-matching equalizer.

where k represents the number of transmission zeros at $\omega = 0$ and with $k \le n$ leading to

$$g(s)g(-s) = h(s)h(-s) + (-1)^k s^{2k}$$

For a given equalizer, the problem consists of defining the unknown polynomial coefficients h_0, $h_1, h_2, \ldots, h_n$ that optimize the response of the system. Once $h(s)$ is known, $g(s)g(-s)$ can be computed using

$$G(s^2) = g(s)g(-s) = G_0 + G_1 s^2 + \cdots + G_n s^{2n}$$

where the polynomial coefficients of $G(s^2)$ are given by

$$\begin{aligned} G_0 &= h_0^2 \\ G_1 &= -h_1^2 + 2h_2 h_0 \\ &\vdots \\ G_i &= (-1)^i h_i^2 + 2h_{2i} h_0 + 2\sum_{j=2}^{i} (-1)^{j-1} h_{j-1} h_{2j-i+1} \\ &\vdots \\ G_k &= G_i\big|_{i=k} + (-1)^k \\ &\vdots \\ G_n &= (-1)^n h_n^2 \end{aligned} \tag{9.5}$$

The left-half roots of $g(s)g(-s)$ are retained to form a Hurwitz polynomial $g(s)$. At this point, the equalizer scattering parameters $e_{11}(s)$, $e_{12}(s)$, $e_{21}(s)$, and $e_{22}(s)$ can be computed. During optimization, they will be needed for computing the error function to be optimized, and after optimization, for example, $e_{11}(s)$ can be used to realize the equalizer.

9.4 Optimization Process

The optimization algorithm is used to define polynomial coefficients $h_0, h_1, \ldots, h_n$ of $h(s)$ for the equalizer that minimize a least mean square error function that includes desired target functions with different weight factors.

For the passive equalizer of this chapter, the error function consists of the forward transmission coefficient and the input refection coefficient:

$$E = \sum_{i=1}^{m} \left[W_1 \left(\frac{S_{21}(\omega_i, \boldsymbol{H})}{S_{21,0}(\omega_i, \boldsymbol{H})} - 1 \right)^2 + W_2 \left(\frac{S_{11}(\omega_i, \boldsymbol{H})}{S_{11,0}(\omega_i, \boldsymbol{H})} - 1 \right)^2 \right] \tag{9.6}$$

where

ω_i is the frequencies used for optimization.

$S_{21}(\omega_i, \boldsymbol{H})$ is the forward transmission coefficient of the equalizer when loaded on the antenna impedance. Note that the antenna impedance is specified as a data file. The target forward transmission coefficient $S_{21,0}(\omega_i, \boldsymbol{H})$ is assumed to be available as a data file.

$S_{11}(\omega_i, \boldsymbol{H})$ is the input reflection coefficient of the equalizer when loaded on the antenna impedance. Note that the antenna impedance is specified as a data file. The target input refection coefficient $S_{11,0}(\omega_i, \boldsymbol{H})$ is assumed to be available as a data file.

The vector $\boldsymbol{H}$ includes the polynomial coefficients $h_0, h_1, \ldots, h_n$ and is used as the $\boldsymbol{X}$ vector in the Levenberg–More optimization algorithms of Chapter 2 and Appendix B.

The main computation in the optimization routine is to define the step $\Delta\boldsymbol{H}$ to add to the elements of vector $\boldsymbol{H}$ at the next iteration given by

$$\Delta\boldsymbol{H} = -\left[\boldsymbol{J}^T\boldsymbol{J} + \lambda\boldsymbol{D}^T\boldsymbol{D}\right]^{-1}\boldsymbol{J}^T e(\boldsymbol{H}) \tag{9.7}$$

where

$e(\boldsymbol{H})$ is the error vector
J is the Jacobian matrix of the error with elements $\partial e_j/\partial h_i$
D is a diagonal matrix
λ is the Levenberg–Marquardt–More parameter described in Appendix B

An important element is the Jacobian matrix. It is based on derivatives of the error functions. For the input reflection coefficient when the equalizer is loaded by the load impedance defined by the measured values of the antenna, analytical expressions would be too lengthy to define and one must use numerical derivatives. For example,

$$\frac{T(\omega_j,\boldsymbol{H})}{\partial h_i} = \frac{T(\omega_j,\boldsymbol{H}+\Delta h_i) - T(\omega_j,\boldsymbol{H})}{\Delta h_i} \tag{9.8}$$

with

$$T(\omega_i,\boldsymbol{H}) = \frac{S_{11}(\omega_i,\boldsymbol{H})}{S_{11,0}(\omega_i,\boldsymbol{H})} - 1 \tag{9.9}$$

where a step of $\Delta h_i = 10^{-5}$ was found satisfactory when starting with $h(s)$ polynomial coefficients initialized at $h_i = \pm 1$.

9.5 Examples of Antenna-Matching Network Designs

9.5.1 Mid-Band Star Antenna

The procedures in this chapter were implemented in a program called ACAD. In this first example, the goal was to improve the return loss of a star antenna using an equalizer and the RFT [3]. The dashed line in Figure 9.4 shows the return loss of the star antenna without the equalizer. The equalizer is taken as an order 7, and the load impedance in the RFT used the measured characteristics of the star antenna. The optimization was done in the 250–350 MHz frequency band using a constant $|S_{11}(j\omega)|$ as target. The solid line in Figure 9.4 shows the simulated return loss at the input of the equalizer when loaded by the measured star antenna. The return loss has been significantly improved with an −17 dB equiripple behavior in the 250–350 MHz band.

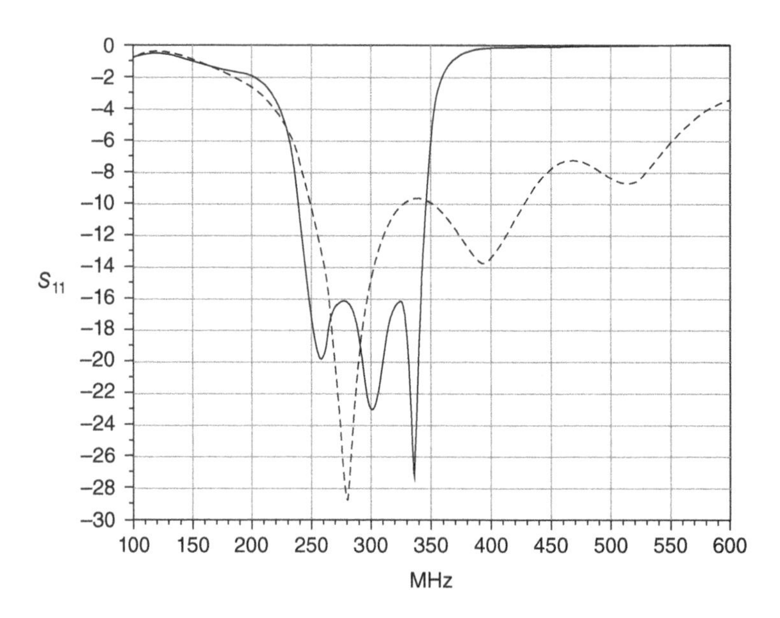

Figure 9.4 Return loss without equalizer (dashed line) and with equalizer (solid line).

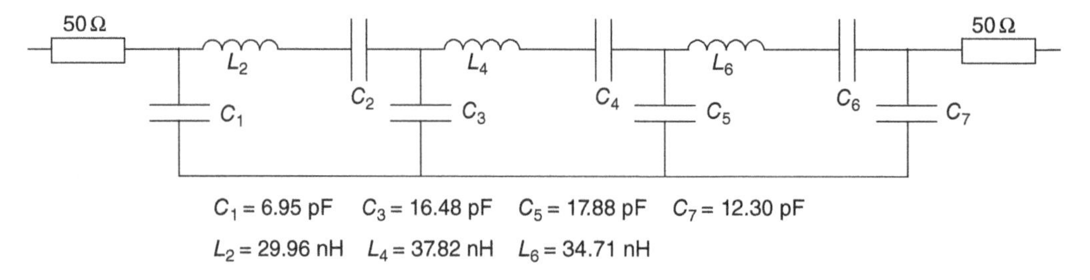

Figure 9.5 Star antenna equalizer circuit for the 250–350 MHz band.

The equalizer circuit was synthesized using the technique in Chapter 2 and is shown in Figure 9.5. The equalizer was then realized on a glass-teflon substrate ($\varepsilon_r = 2.53$), and the measured return loss when loaded with the actual star antenna is given in Figure 9.6.

The measured return loss was nearly equiripple at −13.5 dB and validates the RFT technique for designing antenna-matching networks.

9.5.2 *Broadband Horn Antenna*

In this second example, the goal was to bring the return loss to 30 dB over a very large frequency band extending from 100 to 900 MHz [3]. The return loss of the horn antenna without the equalizer is shown in Figure 9.7. It is barely better than 10 dB in this frequency range.

It is clear that a single equalizer cannot improve the return loss to 30 dB across the entire 100–900 MHz frequency band. Therefore, the idea is to design a set of equalizers placed in a multiplexing network, each covering its own fraction of the 100–900 MHz frequency band. In the case of the horn antenna example, a set of 21 equalizers was chosen.

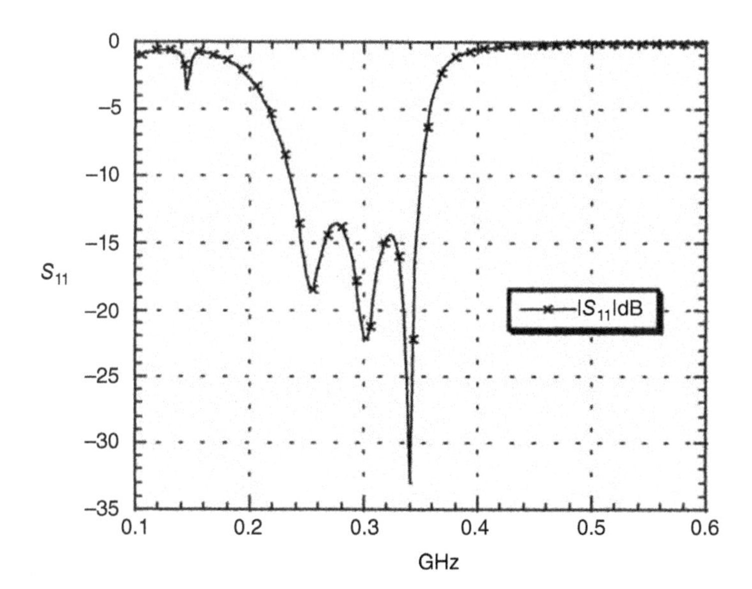

Figure 9.6 Measured return loss of the star antenna with equalizer circuit in the 250–350 MHz band.

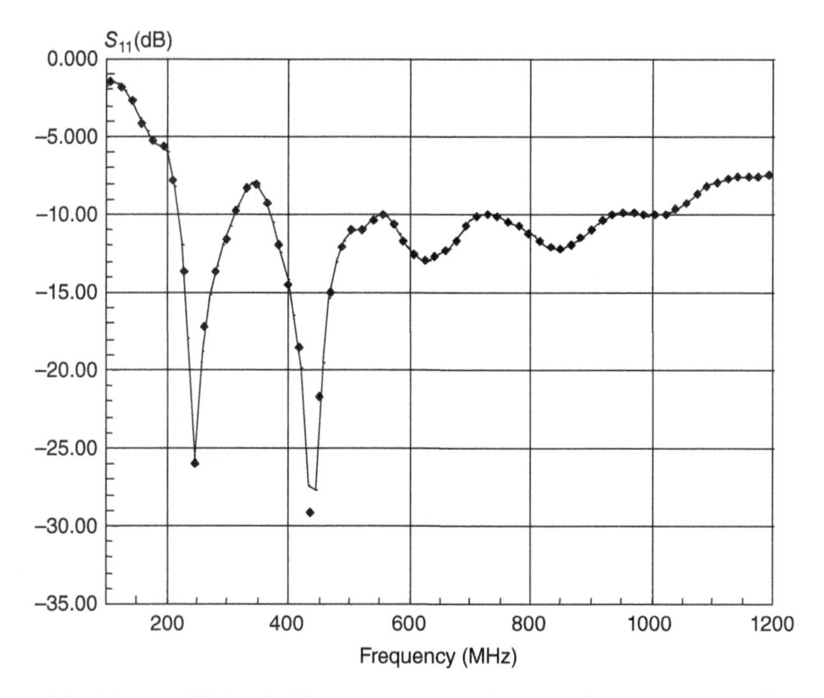

Figure 9.7 Measured $|S_{11}|_{dB}$ for the horn antenna without equalizer from 100 to 1200 MHz.

Each equalizer is optimized in its sub-band according to the RFT technique of this chapter. We give the example of one of these equalizers below. This equalizer was optimized to improve the horn antenna return loss to 30 dB in the sub-band 390–440 MHz. The circuit of this sub-band equalizer is shown in Figure 9.8. It was made of seven elements like in the first example.

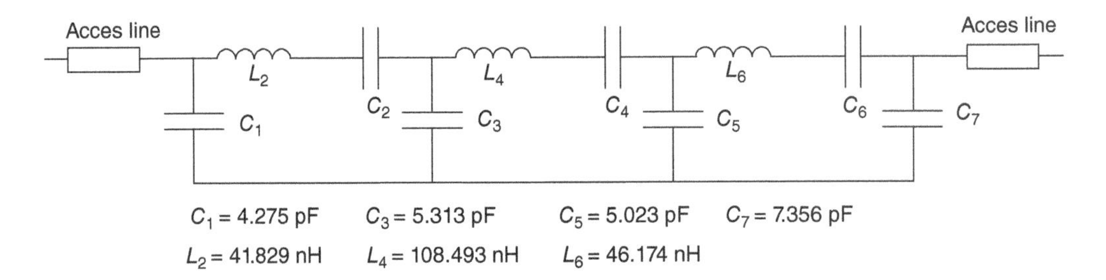

Figure 9.8 Horn antenna equalizer circuit for the 390–440 MHz sub-band.

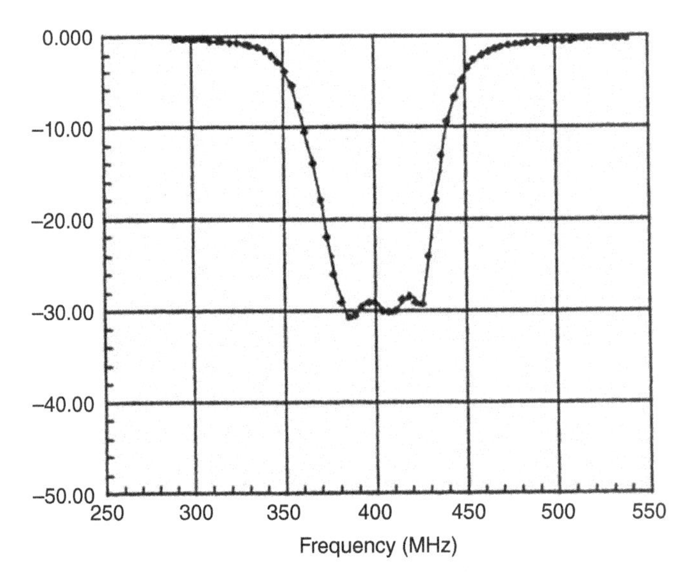

Figure 9.9 Measured return loss of horn antenna with equalizer circuit for 390–440 MHz sub-band.

It was then realized on a glass-teflon substrate ($\varepsilon_r = 2.53$), and the measured return loss when connecting the horn antenna to the actual horn antenna is shown in Figure 9.9. The measured return loss was better than 28.3 dB in the sub-band 390–440 MHz. Therefore, the concept of optimizing separately the equalizers but using the same single actual antenna is valid.

Figure 9.10 shows the simulated return losses of the horn antenna for all 21 equalizer circuits in the 21 sub-bands. Each equalizer was an order 7 and optimized using the same procedures as for the 390–440 MHz sub-band case. Figure 9.10 shows that the return loss is better than 27 dB across the entire 100–900 MHz band showing the feasibility of this solution.

In conclusion, the antenna for RCS measurements needs an improved reflection coefficient. A solution that can solve this antenna-matching problem was presented in this chapter under the form of the ACAD RFT program. It was also shown that a possible solution for improved returned loss over a very large frequency band is the use of a set of equalizer optimized in a given sub-band using the measured performance of the antenna over the entire band. The equalizers are then placed in parallel in a multiplexing system fashion.

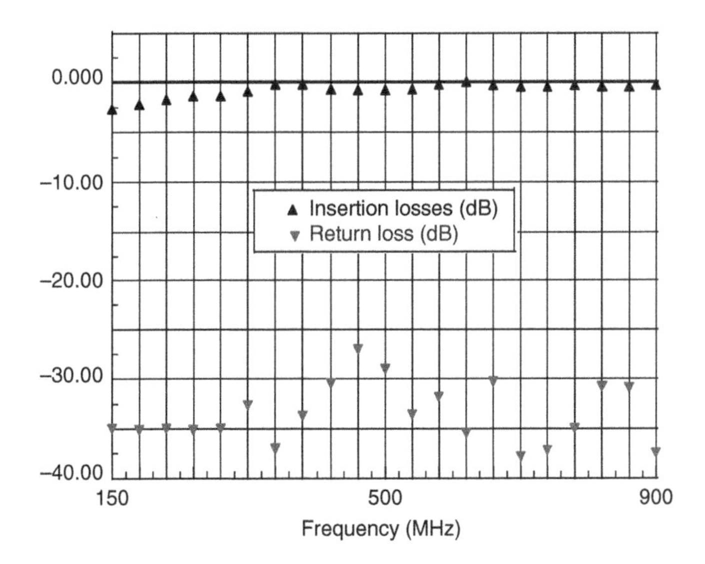

Figure 9.10 Simulated insertion and return losses for each of the 21 sub-bands using seventh-order equalizers.

A limitation of the technique is the large number of equalizers depending on the number of sub-bands needed to cover the entire band. This study was done under contract with CEA (French Office Electronic Atomic).

References

[1] J. Garat, "Microwave techniques for radar cross section measurement, a review," Proceedings of MELECOM'95, Italia, pp. 80–86, May 1996.

[2] O. M. Ramahi, R. Mittra, "Design of a matching network for an HF antenna using the real frequency method," IEEE Trans. Antennas Propag., vol. 37, no. 4, pp. 506–509, April 1989.

[3] E. Kerherve, Ph. Naud, G. Germain, P. Jarry, "RF antenna matching methods for radar cross section measurement," 9th IEEE International Conference on Electronics Circuits & Systems ICECS, Dubrovnik, vol. 1, pp. 85–88, September 15–18, 2002.

[4] V. Belevitch, "Elementary application of the scattering formalism to network design," IRE Trans. Circ. Theory, vol. CT 3, pp. 97–104, June 1956.

Appendix A

Multistage Transducer Gain

A.1 Multistage Transducer Gain Demonstration

To define the expression for the multistage transducer gain, one starts with scattering matrix T_{k-1} as shown in Figure A.1. The scattering matrix T_{k-1} represents the cascade of equalizer–transistor pairs 1 through $k-1$ and assumes that the two-port is doubly terminated by Z_0. The effects of arbitrary source Z_G and load Z_L impedances will be treated later.

One notes that since the two-port is terminated on Z_0 one has $\rho_G = \rho_L = 0$ and the transducer gain of the $k-1$ first stages reduces to

$$G_{T,k-1} = |T_{21,k-1}|^2$$

At this point, the kth equalizer–transistor stage is added as shown in Figure A.2.

Since the system is again terminated by Z_0, the transducer gain of the k first stages reduces to

$$G_{T,k} = |T_{21,k}|^2$$

To simplify the explanation, equations for defining the scattering parameters of the cascade of two scattering matrices are used and are given below. Figure A.3 shows that the cascade of two scattering matrices S_A and S_B can be modeled as a single scattering matrix S_C.

Using the multiplication properties of the transfer matrix and the conversion formulas from scattering matrix to transfer matrix and vice versa (see Appendix D), one can show that

$$S_{C11} = S_{A11} + \frac{S_{A12}S_{A21}S_{B11}}{1 - S_{A22}S_{B11}} \qquad S_{C21} = \frac{S_{A21}S_{B21}}{1 - S_{A22}S_{B11}}$$

$$S_{C12} = \frac{S_{A12}S_{B12}}{1 - S_{A22}S_{B11}} \qquad S_{C22} = S_{B22} + \frac{S_{A22}S_{B12}S_{B21}}{1 - S_{A22}S_{B11}}$$

Microwave Amplifier and Active Circuit Design Using the Real Frequency Technique, First Edition.
Pierre Jarry and Jacques N. Beneat.

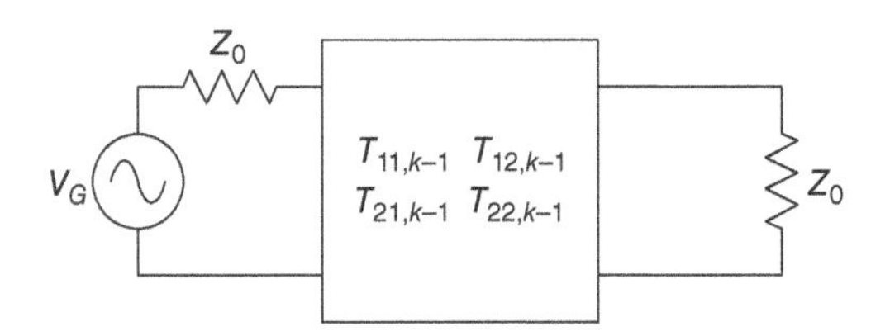

Figure A.1 T_{k-1} represents the first $k-1$ equalizer–transistor stages and Z_G.

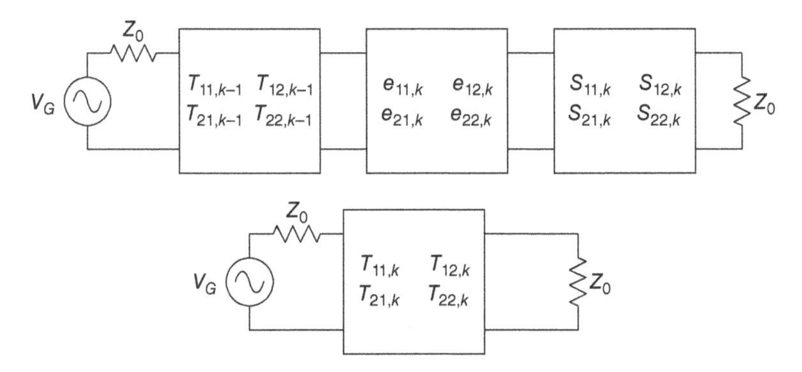

Figure A.2 Adding the kth equalizer–transistor stage to find the gain of the k first stages.

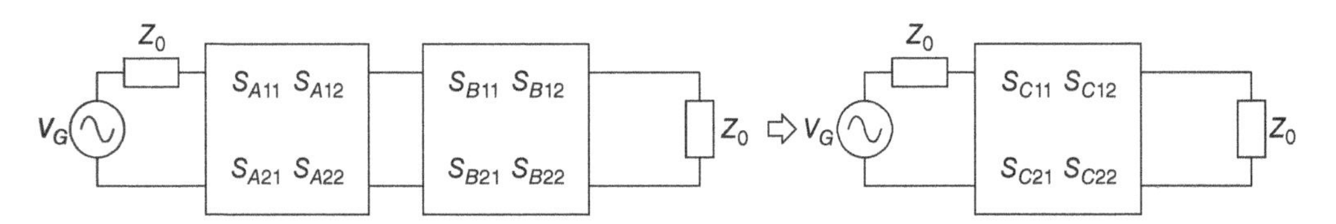

Figure A.3 Scattering parameters for the cascade of two scattering matrices.

Starting from Figure A.1, one adds the kth equalizer as shown in Figure A.4.

Using the formulas above when S_A represents the $k-1$ first equalizer–transistor stages and Z_G, and S_B represents the kth equalizer, the scattering parameters of S_C are such that

$$S_{C21} = \frac{T_{21,k-1}e_{21,k}}{1-T_{22,k-1}e_{11,k}}$$

$$S_{C22} = e_{22,k} + \frac{T_{22,k-1}e_{12,k}e_{21,k}}{1-T_{22,k-1}e_{11,k}}$$

At this point, the kth transistor is added as shown in Figure A.5. In this figure the $k-1$ first stages and the kth equalizer are modeled as scattering matrix S_C.

In this case, the cascade of the two scattering matrices of Figure A.5 represents the scattering matrix T_k of the k first stages as shown in Figure A.2. Using the cascade of scattering matrix formulas above when S_A represents S_C of Figure A.5 and S_B represents the kth transistor, the scattering parameters of T_k are such that

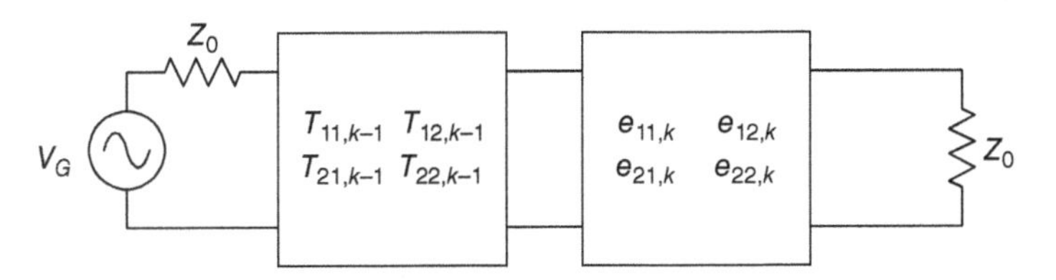

Figure A.4 First adding the kth equalizer.

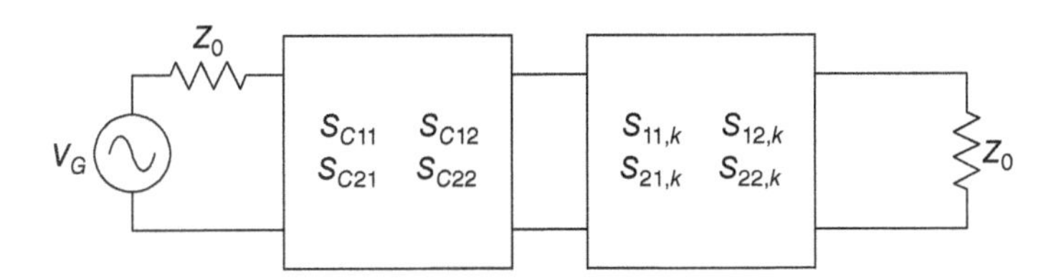

Figure A.5 Adding the kth transistor.

$$T_{21,k} = \frac{S_{C21}S_{21,k}}{1 - S_{C22}S_{11,k}} = \frac{T_{21,k-1}e_{21,k}}{1 - T_{22,k-1}e_{11,k}} \frac{S_{21,k}}{1 - S_{C22}S_{11,k}}$$

and taking the magnitude square

$$|T_{21,k}|^2 = |T_{21,k-1}|^2 \frac{|e_{21,k}|^2}{|1 - T_{22,k-1}e_{11,k}|^2} \frac{|S_{21,k}|^2}{|1 - S_{C22}S_{11,k}|^2}$$

with $S_{C22} = e_{22,k} + \dfrac{T_{22,k-1}e_{12,k}e_{21,k}}{1 - T_{22,k-1}e_{11,k}}$

At this stage, some terms are renamed:

$$T_{22,k-1} \Rightarrow S_{G,k}$$
$$S_{C22} \Rightarrow \hat{e}_{22,k}$$

In the case of a reciprocal equalizer one has $e_{12,k} = e_{21,k}$. In that case, the transducer gain $G_{T,k}$ representing the k first stages can be expressed in terms of the transducer gain $G_{T,k-1}$ representing the $k-1$ first stages by

$$G_{T,k} = G_{T,k-1} \frac{|e_{21,k}|^2}{|1 - S_{G,k}e_{11,k}|^2} \frac{|S_{21,k}|^2}{|1 - \hat{e}_{22,k}S_{11,k}|^2}$$

with

$$\hat{e}_{22,k} = e_{22,k} + \frac{S_{G,k}e_{21,k}^2}{1 - S_{G,k}e_{11,k}}$$

and

$$S_{G,k} = S_{22,k-1} + \frac{\hat{e}_{22,k-1}S_{12,k-1}S_{21,k-1}}{1-\hat{e}_{22,k-1}S_{11,k-1}}$$

Note that $S_{G,k}$ represents the output reflection coefficient of the $k-1$ first stages $(T_{22,k-1})$ when it is loaded on the left. Therefore, the expression of $S_{G,k}$ directly uses the scattering parameters of transistor $k-1$ but uses the output reflection coefficient of the $k-1$ equalizer when loaded on the left noted as $\hat{e}_{22,k-1}$. It is loaded by the $k-2$ first equalizer–transistor stages and the effect of Z_G.

Since the equation for the transducer gain is recursive, it is important to define the starting point. This is the first equalizer–transistor loaded on the arbitrary load Z_G as shown in Figure A.6.

Referring to (1.22), the transducer power gain is given by

$$G_T = |S_{C21}|^2 \frac{\left(1-|\rho_G|^2\right)\left(1-|\rho_L|^2\right)}{|(1-S_{C11}\rho_G)(1-S_{C22}\rho_L)-S_{C12}S_{C21}\rho_G\rho_L|^2}$$

where scattering matrix S_C represents the cascade of the first equalizer and transistor.

As seen in Figure A.6, $Z_L = Z_0$ so that $\rho_L = 0$ and the transducer gain reduces to

$$G_T = \frac{\left(1-|\rho_G|^2\right)}{|1-S_{C11}\rho_G|^2}|S_{C21}|^2$$

where

$$S_{C11} = e_{11,1} + \frac{e_{21,1}^2 S_{11,1}}{1-e_{22,1}S_{11,1}} \quad \text{and} \quad S_{C21} = \frac{e_{21,1}S_{21,1}}{1-e_{22,1}S_{11,1}}$$

so that

$$G_T = \frac{\left(1-|\rho_G|^2\right)}{|1-S_{C11}\rho_G|^2} \frac{|e_{21,1}|^2|S_{21,1}|^2}{|1-e_{22,1}S_{11,1}|^2}$$

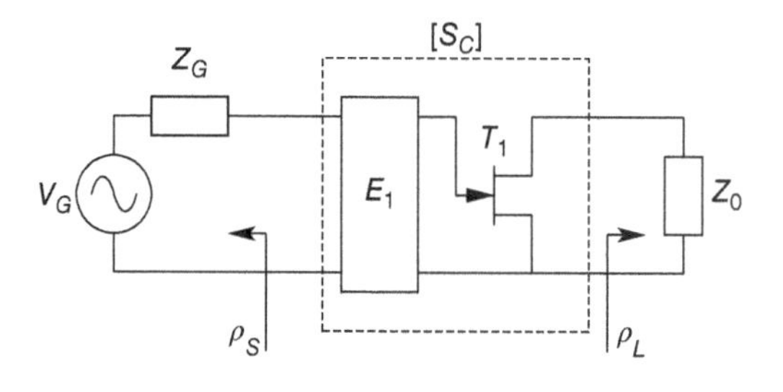

Figure A.6 Starting point for the recursive formula: The first equalizer–transistor pair loaded on Z_G.

Now when using the recursive gain formula for the first stage ($k = 1$) one gets

$$G_{T,1} = G_{T,0} \frac{|e_{21,1}|^2}{|1 - S_{G,1} e_{11,1}|^2} \frac{|S_{21,1}|^2}{|1 - \hat{e}_{22,1} S_{11,1}|^2}$$

with $\hat{e}_{22,1} = e_{22,1} + \dfrac{S_{G,1} e_{21,1}^2}{1 - S_{G,1} e_{11,1}}$

By developing and refactorizing the denominators, it can be shown that two transducer gain expressions are the same when taking

$$S_{G,1} = \rho_G \quad \text{and} \quad G_{T,0} = \left(1 - |\rho_G|^2\right)$$

In the case when the system is terminated on an arbitrary load Z_L, one adds a $(k+1)$th equalizer as shown in Figure A.7.

In this case the scattering matrix S_C represents the cascade of the k first stages and the $(k+1)$th equalizer and has for scattering parameters

$$S_{C21} = \frac{T_{21,k} e_{21,k+1}}{1 - T_{22,k} e_{11,k+1}}$$

$$S_{C22} = e_{22,k+1} + \frac{T_{22,k} e_{12,k+1} e_{21,k+1}}{1 - T_{22,k} e_{11,k+1}}$$

This case is modeled in Figure A.8 and is used to define the transducer power gain.

From (1.22), the transducer power gain is given by

$$G_T = |S_{C21}|^2 \frac{\left(1 - |\rho_G|^2\right)\left(1 - |\rho_L|^2\right)}{|(1 - S_{C11}\rho_G)(1 - S_{C22}\rho_L) - S_{C12} S_{C21} \rho_G \rho_L|^2}$$

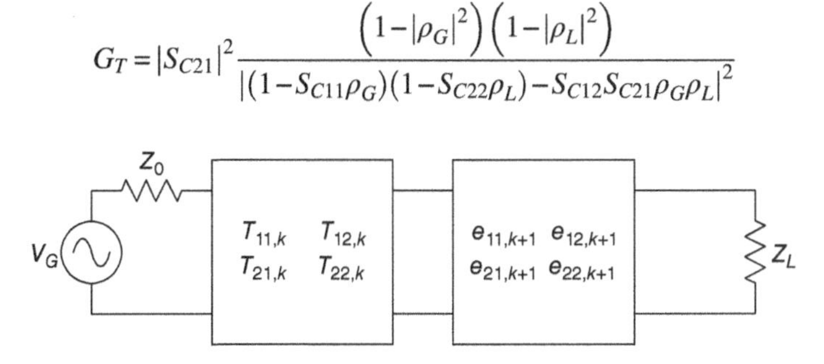

Figure A.7 Adding a last $(k+1)$th equalizer to compensate for an arbitrary load Z_L.

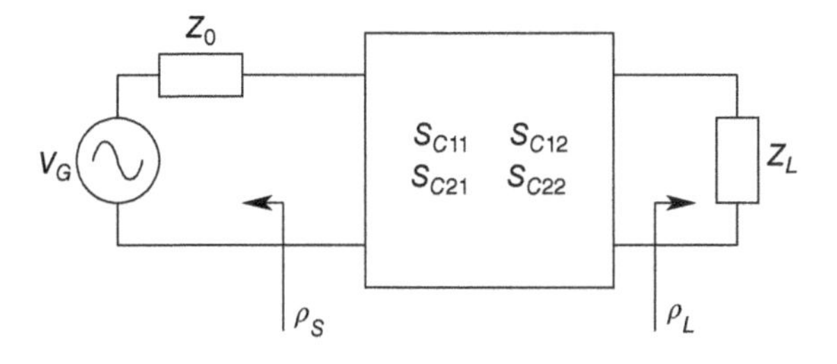

Figure A.8 Reflection coefficients used in the transducer power gain.

and in the case of Figure A.8, $Z_G = Z_0$ so that $\rho_G = 0$ and the transducer gain reduces to

$$G_T = |S_{C21}|^2 \frac{\left(1-|\rho_L|^2\right)}{|1-S_{C22}\rho_L|^2}$$

And replacing the scattering parameters using the expressions from the cascade of T_k and $(k+1)$th equalizer, one gets

$$G_T = \frac{|T_{21,k}|^2 |e_{21,k+1}|^2}{|1-T_{22,k}e_{11,k+1}|^2} \frac{\left(1-|\rho_L|^2\right)}{|1-S_{C22}\rho_L|^2}$$

Using the same renaming as before

$$T_{22,k} \Rightarrow S_{G,k+1}$$

$$S_{C22} \Rightarrow \hat{e}_{22,k+1}$$

and for a reciprocal equalizer $e_{12,k} = e_{21,k}$, the transducer gain can be written as

$$G_{T,k+1} = |G_{T,k}|^2 \frac{|e_{21,k+1}|^2}{|1-S_{G,k+1}e_{11,k+1}|^2} \frac{\left(1-|\rho_L|^2\right)}{|1-\hat{e}_{22,k+1}\rho_L|^2}$$

with

$$\hat{e}_{22,k+1} = e_{22,k+1} + \frac{S_{G,k+1}e_{21,k+1}^2}{1-S_{G,k+1}e_{11,k+1}}$$

and

$$S_{G,k+1} = S_{22,k} + \frac{\hat{e}_{22,k}S_{12,k}S_{21,k}}{1-\hat{e}_{22,k}S_{11,k}}$$

A.2 Target Multistage Transducer Gain

This section defines a suitable target value for the transducer gain. It is based on the maximum available gain possible combined with stability considerations.

One starts with the first equalizer–transistor stage as shown in Figure A.9.

The general transducer gain formula when terminated on complex loads is

$$T = \frac{\left(1-|\rho_1|^2\right)|S_{21}|^2\left(1-|\rho_2|^2\right)}{|(1-\rho_1 S_{11})(1-\rho_2 S_{22}) - S_{12}S_{21}\rho_1\rho_2|^2}$$

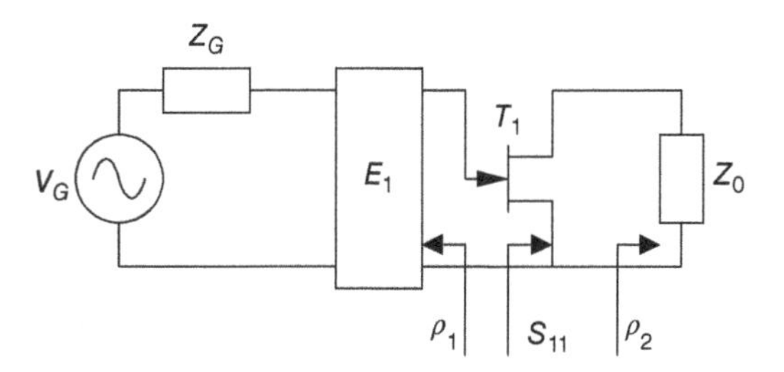

Figure A.9 Defining the maximum gain for the first equalizer–transistor stage.

where in this case

$$\rho_1 = \frac{Z_{\text{left}} - Z_0}{Z_{\text{left}} + Z_0} \text{ and where } Z_{\text{left}} \text{ represents the load made of } Z_G \text{ and } E_1$$

$$\rho_2 = \frac{Z_L - Z_0}{Z_L + Z_0} \text{ but where } Z_L = Z_0 \text{ so that } \rho_2 = 0$$

S_{ij} correspond to the scattering parameters of the transistor, so that

$$T = \frac{\left(1 - |\rho_1|^2\right)|S_{21}|^2}{|(1 - \rho_1 S_{11})|^2}$$

To adapt the input, one takes $\rho_1 = S_{11}^*$ so that the maximum gain is given by

$$T_{MAX} = \frac{\left(1 - |\rho_1|^2\right)|S_{21}|^2}{\left|\left(1 - \rho_1 \rho_1^*\right)\right|^2} = \frac{\left(1 - |\rho_1|^2\right)|S_{21}|^2}{\left|1 - |\rho_1|^2\right|^2} = \frac{|S_{21}|^2}{\left|1 - |\rho_1|^2\right|} \quad \text{or} \quad T_{MAX} = \frac{|S_{21}|^2}{\left|1 - |S_{11}|^2\right|}$$

This would represent a maximum gain when terminated on Z_0. Note that Z_0 might not provide the maximum overall gain (only the input side is truly adapted). Note that T_{MAX} is a function of frequency because of the frequency dependence of the $S'_{ij}s$.

To define a simple constant for the target gain, one can take the minimum value T_{MAX} over the frequency band of interest. The target function $T_{0,1}$ for the transducer gain for the first equalizer is given below

$$T_{0,1} = \min\left\{\frac{|S_{21,1}|^2}{\left|1 - |S_{11,1}|^2\right|}\right\}$$

When the second equalizer–transistor pair is added as shown in Figure A.10, the gain expression becomes more complicated. Only an estimate for the maximum gain based on the unilateral gain ($S_{12} = 0$) is proposed.

The general transducer gain formula when terminated on complex loads is

$$T = \frac{\left(1 - |\rho_1|^2\right)|S_{21}|^2\left(1 - |\rho_2|^2\right)}{|(1 - \rho_1 S_{11})(1 - \rho_2 S_{22}) - S_{12} S_{21} \rho_1 \rho_2|^2}$$

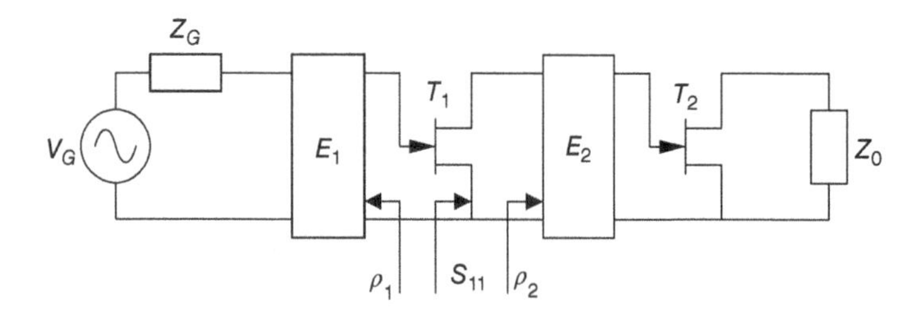

Figure A.10 Defining the maximum gain when adding the second equalizer–transistor pair.

In this case, the maximum gain occurs for $\rho_1 = S_1^*$ and $\rho_2 = S_2^*$ where

$$S_1 = S_{11} + \frac{S_{12}S_{21}\rho_2}{1 - S_{22}\rho_2}$$

$$S_2 = S_{22} + \frac{S_{12}S_{21}\rho_1}{1 - S_{11}\rho_1}$$

This leads to rather difficult equations. An approximate solution would be to take $\rho_1 = S_{11}^*$ and $\rho_2 = S_{22}^*(S_{12} \approx 0)$. In this case, the expression for the maximum gain becomes

$$T_{MAX} = \frac{\left(1-|\rho_1|^2\right)|S_{21}|^2\left(1-|\rho_2|^2\right)}{\left|\left(1-|\rho_1|^2\right)\left(1-|\rho_2|^2\right) - S_{12}S_{21}\rho_1\rho_2\right|^2} = \frac{\left(1-|\rho_1|^2\right)\left(1-|\rho_2|^2\right)|S_{21}|^2}{\left|\left(1-|\rho_1|^2\right)\right|^2\left|\left(1-|\rho_2|^2\right)\right|^2\left|1 - \dfrac{S_{12}S_{21}\rho_1\rho_2}{\left(1-|\rho_1|^2\right)\left(1-|\rho_2|^2\right)}\right|^2}$$

$$T_{MAX} = \frac{|S_{21}|^2}{\left|\left(1-|\rho_1|^2\right)\right|\left|\left(1-|\rho_2|^2\right)\right|}\frac{1}{\left|1 - \dfrac{S_{12}S_{21}\rho_1\rho_2}{\left(1-|\rho_1|^2\right)\left(1-|\rho_2|^2\right)}\right|^2}$$

$$T_{MAX} = \frac{|S_{21}|^2}{\left|1-|S_{11}|^2\right|\left|1-|S_{22}|^2\right|}\frac{1}{\left|1 - \dfrac{S_{12}S_{21}S_{11}^*S_{22}^*}{\left(1-|S_{11}|^2\right)\left(1-|S_{22}|^2\right)}\right|^2}$$

Referring (1.25), the unilateral gain T_U happens for $S_{12} = 0$ and

$$T_U = \frac{\left(1-|\rho_1|^2\right)|S_{21}|^2\left(1-|\rho_2|^2\right)}{|(1-\rho_1 S_{11})(1-\rho_2 S_{22})|^2} = \frac{|S_{21}|^2}{\left|\left(1-|\rho_1|^2\right)\right|\left|\left(1-|\rho_2|^2\right)\right|}$$

and the maximum unilateral gain $T_{U\ MAX}$ happens for $\rho_1 = S_{11}^*$ and $\rho_2 = S_{22}^*$ so that

$$T_{U\ MAX} = \frac{|S_{21}|^2}{\left|1-|S_{11}|^2\right|\left|1-|S_{22}|^2\right|}$$

and

$$T_{MAX} = T_{U\ MAX} \frac{1}{\left|1 - \dfrac{S_{12}S_{21}S_{11}^{*}S_{22}^{*}}{\left(1-|S_{11}|^2\right)\left(1-|S_{22}|^2\right)}\right|^2}$$

The target function for the second stage is chosen to be a constant such that

$$T_{0,2} = \min\left\{T_{MAX,1}\frac{|S_{21,2}|^2}{\left|1-|S_{11,2}|^2\right|}\right\}$$

where

$$T_{MAX,1} = \frac{|S_{21,1}|^2}{\left|1-|S_{11,1}|^2\right|\left|1-|S_{22,1}|^2\right|} \frac{1}{\left|1 - \dfrac{S_{12,1}S_{21,1}S_{11,1}^{*}S_{22,1}^{*}}{\left(1-|S_{11,1}|^2\right)\left(1-|S_{22,1}|^2\right)}\right|^2}$$

Using a similar approach, the target function for the kth stage will be a constant such that

$$T_{0,k} = \min\left\{T_{MAX,1} \times T_{MAX,2} \times \cdots \times T_{MAX,k-1}\frac{|S_{21,k}|^2}{\left|1-|S_{11,k}|^2\right|}\right\}$$

where

$$T_{MAX,i} = \frac{|S_{21,i}|^2}{\left|1-|S_{11,i}|^2\right|\left|1-|S_{22,i}|^2\right|} \frac{1}{\left|1 - \dfrac{S_{12,i}S_{21,i}S_{11,i}^{*}S_{22,i}^{*}}{\left(1-|S_{11,i}|^2\right)\left(1-|S_{22,i}|^2\right)}\right|^2}$$

and finally since there is no transistor added for the $(k+1)$th equalizer, the target function for the gain will be a constant given by

$$T_{0,k+1} = \min\{T_{MAX,1} \times T_{MAX,2} \times \cdots \times T_{MAX,k-1} \times T_{MAX,k}\}$$

Appendix B

Levenberg–Marquardt–More Optimization Algorithm

B.1 Levenberg–Marquardt–More Optimization Algorithm

This algorithm assumes the optimization of a single-valued function $E(X)$ made of a sum of squares:

$$E(X)=\sum_{i=1}^{m}[e_i(X)]^2 \quad e_i(X)=\frac{T(\omega_i,X)}{T_0}-1 \quad X=\begin{bmatrix}x_1\\x_2\\\vdots\\x_n\end{bmatrix}$$

where

X is the parameter vector made of the n parameters to optimize
$e(X)$ is the error vector made of m single errors between target and actual function at m frequencies:

$$e^T(X)=[e_1(\omega_1,X),e_2(\omega_2,X),e_3(\omega_3,X),\ldots,e_m(\omega_m,X)]$$

The single-valued function $E(X)$ can be written as the norm squared of vector $e(X)$:

$$E(X)=e^T(X)e(X)=\|e(X)\|^2$$

and the gradient of $E(X)$ is given by

$$\nabla E(X)=\nabla(E(X))=\nabla\left(\|e(X)\|^2\right)=2J^Te(X)$$

Microwave Amplifier and Active Circuit Design Using the Real Frequency Technique, First Edition.
Pierre Jarry and Jacques N. Beneat.

$$\nabla E(\boldsymbol{X}) = \begin{bmatrix} \frac{\partial E}{\partial x_1} \\ \frac{\partial E}{\partial x_2} \\ \vdots \\ \frac{\partial E}{\partial x_n} \end{bmatrix}$$

where $\boldsymbol{J}$ is the Jacobian matrix defined by

$$\boldsymbol{J} = \begin{bmatrix} \frac{\partial e_1(\boldsymbol{X})}{\partial x_1} & \frac{\partial e_1(\boldsymbol{X})}{\partial x_2} & \cdots & \frac{\partial e_1(\boldsymbol{X})}{\partial x_n} \\ \frac{\partial e_2(\boldsymbol{X})}{\partial x_1} & \frac{\partial e_2(\boldsymbol{X})}{\partial x_2} & \cdots & \frac{\partial e_2(\boldsymbol{X})}{\partial x_n} \\ \frac{\partial e_3(\boldsymbol{X})}{\partial x_1} & \frac{\partial e_3(\boldsymbol{X})}{\partial x_2} & \cdots & \frac{\partial e_3(\boldsymbol{X})}{\partial x_n} \\ \frac{\partial e_4(\boldsymbol{X})}{\partial x_1} & \frac{\partial e_4(\boldsymbol{X})}{\partial x_2} & \cdots & \frac{\partial e_4(\boldsymbol{X})}{\partial x_n} \\ \vdots & \vdots & \vdots & \vdots \\ \frac{\partial e_m(\boldsymbol{X})}{\partial x_1} & \frac{\partial e_m(\boldsymbol{X})}{\partial x_2} & \cdots & \frac{\partial e_m(\boldsymbol{X})}{\partial x_n} \end{bmatrix}$$

The error vector at $X + \Delta X$ can be approximated using a first-order Taylor series:

$$e(\boldsymbol{X} + \Delta \boldsymbol{X}) \approx e(\boldsymbol{X}) + \boldsymbol{J}\Delta \boldsymbol{X}$$

The gradient of $E(X + \Delta X)$ becomes in this case

$$\nabla E(\boldsymbol{X} + \Delta \boldsymbol{X}) \approx 2\boldsymbol{J}^T[e(\boldsymbol{X}) + \boldsymbol{J}\Delta \boldsymbol{X}]$$

To minimize the single-valued function $E(\boldsymbol{X})$ one makes the gradient $E(X + \Delta X)$ equal to zero:

$$2\boldsymbol{J}^T[e(\boldsymbol{X}) + \boldsymbol{J}\Delta \boldsymbol{X}] = \boldsymbol{J}^T e(\boldsymbol{X}) + \boldsymbol{J}^T\boldsymbol{J}\Delta \boldsymbol{X} = 0$$

This provides conditions on the incremental change vector $\boldsymbol{X}$ that minimizes the single-valued function:

$$\Delta \boldsymbol{X} = -[\boldsymbol{J}^T\boldsymbol{J}]^{-1}\boldsymbol{J}^T e(\boldsymbol{X})$$

The optimization algorithm is then an iterative process. Starting with an initial parameter vector $\boldsymbol{X}$, one computes the appropriate incremental change vector $\Delta \boldsymbol{X}$ based on the Jacobian matrix. Note that the derivatives in the Jacobian matrix are often computed numerically since it is difficult to define closed-form expressions for these derivatives in many of the microwave problems.

The incremental change vector $\Delta \boldsymbol{X}$ is then added to the current parameter vector $\boldsymbol{X}$ to form the parameter vector for the next iteration of the program. The process is then repeated until the algorithm converges to a solution. This happens when $\boldsymbol{X}$ no longer varies between iterations.

Levenberg [1] and Marquardt [2] proposed the addition of a parameter λ in the expression of the incremental change vector ΔX to improve convergence issues of the iterative process. In this book, the practical implementation of the technique proposed by More [3] is used. In this case, the expression for the incremental change vector is given by

$$\Delta X = -\left[J^T J + \lambda D^T D\right]^{-1} J^T e(X)$$

where D is a $n \times n$ diagonal matrix and the parameter λ is computed to satisfy the Taylor series approximation assumption so that

$$\|D\,\Delta X\| \leq \Delta$$

where Δ is a constant greater than 0.

In other words $D\,\Delta X$ is a n column vector and its norm should be kept small at every iteration. Note that at every iteration the Jacobian and D matrices should be computed and a suitable value for Δ must be defined so that the parameter λ can be computed.

The main steps of the optimization algorithm are provided next. They provide the steps to define the incremental change vector ΔX at the kth iteration so that the new parameter vector $X + \Delta X$ can be computed for the next iteration.

B.1.1 Overall Algorithm

1. Given that $\Delta_k > 0$, find $\lambda_k \geq 0$ such that
   $\|D_k\,\Delta X_k\| \leq \Delta_k$ where $\Delta X_k = -\left[J_k^T J_k + \lambda_k D_k^T D_k\right]^{-1} J_k^T e(X_k)$
   if $\lambda_k = 0$ and $\|D_k\,\Delta X_k\| \leq \Delta_k$ then directly take $\Delta X_k = -\left[J_k^T J_k\right]^{-1} J_k^T e(X_k)$
   If not, find λ_k such that $\left\| -D_k \left[J_k^T J_k + \lambda_k D_k^T D_k\right]^{-1} J_k^T e(X_k)\right\| = \Delta_k$
   then take $\Delta X_k = -\left[J_k^T J_k + \lambda_k D_k^T D_k\right]^{-1} J_k^T e(X_k)$
2. If $e(X_k + \Delta X_k) < e(X_k)$ then take $X_{k+1} = X_k + \Delta X_k$ and compute J_{k+1}
   If not then take $X_{k+1} = X_k$ and $J_{k+1} = J_k$
3. Choose Δ_{k+1} and D_{k+1}

The overall algorithm requires three things: How to find λ_k once D_k and Δ_k are given, how to compute D_k, and how to define a suitable value for Δ_k. These procedures are described later.

B.1.2 Finding λ_k When D_k and Δ_k Have Been Defined

The goal is to solve $\left\| D_k \left[J_k^T J_k + \lambda_k D_k^T D_k\right]^{-1} J_k^T e(X_k)\right\| = \Delta_k$ when D_k and Δ_k are known.

First, to simply the notations, the following will be used:

$$D = D_k, J^T = J_k^T, X = X_k, \Delta X = \Delta X_k, \lambda = \lambda_k, \Delta = \Delta_k$$

The problem consists of finding the value of λ that makes the $\phi(\lambda)$ function equal to zero and where $\phi(\lambda)$ is given by:

$$\phi(\lambda) = \|D\,\Delta X\| - \Delta = \left\| D\left[J^T J + \lambda D^T D\right]^{-1} J^T e(X)\right\| - \Delta$$

This function is sketched in Figure B.1.

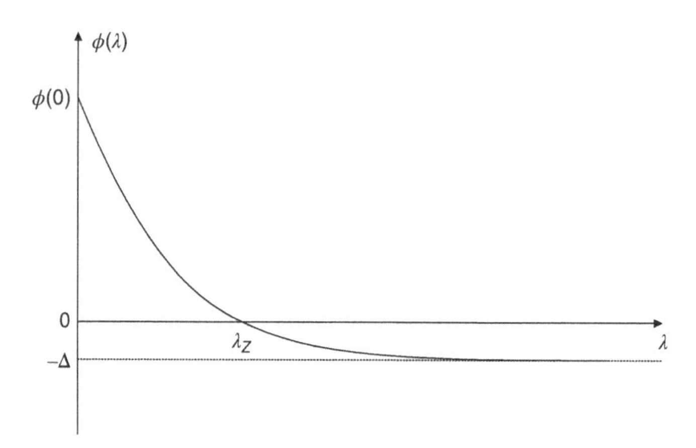

Figure B.1 Finding the zero crossing of the $\phi(\lambda)$ function.

Note that $\phi(\lambda) = \|\boldsymbol{D}\Delta \boldsymbol{X}\| - \Delta$ will be greater than 0 at $\lambda = 0$ since $\|\boldsymbol{D}\Delta \boldsymbol{X}\| > \Delta$ at $\lambda = 0$ according to step 1 of the overall algorithm and that the norm of a vector is 0 or greater. This means that the minimum value of $\phi(\lambda)$ is $-\Delta$. In Figure B.1, the function is shown to be equal to zero at $\lambda = \lambda_Z$ and is the solution that is wanted.

At this stage, one could use a typical single-parameter Newton method to find the zero crossing. This would be an iterative process using an increment such that

$$\lambda_{i+1} = \lambda_i - \frac{\phi(\lambda_i)}{\phi'(\lambda_i)}$$

where $\phi'(\lambda)$ is the first derivative of $\phi(\lambda)$.

However, in Ref. [3], a more robust technique is proposed that can avoid convergence issues of the Newton method for convex functions. More [3] proposes an iterative process using an increment such that

$$\lambda_{i+1} = \lambda_i - \left(\frac{\phi(\lambda_i) + \Delta}{\Delta}\right) \times \frac{\phi(\lambda_i)}{\phi'(\lambda_i)}$$

along with defining a limit interval $[l_i, u_i]$.

In this technique, one starts with $\lambda_0 = 0$ and a starting limit interval $[l_0, u_0]$ such that

$$l_0 = -\frac{\phi(0)}{\phi'(0)} \quad \text{and} \quad u_0 = \frac{\|\boldsymbol{J}\boldsymbol{D}^{-1}e(\boldsymbol{X})\|}{\Delta}$$

The algorithm steps are given below.

B.1.3 Modified Newton Method Algorithm for Computing λ_k

a. If $\lambda_i \notin [l_i, u_i]$ then take $\lambda_i = \max\left(0.001 \times u_i, \sqrt{l_i u_i}\right)$
b. evaluate $\phi(\lambda_i)$ and $\phi'(\lambda_i)$

If $\phi(\lambda_i) < 0$ then take $u_{i+1} = \lambda_i$, otherwise take $u_{i+1} = u_i$

Take $l_{i+1} = \max\left(l_i, \lambda_i - \frac{\phi(\lambda_i)}{\phi'(\lambda_i)}\right)$

c. compute $\lambda_{i+1} = \lambda_i - \left(\frac{\phi(\lambda_i) + \Delta}{\Delta} \right) \times \frac{\phi(\lambda_i)}{\phi'(\lambda_i)}$

Modified step (b):

(b.1) evaluate $\phi(\lambda_i)$ and $\phi'(\lambda_i)$
if $\phi(\lambda_i) < 0$ then take $u_{i+1} = \alpha_i$, otherwise take $u_{i+1} = u_i$

(b.2) compute $\beta_i = \lambda_i - \frac{\phi(\lambda_i)}{\phi'(\lambda_i)}$

(b.2.1) if $\beta_i > 0$ then
if $l_i > \beta_i$ and $\phi(\lambda_i) > 0$ then take $l_{i+1} = l_i$
otherwise take $l_{i+1} = \beta_i$

(b.2.2) if $\beta_i < 0$ then take $l_i = 2l_{i-1}$ and $\lambda_i = 2\lambda_{i-1}$ and go back to step (b.1)

(b.2.3) if $l_i > u_i$ then take $l_i = u_i$ and $u_i = l_i$ (swap them)

The derivative of $\phi(\lambda)$ used in the algorithm can be computed using

$$\phi'(\lambda) = -\frac{[\boldsymbol{D}^T\boldsymbol{q}]^T [\boldsymbol{J}^T\boldsymbol{J} + \lambda\boldsymbol{D}^T\boldsymbol{D}]^{-1} [\boldsymbol{D}^T\boldsymbol{q}]}{\|\boldsymbol{q}\|}$$

where $\boldsymbol{q} = \boldsymbol{D}\Delta\boldsymbol{X} = \boldsymbol{D}[\boldsymbol{J}^T\boldsymbol{J} + \lambda\boldsymbol{D}^T\boldsymbol{D}]^{-1}\boldsymbol{J}^T e(\boldsymbol{X})$

B.1.4 Computing the $\boldsymbol{D_k}$ Matrix

$\boldsymbol{D_k}$ is a $n \times n$ diagonal matrix where n is the number of parameters to optimize.

$$\boldsymbol{D_k} = \begin{bmatrix} d_1^k & & & 0 \\ & d_2^k & & \\ & & \ddots & \\ 0 & & & d_n^k \end{bmatrix}$$

More [3] proposed that the diagonal elements be given by

$$d_i^k = \max\left\{ d_i^{k-1}, \left\| \frac{\partial e(\boldsymbol{X_k})}{\partial x_i} \right\| \right\} k \geq 1$$

where d_i^{k-1} is the ith diagonal element from the previous iteration (i.e., $k-1$). One takes the maximum between d_i^{k-1} and the norm of the ith column of the Jacobian matrix.

B.1.4.1 Defining the Value of Δ

In order to define a suitable value for Δ, one considers the quantity:

$$\rho(\boldsymbol{X}) = \frac{\|e(\boldsymbol{X})\|^2 - \|e(\boldsymbol{X} + \Delta\boldsymbol{X})\|^2}{\|e(\boldsymbol{X})\|^2 - \|e(\boldsymbol{X}) + \boldsymbol{J}\Delta\boldsymbol{X}\|^2}$$

This quantity can show the Taylor series assumption

$$e(X+\Delta X) \approx e(X) + J\Delta X$$

since $\rho(X)=1$ when the Taylor series is an exact approximation. In that case, there is an exact equality:

$$e(X+\Delta X) = e(X) + J\Delta X$$

This quantity can show if the overall error will increase or decrease between iterations since $\rho(X) \leq 0$ when $\|e(X+\Delta X)\| \geq \|e(X)\|$ so that the values taken by $\rho(X)$ can be used to define a suitable value for Δ when evaluating $\|D\Delta X\| < \Delta$.

The algorithm and rule of thumb numbers proposed by More [3] are given next.

B.1.4.2 Algorithm for Defining Δ_{k+1}

a. If $\rho(X) < 10^{-4}$ then take $X_{k+1} = X_k$ and $J_{k+1} = J_k$

   if $\rho(X) > 10^{-4}$ then take $X_{k+1} = X_k + \Delta X_k$ and compute J_{k+1}

b. if $\rho(X) < 0.25$ then take $\Delta_{k+1} \in [0.1\Delta_k, 0.5\Delta_k]$

   if $0.25 < \rho(X) < 0.75$ and $\lambda_k = 0$ or if $0.25 < \rho(X)$ and $\lambda_k = 0$ then take $\Delta_{k+1} = 2\|D_k \Delta X_k\|$

References

[1] K. Levenberg, "A method for the solution of certain non-linear problems in least squares," Q. Appl. Math., vol. 2, pp. 164–168, 1944.

[2] D. Marquardt, "An algorithm for least-squares estimation of nonlinear parameters," J. Soc. Ind. Appl. Math., vol. 11, no. 2, pp. 431–441, June 1963.

[3] J. J. More, The Levenberg-Marquardt Algorithm: Implementation and Theory, Lecture Notes in Mathematics 630, Springer Verlag, New York, 1978.

Appendix C

Noise Correlation Matrix

C.1 Definition

A two-port that has internal sources of noise can be modeled as a noiseless two-port with external voltage and current noise sources as shown in Figure C.1.

The noise in the two-port is characterized by its noise correlation matrix such that

$$N = \begin{vmatrix} \overline{ee*} & \overline{ei*} \\ \overline{ie*} & \overline{ii*} \end{vmatrix} \tag{C.1}$$

where

$$\overline{ee*} = 4kTR_n\Delta f \tag{C.2}$$

$$\overline{ei*} = 4kT\rho\Delta f \tag{C.3}$$

$$\overline{ie*} = 4kT\rho*\Delta f \tag{C.4}$$

$$\overline{ee*} = 4kTG_n\Delta f \tag{C.5}$$

R_n is the noise resistance
G_n is the noise conductance
ρ is the correlation coefficient (complex valued)
T is the temperature
k is the Boltzmann constant
Δf is the frequency band

Microwave Amplifier and Active Circuit Design Using the Real Frequency Technique, First Edition.
Pierre Jarry and Jacques N. Beneat.

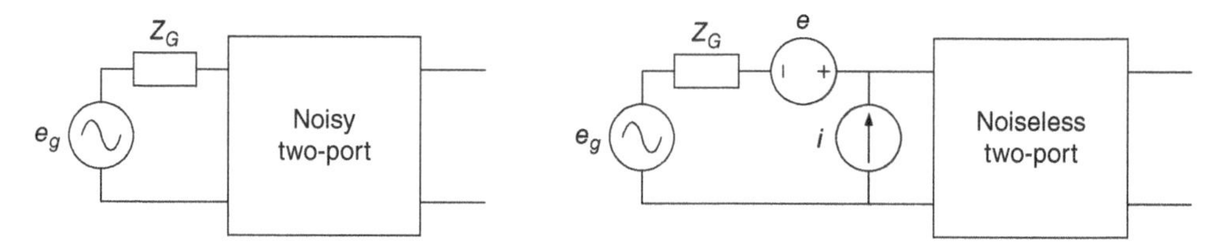

Figure C.1 Modeling a noisy two-port as a noiseless two-port with external noise sources.

In this case, the noise correlation matrix can be expressed as

$$N = 4kT\Delta f \begin{vmatrix} R_n & \rho \\ \rho* & G_n \end{vmatrix} \tag{C.6}$$

In the case of a passive two-port, the elements in the noise correlation matrix are given in terms of the *ABCD* parameters.

$$R_n = \frac{BA* + AB*}{2} \tag{C.7}$$

$$G_n = \frac{DC* + CD*}{2} \tag{C.8}$$

$$\rho = \frac{DA* + CB* - 1}{2} \tag{C.9}$$

These equations assume a noise power at the output of the two-port such that

$$N_{OUT} = \frac{kT\Delta f}{2} \tag{C.10}$$

In the case of a transistor defined by its noise factor

$$F = F_{\min} + \frac{R_n}{G_G} |Y_G - Y_{G\min}|^2 \tag{C.11}$$

where $Y_G = G_G + jB_G$ and $Y_{G\min} = G_{G\min} + jB_{G\min}$ so that

$$F = F_{\min} + \frac{R_n}{G_G}\left[(G_G - G_{G\min})^2 + (B_G - B_{G\min})^2\right] \tag{C.12}$$

Then the elements in the noise correlation matrix are given by

$$R_n = R_n \tag{C.13}$$

$$G_n = \left(G_{G\min}{}^2 + B_{G\min}{}^2\right) R_n \tag{C.14}$$

$$\rho = \left(\frac{F_{\min} - 1}{2} - G_{G\min} R_n\right) + jB_{G\min} R_n \tag{C.15}$$

and where

$$G_{G\min} = Y_0 \frac{1 - |\rho_{\min}|^2}{1 + 2\text{Re}\{\rho_{\min}\} + |\rho_{\min}|^2} \tag{C.16}$$

$$B_{G\min} = -Y_0 \frac{2\text{Im}\{\rho_{\min}\}}{1 + 2\text{Re}\{\rho_{\min}\} + |\rho_{\min}|^2} \tag{C.17}$$

when $\rho_{\min}$ is the reflection coefficient

$$\rho_{\min} = \frac{Y_0 - Y_{G\min}}{Y_0 + Y_{G\min}} \tag{C.18}$$

C.2 Noise Correlation Matrix for Interconnection of Two Two-ports

When interconnecting two systems as shown in Figure C.2 characterized by their noise correlation matrix N_1 and N_2, we are interested in defining the noise correlation matrix N of a single two-port representing this interconnection as seen in Figure C.3.

The first two-port has a noise correlation matrix

$$N_1 = \begin{vmatrix} \overline{e_1 e_1*} & \overline{e_1 i_1*} \\ \overline{i_1 e_1*} & \overline{i_1 i_1*} \end{vmatrix}$$

The second two-port has a noise correlation matrix

$$N_2 = \begin{vmatrix} \overline{e_2 e_2*} & \overline{e_2 i_2*} \\ \overline{i_2 e_2*} & \overline{i_2 i_2*} \end{vmatrix}$$

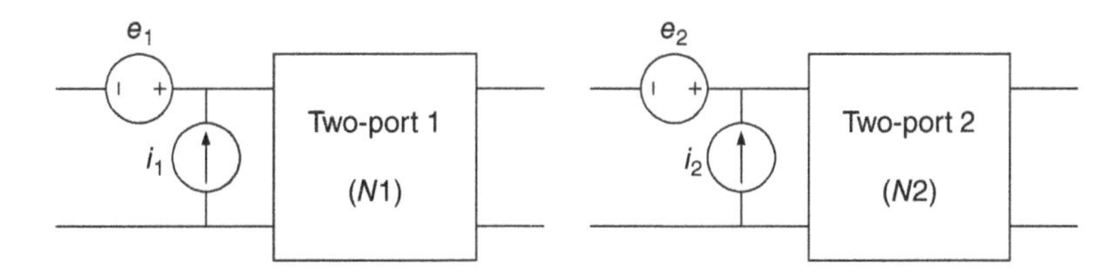

Figure C.2 Two noisy two-ports.

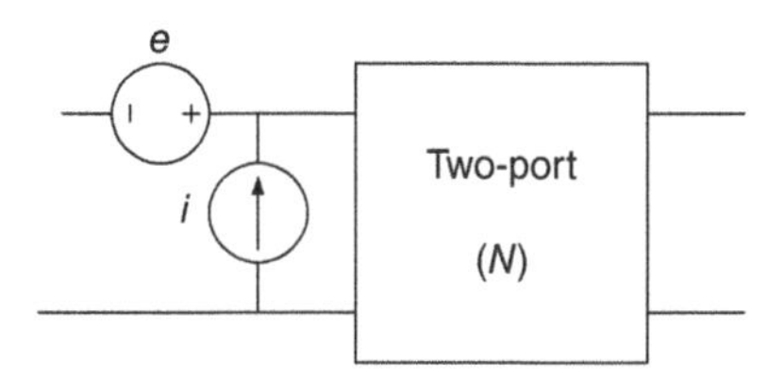

Figure C.3 Equivalent noisy two-port.

Since the expressions can become very tedious we define general interconnection parameters a_1, b_1, c_1, d_1 and a_2, b_2, c_2, d_2 such that

$$e = a_1 e_1 + b_1 i_1 + a_2 e_2 + b_2 i_2 \tag{C.19}$$

$$i = c_1 e_1 + d_1 i_1 + c_2 e_2 + d_2 i_2 \tag{C.20}$$

In the case when the noise sources of the first two-port are not correlated with the noise sources of the second two-port, the noise correlation matrix of the interconnection becomes

$$N = \begin{vmatrix} \overline{ee*} & \overline{ei*} \\ \overline{ie*} & \overline{ii*} \end{vmatrix} \tag{C.21}$$

where

$$\begin{aligned} \overline{ee^*} &= a_1 a_1^* \overline{e_1 e_1^*} + b_1 b_1^* \overline{i_1 i_1^*} + a_2 a_2^* \overline{e_2 e_2^*} + b_2 b_2^* \overline{i_2 i_2^*} \\ &+ a_1 b_1^* \overline{e_1 i_1^*} + b_1 a_1^* \overline{i_1 e_1^*} + a_2 b_2^* \overline{e_2 i_2^*} + b_2 a_2^* \overline{i_2 e_2^*} \end{aligned} \tag{C.22}$$

$$\begin{aligned} \overline{ii^*} &= c_1 c_1^* \overline{e_1 e_1^*} + d_1 d_1^* \overline{i_1 i_1^*} + c_2 c_2^* \overline{e_2 e_2^*} + d_2 d_2^* \overline{i_2 i_2^*} \\ &+ c_1 d_1^* \overline{e_1 i_1^*} + d_1 c_1^* \overline{i_1 e_1^*} + c_2 d_2^* \overline{e_2 i_2^*} + d_2 c_2^* \overline{i_2 e_2^*} \end{aligned} \tag{C.23}$$

$$\begin{aligned} \overline{ei^*} &= a_1 c_1^* \overline{e_1 e_1^*} + b_1 d_1^* \overline{i_1 i_1^*} + a_2 c_2^* \overline{e_2 e_2^*} + b_2 d_2^* \overline{i_2 i_2^*} \\ &+ a_1 d_1^* \overline{e_1 i_1^*} + b_1 c_1^* \overline{i_1 e_1^*} + a_2 d_2^* \overline{e_2 i_2^*} + b_2 c_2^* \overline{i_2 e_2^*} = \left(\overline{ie^*}\right)^* \end{aligned} \tag{C.24}$$

To follow, we provide the general interconnection parameters a_1, b_1, c_1, d_1 and a_2, b_2, c_2, d_2 for a few typical interconnection situations encountered in practice.

C.2.1 Noise Correlation Matrix for Two Two-Ports in Cascade

Figure C.4 shows the case of two systems connected in cascade.

The system equations are

$$e = e_1 + A_1 e_2 + B_1 i_2 \tag{C.25}$$

$$i = i_1 + C_1 e_2 + D_1 i_2 \tag{C.26}$$

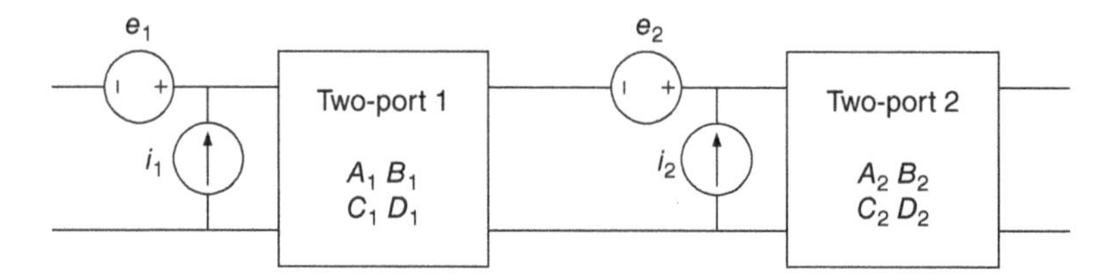

Figure C.4 Two systems connected in cascade.

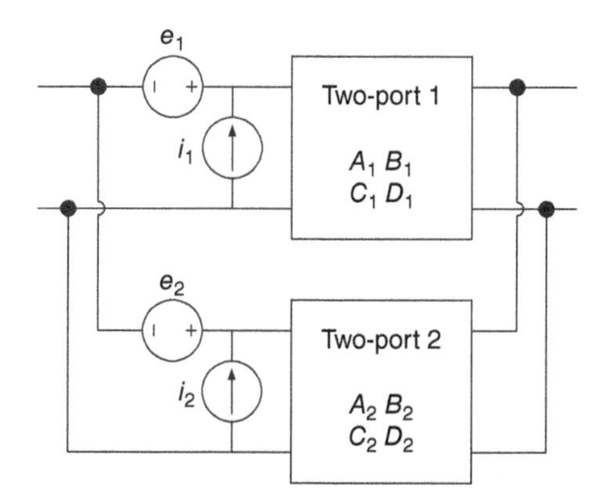

Figure C.5 Two systems connected in parallel.

and the interconnection parameters are given by

$a_1 = 1$	$b_1 = 0$	$c_1 = 0$	$d_1 = 1$
$a_2 = A_1$	$b_2 = B_1$	$c_2 = C_1$	$d_2 = D_1$

C.2.2 Noise Correlation Matrix for Two Two-Ports in Parallel

Figure C.5 shows the case of two systems connected in parallel.

The system equations are

$$e = \frac{B_2}{B_1 + B_2} e_1 + \frac{B_1}{B_1 + B_2} e_2 \tag{C.27}$$

$$i = \frac{D_2 - D_1}{B_1 + B_2} e_1 + i_1 + \frac{D_1 - D_2}{B_1 + B_2} e_2 + i_2 \tag{C.28}$$

and the interconnection parameters are given by

$a_1 = \frac{B_2}{B_1 + B_2}$	$b_1 = 0$	$c_1 = \frac{D_2 - D_1}{B_1 + B_2}$	$d_1 = 1$
$a_2 = \frac{B_1}{B_1 + B_2}$	$b_2 = 0$	$c_2 = \frac{D_1 - D_2}{B_1 + B_2}$	$d_2 = 1$

C.2.3 Noise Correlation Matrix for Two Two-Ports in Series

Figure C.6 shows the case of two systems connected in series.

The system equations are

$$e = e_1 + \frac{A_2 - A_1}{C_1 + C_2} i_1 + e_2 + \frac{A_1 - A_2}{C_1 + C_2} i_2 \tag{C.29}$$

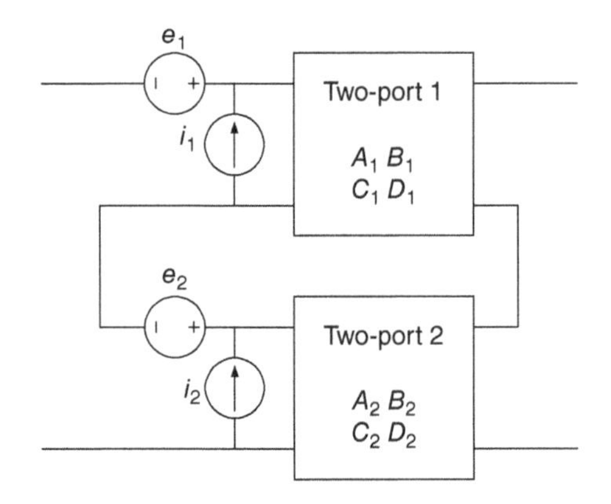

Figure C.6 Two systems connected in series.

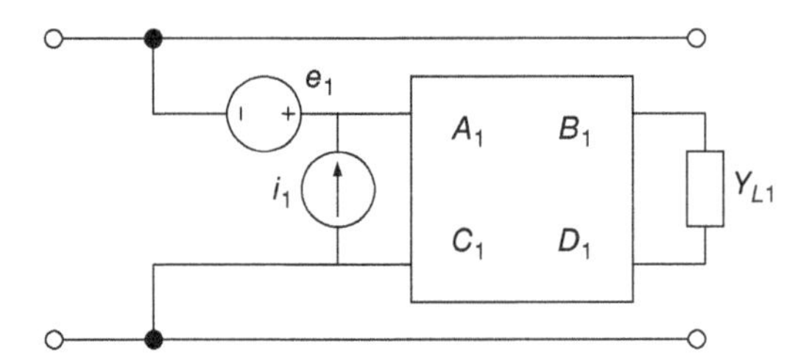

Figure C.7 Two-port system made of a loaded two-port connected in parallel.

$$i = \frac{C_2}{C_1 + C_2} i_1 + \frac{C_1}{C_1 + C_2} i_2 \tag{C.30}$$

and the interconnection parameters are given by

$a_1 = 1$	$b_1 = \frac{A_2 - A_1}{C_1 + C_2}$	$c_1 = 0$	$d_1 = \frac{C_2}{C_1 + C_2}$
$a_2 = 1$	$b_2 = \frac{A_1 - A_2}{C_1 + C_2}$	$c_2 = 0$	$d_2 = \frac{C_1}{C_1 + C_2}$

C.3 Two-port System Made of a Loaded Two-Port Connected in Parallel

Figure C.7 shows the case of a two-port system made of a loaded two-port connected in parallel.

The system equations are

$$\begin{cases} e = 0 \\ i = -\frac{C_1 + D_1 Y_{L1}}{A_1 + B_1 Y_{L1}} e_1 + i_1 + \frac{A_1 D_1 - B_1 C_1}{A_1 + B_1 Y_{L1}} i_{L1} \\ \overline{i_{L1} i_{L1}^*} = 4kT\Delta f \mathrm{Re}\{Y_{L1}\} \end{cases} \tag{C.31}$$

and the interconnection parameters are given by

$a_1 = 0$	$b_1 = 0$	$c_1 = -\dfrac{C_1 + D_1 Y_{L1}}{A_1 + B_1 Y_{L1}}$	$d_1 = 1$
$a_2 = 0$	$b_2 = 0$	$c_2 = 0$	$d_2 = 0$

but one should add the term below to the noise conductance.

$$G_{nL1} = \mathrm{Re}\{Y_{L1}\}\left|\frac{A_1 D_1 - B_1 C_1}{A_1 + B_1 Y_{L1}}\right|^2 \tag{C.32}$$

and

$$G_n' = G_n + G_{nL1}$$

C.4 Two-Port System Made of a Loaded Two-port Connected in Series

Figure C.8 shows the case of a two-port system made of a loaded two-port connected in series.

The system equations are

$$\begin{cases} e = e_1 - \dfrac{A_1 Z_{L1} + B_1}{C_1 Z_{L1} + D_1} i_1 + \dfrac{A_1 D_1 - B_1 C_1}{C_1 Z_{L1} + D_1} e_{L1} \\ i = 0 \\ \overline{e_{L1} e_{L1}^*} = 4kT\Delta f \mathrm{Re}\{Z_{L1}\} \end{cases} \tag{C.33}$$

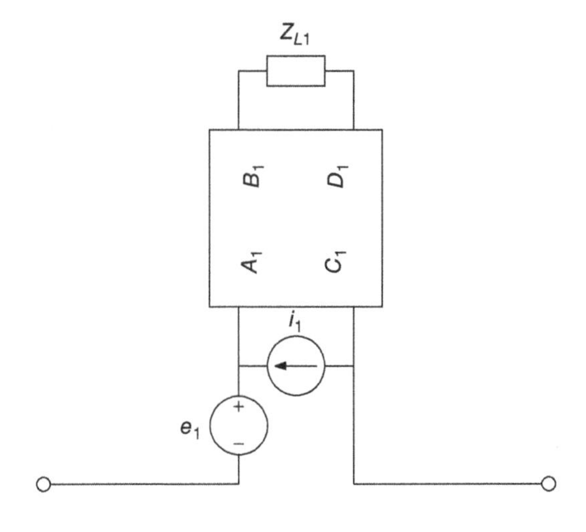

Figure C.8 Two-port system made of a loaded two-port connected in series.

and the interconnection parameters are given by

$a_1 = 1$	$b_1 = -\dfrac{A_1 Z_{L1} + B_1}{C_1 Z_{L1} + D_1}$	$c_1 = 0$	$d_1 = 0$
$a_2 = 0$	$b_2 = 0$	$c_2 = 0$	$d_2 = 0$

but one should add the term below to the noise resistance.

$$R_{nL1} = \mathrm{Re}\{Z_{L1}\}\left|\frac{A_1 D_1 - B_1 C_1}{C_1 Z_{L1} + D_1}\right|^2 \tag{C.34}$$

and

$$R_n' = R_n + R_{nL1}$$

Appendix D

Network Synthesis Using the Transfer Matrix

D.1 Definitions

D.1.1 Transfer Matrix

We first provide the definition of the transfer matrix of a lossless two-port as shown in Figure D.1.

The lossless two-port can be characterized by its scattering matrix such that

$$\begin{pmatrix} b_1 \\ b_2 \end{pmatrix} = \begin{pmatrix} S_{11} & S_{12} \\ S_{21} & S_{22} \end{pmatrix} \begin{pmatrix} a_1 \\ a_2 \end{pmatrix}$$

However, when placing two lossless two-ports in cascade, one cannot simply use matrix multiplication to define the scattering matrix of the cascade. It is then advantageous to use what is called the transfer matrix such that the transfer matrix of two lossless two-ports in cascade will be equal to the multiplication of the two transfer matrices. We define the transfer matrix as

$$\begin{pmatrix} b_1 \\ a_1 \end{pmatrix} = \begin{pmatrix} T_{11} & T_{12} \\ T_{21} & T_{22} \end{pmatrix} \begin{pmatrix} a_2 \\ b_2 \end{pmatrix}$$

The transfer matrix can be constructed from the scattering matrix using the conversion formulas below.

$$T = \begin{pmatrix} \dfrac{S_{12}S_{21} - S_{11}S_{22}}{S_{21}} & \dfrac{S_{11}}{S_{21}} \\ -\dfrac{S_{22}}{S_{21}} & \dfrac{1}{S_{21}} \end{pmatrix} \quad \text{or} \quad T = \frac{1}{S_{21}} \begin{pmatrix} S_{12}S_{21} - S_{11}S_{22} & S_{11} \\ -S_{22} & 1 \end{pmatrix}$$

Microwave Amplifier and Active Circuit Design Using the Real Frequency Technique, First Edition.
Pierre Jarry and Jacques N. Beneat.

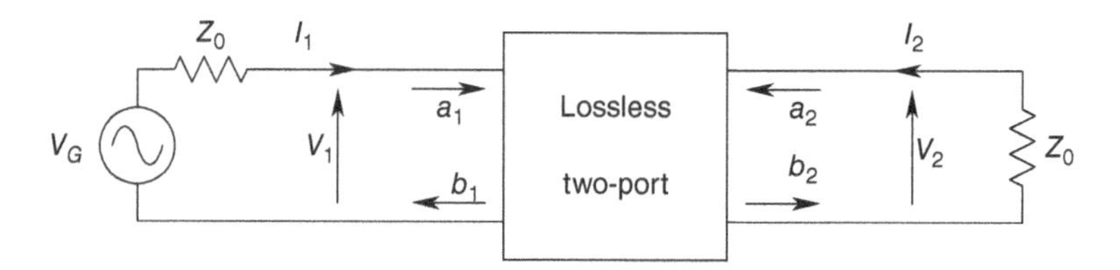

Figure D.1 Lossless two-port and incident and reflected waves.

The scattering matrix can be constructed from the transfer matrix using the conversion formulas below.

$$S=\begin{pmatrix} \dfrac{T_{12}}{T_{22}} & \dfrac{T_{11}T_{22}-T_{12}T_{21}}{T_{22}} \\ \dfrac{1}{T_{22}} & -\dfrac{T_{21}}{T_{22}} \end{pmatrix}$$

D.1.2 *Transfer Matrix of a Lossless Reciprocal Two-Port*

For a lossless and reciprocal two-port, the scattering matrix can be represented using three polynomials $f(s)$, $g(s)$, and $h(s)$ such that

$$S=\begin{pmatrix} \dfrac{h(s)}{g(s)} & \dfrac{f(s)}{g(s)} \\ \dfrac{f(s)}{g(s)} & \dfrac{-\varepsilon h(-s)}{g(s)} \end{pmatrix}$$

where $f(s)$ is an even or odd function. If $f(s)$ is even then $\varepsilon=1$ and if $f(s)$ is odd then $\varepsilon=-1$. In addition, the lossless condition $|e_{11}|^2+|e_{12}|^2=1$ leads to

$$g(s)g(-s)=h(s)h(-s)+f(s)f(-s)$$

Using the conversion formulas above, we find that the transfer matrix of this two-port is given by

$$T=\frac{g(s)}{f(s)}\begin{pmatrix} \dfrac{f(s)f(s)}{g(s)g(s)}-\dfrac{-\varepsilon h(-s)}{g(s)}\dfrac{h(s)}{g(s)} & \dfrac{h(s)}{g(s)} \\ \dfrac{\varepsilon h(-s)}{g(s)} & 1 \end{pmatrix}$$

or

$$T=\frac{1}{f(s)}\begin{pmatrix} \dfrac{f(s)f(s)+\varepsilon h(s)h(-s)}{g(s)} & h(s) \\ \varepsilon h(-s) & g(s) \end{pmatrix}$$

and from the lossless condition

$$g(-s)=\frac{f(s)f(-s)+h(s)h(-s)}{g(s)}$$

In the case $f(s)=f(-s)$ is even so that $\varepsilon=1$, we have

$$\frac{f(s)f(s)+\varepsilon h(s)h(-s)}{g(s)}=\frac{f(s)f(-s)+h(s)h(-s)}{g(s)} \quad \text{and} \quad T=\frac{1}{f(s)}\begin{pmatrix} g(-s) & h(s) \\ \varepsilon h(-s) & g(s) \end{pmatrix}$$

In the case $f(s)=-f(-s)$ is odd so that $\varepsilon=-1$, we have

$$\frac{f(s)f(s)+\varepsilon h(s)h(-s)}{g(s)}=\frac{-f(s)f(-s)-h(s)h(-s)}{g(s)} \quad \text{and} \quad T=\frac{1}{f(s)}\begin{pmatrix} -g(-s) & h(s) \\ \varepsilon h(-s) & g(s) \end{pmatrix}$$

Since the sign of $g(-s)$ follows ε, the transfer matrix of the reciprocal lossless two-port can be expressed as

$$T=\frac{1}{f(s)}\begin{pmatrix} \varepsilon g(-s) & h(s) \\ \varepsilon h(-s) & g(s) \end{pmatrix}$$

Next, we define the transfer matrix of the some common elements found in practice. The technique is based on converting the *ABCD* matrix of the element to the scattering matrix and then to the transfer matrix.

D.1.3 Transfer Matrix of Common Elements

The scattering matrix can be constructed from the *ABCD* matrix for terminations Z_0 as shown in Figure D.1 using the conversion formulas:

$$\begin{pmatrix} S_{11} & S_{12} \\ S_{21} & S_{22} \end{pmatrix}=\begin{pmatrix} \dfrac{A+BY_0-CZ_0-D}{A+BY_0+CZ_0+D} & \dfrac{2(AD-BC)}{A+BY_0+CZ_0+D} \\ \dfrac{2}{A+BY_0+CZ_0+D} & \dfrac{-A+BY_0-CZ_0+D}{A+BY_0+CZ_0+D} \end{pmatrix}$$

In the next section, we provide the results for the normalized to 1 Ω case so that $Z_0=1$ and $Y_0=1$. In this case, the scattering matrix is given in terms of the *ABCD* matrix by:

$$\begin{pmatrix} S_{11} & S_{12} \\ S_{21} & S_{22} \end{pmatrix}=\begin{pmatrix} \dfrac{A+B-C-D}{A+B+C+D} & \dfrac{2(AD-BC)}{A+B+C+D} \\ \dfrac{2}{A+B+C+D} & \dfrac{-A+B-C+D}{A+B+C+D} \end{pmatrix}$$

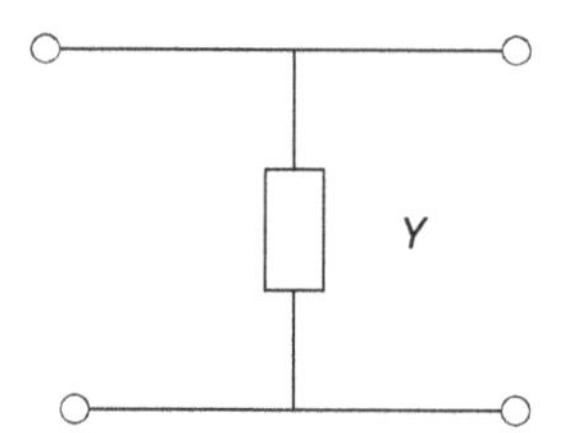

Figure D.2 Parallel admittance.

D.1.3.1 Transfer Matrix of a Parallel Admittance

We are interested in defining the transfer matrix of a parallel admittance as shown in Figure D.2.

In this case we have

$$\begin{pmatrix} A & B \\ C & D \end{pmatrix} = \begin{pmatrix} 1 & 0 \\ Y & 1 \end{pmatrix}$$

For the case of a parallel admittance Y we use $A=1$, $B=0$, $D=1$, and $C=y$, where y is the normalized to 1 Ω admittance (i.e., $y=Y/Y_0$) where Y_0 is the characteristic admittance (e.g., $Z_0=50\Omega$) for applying the $ABCD$ to scattering matrix conversion formulas. In this case, the scattering matrix normalized to 1 Ω is given by

$$\begin{pmatrix} S_{11} & S_{12} \\ S_{21} & S_{22} \end{pmatrix} = \begin{pmatrix} \dfrac{-y}{2+y} & \dfrac{2}{2+y} \\ \dfrac{2}{2+y} & \dfrac{-y}{2+y} \end{pmatrix}$$

and using the scattering matrix to transfer matrix conversion formulas, the transfer matrix normalized to 1 Ω is given by

$$T = \begin{pmatrix} \dfrac{S_{12}S_{21}-S_{11}S_{22}}{S_{21}} & \dfrac{S_{11}}{S_{21}} \\ -\dfrac{S_{22}}{S_{21}} & \dfrac{1}{S_{21}} \end{pmatrix} = \begin{pmatrix} \dfrac{\dfrac{2}{2+y}\dfrac{2}{2+y}-\dfrac{-y}{2+y}\dfrac{-y}{2+y}}{\dfrac{2}{2+y}} & \dfrac{\dfrac{-y}{2+y}}{\dfrac{2}{2+y}} \\ -\dfrac{\dfrac{-y}{2+y}}{\dfrac{2}{2+y}} & \dfrac{1}{\dfrac{2}{2+y}} \end{pmatrix} = \begin{pmatrix} \dfrac{2-y}{2} & \dfrac{-y}{2} \\ \dfrac{y}{2} & \dfrac{2+y}{2} \end{pmatrix}$$

D.1.3.2 Transfer Matrix of a Series Impedance

We are interested in defining the transfer matrix of a series impedance as shown in Figure D.3.

In this case we have

$$\begin{pmatrix} A & B \\ C & D \end{pmatrix} = \begin{pmatrix} 1 & Z \\ 0 & 1 \end{pmatrix}$$

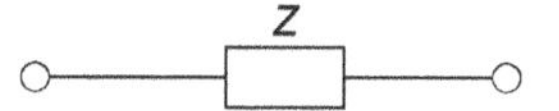

Figure D.3 Series impedance.

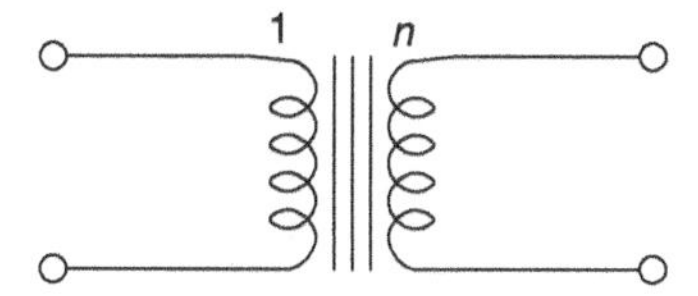

Figure D.4 Ideal transformer.

For the case of a series impedance Z we use $A=1$, $C=0$, $D=1$, and $B=z$, where z is the normalized to 1 Ω admittance (i.e., $z=Z/Z_0$) where Z_0 is the characteristic impedance (e.g., $Z_0=50\Omega$) for applying the *ABCD* to scattering matrix conversion formulas. In this case the scattering matrix normalized to 1 Ω is given by

$$\begin{pmatrix} S_{11} & S_{12} \\ S_{21} & S_{22} \end{pmatrix} = \begin{pmatrix} \dfrac{z}{2+z} & \dfrac{2}{2+z} \\ \dfrac{2}{2+z} & \dfrac{z}{2+z} \end{pmatrix}$$

and using the scattering matrix to transfer matrix conversion formulas, the transfer matrix normalized to 1 Ω is given by

$$T = \begin{pmatrix} \dfrac{S_{12}S_{21}-S_{11}S_{22}}{S_{21}} & \dfrac{S_{11}}{S_{21}} \\ -\dfrac{S_{22}}{S_{21}} & \dfrac{1}{S_{21}} \end{pmatrix} = \begin{pmatrix} \dfrac{\dfrac{2}{2+z}\dfrac{2}{2+z}-\dfrac{z}{2+z}\dfrac{z}{2+z}}{\dfrac{2}{2+z}} & \dfrac{\dfrac{z}{2+z}}{\dfrac{2}{2+z}} \\ -\dfrac{\dfrac{z}{2+z}}{\dfrac{2}{2+z}} & \dfrac{1}{\dfrac{2}{2+z}} \end{pmatrix} = \begin{pmatrix} \dfrac{2-z}{2} & \dfrac{z}{2} \\ \dfrac{-z}{2} & \dfrac{2+z}{2} \end{pmatrix}$$

D.1.3.3 Transfer Matrix of an Ideal Transformer

We are interested in defining the transfer matrix of an ideal transformer as shown in Figure D.4.

In this case we have

$$\begin{pmatrix} A & B \\ C & D \end{pmatrix} = \begin{pmatrix} \dfrac{1}{n} & 0 \\ 0 & n \end{pmatrix}$$

For the case of an ideal transformer we use $A = 1/n$, $B = 0$, $C = 0$, $D = n$ for applying the *ABCD* to scattering matrix conversion formulas. In this case the scattering matrix normalized to $1\,\Omega$ is given by

$$\begin{pmatrix} S_{11} & S_{12} \\ S_{21} & S_{22} \end{pmatrix} = \begin{pmatrix} \dfrac{1/n-n}{1/n+n} & \dfrac{2}{1/n+n} \\ \dfrac{2}{1/n+n} & \dfrac{-1/n+n}{1/n+n} \end{pmatrix} = \begin{pmatrix} \dfrac{1-n^2}{1+n^2} & \dfrac{2n}{1+n^2} \\ \dfrac{2n}{1+n^2} & \dfrac{-(1-n^2)}{1+n^2} \end{pmatrix}$$

and using the scattering matrix to transfer matrix conversion formulas, the transfer matrix normalized to $1\,\Omega$ is given by

$$T = \begin{pmatrix} \dfrac{S_{12}S_{21} - S_{11}S_{22}}{S_{21}} & \dfrac{S_{11}}{S_{21}} \\ -\dfrac{S_{22}}{S_{21}} & \dfrac{1}{S_{21}} \end{pmatrix} = \begin{pmatrix} \dfrac{\dfrac{2n}{1+n^2}\dfrac{2n}{1+n^2} - \dfrac{-(1-n^2)}{1+n^2}\dfrac{1-n^2}{1+n^2}}{\dfrac{2n}{1+n^2}} & \dfrac{\dfrac{1-n^2}{1+n^2}}{\dfrac{2n}{1+n^2}} \\ -\dfrac{\dfrac{-(1-n^2)}{1+n^2}}{\dfrac{2n}{1+n^2}} & \dfrac{1}{\dfrac{2n}{1+n^2}} \end{pmatrix} = \begin{pmatrix} \dfrac{1+n^2}{2n} & \dfrac{1-n^2}{2n} \\ \dfrac{1-n^2}{2n} & \dfrac{1+n^2}{2n} \end{pmatrix}$$

D.2 Extracting a Transfer Matrix

We are interested first in defining the transfer matrix of the cascade of two systems characterized by their transfer matrices. The transfer matrix of the cascade is then equal to the multiplication of the two transfer matrices.

$$T = T_1 T_2$$

where

$$T_1 = \frac{1}{f_1(s)} \begin{pmatrix} \varepsilon g_1(-s) & h_1(s) \\ \varepsilon h_1(-s) & g_1(s) \end{pmatrix} \quad T_2 = \frac{1}{f_2(s)} \begin{pmatrix} \varepsilon g_2(-s) & h_2(s) \\ \varepsilon h_2(-s) & g_2(s) \end{pmatrix}$$

Using the multiplication property of two matrices, we find that

$$T = \frac{1}{f_1(s) f_2(s)} \begin{pmatrix} \varepsilon g_1(-s)\varepsilon g_2(-s) + h_1(s)\varepsilon h_2(-s) & g_2(s)h_1(s) + \varepsilon g_1(-s)h_2(s) \\ g_1(s)\varepsilon h_2(-s) + \varepsilon g_2(-s)\varepsilon h_1(-s) & g_1(s)g_2(s) + \varepsilon h_1(-s)h_2(s) \end{pmatrix}$$

and

$$T = \frac{1}{f_1(s) f_2(s)} \begin{pmatrix} g_1(-s)g_2(-s) + \varepsilon h_1(s)h_2(-s) & g_2(s)h_1(s) + \varepsilon g_1(-s)h_2(s) \\ g_2(-s)h_1(-s) + \varepsilon g_1(s)h_2(-s) & g_1(s)g_2(s) + \varepsilon h_1(-s)h_2(s) \end{pmatrix}$$

and

$$T = \frac{1}{f(s)} \begin{pmatrix} \varepsilon g(-s) & h(s) \\ \varepsilon h(-s) & g(s) \end{pmatrix}$$

with

$$f(s) = f_1(s) f_2(s)$$
$$g(s) = g_1(s) g_2(s) + \varepsilon h_1(-s) h_2(s)$$
$$h(s) = g_2(s) h_1(s) + \varepsilon g_1(-s) h_2(s)$$

We are also interested in finding the inverse transfer matrix of a lossless reciprocal two-port. The lossless reciprocal two-port has a transfer matrix of the form

$$T = \frac{1}{f(s)} \begin{pmatrix} \varepsilon g(-s) & h(s) \\ \varepsilon h(-s) & g(s) \end{pmatrix}$$

Using the inverse property of a matrix, we find that

$$T^{-1} = f(s) \begin{pmatrix} \dfrac{g(s)}{\varepsilon g(s) g(-s) - \varepsilon h(s) h(-s)} & \dfrac{-h(s)}{\varepsilon g(s) g(-s) - \varepsilon h(s) h(-s)} \\ \dfrac{-\varepsilon h(-s)}{\varepsilon g(s) g(-s) - \varepsilon h(s) h(-s)} & \dfrac{\varepsilon g(-s)}{\varepsilon g(s) g(-s) - \varepsilon h(s) h(-s)} \end{pmatrix}$$

At this point, we note that from the lossless condition

$$g(s) g(-s) - h(s) h(-s) = f(s) f(-s)$$

and we can express the inverse transfer matrix as

$$T^{-1} = \frac{f(s)}{\varepsilon f(s) f(-s)} \begin{pmatrix} g(s) & -h(s) \\ -\varepsilon h(-s) & \varepsilon g(-s) \end{pmatrix}$$

and since $\varepsilon = 1$ for $f(s) = f(-s)$ even, and $\varepsilon = -1$ for $f(s) = -f(-s)$ odd, we can express the inverse transfer matrix of a lossless reciprocal two-port as

$$T^{-1} = \frac{1}{f(s)} \begin{pmatrix} g(s) & -h(s) \\ -\varepsilon h(-s) & \varepsilon g(-s) \end{pmatrix}$$

We are now interested in extracting a transfer matrix named T_e from a starting transfer matrix named T and also defining the remaining transfer matrix T_r. This means that the matrix T will be equal to the multiplication of matrix T_e by the matrix T_r in that order since matrix multiplication is not commutative. Therefore, we have

$$T = T_e T_r$$

and the remaining transfer matrix T_r will be given by the multiplication of the inverse transfer matrix T_e^{-1} by transfer matrix T such that

$$T_r = T_e^{-1} T$$

and where

$$T_e = \frac{1}{f_e(s)} \begin{pmatrix} \varepsilon g_e(-s) & h_e(s) \\ \varepsilon h_e(-s) & g_e(s) \end{pmatrix} \quad \text{and} \quad T_r = \frac{1}{f_r(s)} \begin{pmatrix} \varepsilon g_r(-s) & h_r(s) \\ \varepsilon h_r(-s) & g_r(s) \end{pmatrix}$$

and

$$T_e^{-1} = \frac{1}{f_e(s)} \begin{pmatrix} g_e(s) & -h_e(s) \\ -\varepsilon h_e(-s) & \varepsilon g_e(-s) \end{pmatrix}$$

and

$$T = \frac{1}{f(s)} \begin{pmatrix} \varepsilon g(-s) & h(s) \\ \varepsilon h(-s) & g(s) \end{pmatrix}$$

applying directly the multiplication of two transfer matrices taking $T_1 = T_e^{-1}$ and $T_2 = T$, we have for transfer matrix T_r

$$f_r(s) = f_1(s) f_2(s) = f_e(s) f(s)$$

$$g_r(s) = g_1(s) g_2(s) + \varepsilon h_1(-s) h_2(s) = \varepsilon g_e(-s) g(s) - \varepsilon h_e(-s) h(s)$$

$$h_r(s) = g_2(s) h_1(s) + \varepsilon g_1(-s) h_2(s) = -g(s) h_e(s) + g_e(s) h(s) = g_e(s) h(s) - h_e(s) g(s)$$

Note that when computing $T = T_e T_r$, one has $f(s) = f_e(s) f_r(s) = f_e^2(s) f(s)$ and not $f(s)$, so we need to scale $f_r(s)$ by $1/f_e^2(s)$ to ensure that $T = T_e T_r$.

The remaining transfer T_r matrix that will be used in rest of this appendix is therefore given by

$$f_r(s) = \frac{f(s)}{f_e(s)}$$

$$g_r(s) = \frac{\varepsilon g_e(-s) g(s) - \varepsilon h_e(-s) h(s)}{f_e^2(s)}$$

$$h_r(s) = \frac{g_e(s) h(s) - h_e(s) g(s)}{f_e^2(s)}$$

In order for the remaining transfer matrix T_r to have its coefficients $f_r(s)$, $g_r(s)$, and $h_r(s)$ remain polynomials, several conditions will need to be satisfied when extracting a circuit element from the starting transfer matrix T.

In Chapter 7, we have limited $f(s)$ of an equalizer to be of the form:

$$f(s) = s^k \prod_{i=1}^{m} \left(s^2 + \omega_i^2\right)$$

and we assume

$$h(s) = h_0 + h_1 s + \cdots + h_{n-1} s^{n-1} + h_n s^n$$

$$g(s) = g_0 + g_1 s + \cdots + g_{n-1} s^{n-1} + g_n s^n$$

The equalizer is also a lossless network so that

$$g(s)g(-s) = h(s)h(-s) + f(s)f(-s)$$

Under these conditions, we can see that

if $n > k + m$ then the terms at s^{2n} provide the equality

$$g_n^2 = h_n^2 \quad \text{or} \quad g_n = \pm h_n$$

if $n > 2 + k + m$ then the terms at s^{2n-2} provide the equality

$$2 g_n g_{n-2} - g_{n-1}^2 = 2 h_n h_{n-2} - h_{n-1}^2$$

if $k = 1$ then the terms at s^0 provide the equality

$$g_0^2 = h_0^2 \quad \text{or} \quad g_0 = \pm h_0$$

if $k = 2$ then the terms at s^2 provide the equality

$$2 g_2 g_0 - g_1^2 = 2 h_2 h_0 - h_1^2$$

These properties will provide some indications on the type of element to be extracted.

D.3 Network Synthesis

D.3.1 Extracting a Series Inductor

We are interested in defining how to extract the series inductor shown in Figure D.5.

The transfer matrix of this case gives

$$T_e = \begin{pmatrix} 1 - \dfrac{Ls}{2} & \dfrac{Ls}{2} \\ -\dfrac{Ls}{2} & 1 + \dfrac{Ls}{2} \end{pmatrix}$$

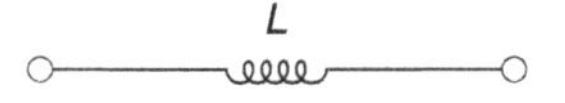

Figure D.5 Series inductor.

so that

$$f_e(s) = 1$$
$$g_e(s) = 1 + \frac{Ls}{2}$$
$$h_e(s) = \frac{Ls}{2}$$
$$\varepsilon = 1$$

and applying the general extraction formulas of Section **D.2**, we find that

$$f_r(s) = \frac{f(s)}{f_e(s)} = f(s)$$

$$g_r(s) = \frac{\varepsilon g_e(-s)g(s) - \varepsilon h_e(-s)h(s)}{f_e^2(s)} = \frac{\left(1 - \frac{Ls}{2}\right)g(s) - \left(-\frac{Ls}{2}\right)h(s)}{1} = g(s) + \frac{Ls}{2}(h(s) - g(s))$$

$$h_r(s) = \frac{g_e(s)h(s) - h_e(s)g(s)}{f_e^2(s)} = \frac{\left(1 + \frac{Ls}{2}\right)h(s) - \frac{Ls}{2}g(s)}{1} = h(s) + \frac{Ls}{2}(h(s) - g(s))$$

In summary

$$f_r(s) = f(s)$$
$$g_r(s) = g(s) + \frac{Ls}{2}(h(s) - g(s))$$
$$h_r(s) = h(s) + \frac{Ls}{2}(h(s) - g(s))$$

Assuming that $g(s)$ and $h(s)$ are polynomials of order n, $g_r(s)$ will have for $(n+1)$th and nth polynomial coefficients

$$g_{rn+1} = \frac{L}{2}(h_n - g_n) \text{ and } g_{rn} = g_n + \frac{L}{2}(h_{n-1} - g_{n-1})$$

In order to reduce the order of the polynomials in the remaining transfer matrix, one must have these two coefficients equal to zero. In the first case, one must have a precondition that $h_n = g_n$, and for the second case, we can take the inductor to be

$$L = \frac{2g_n}{g_{n-1} - h_{n-1}}$$

D.3.2 *Extracting a Parallel Capacitor*

We are interested in defining how to extract the parallel capacitor shown in Figure D.6.

The transfer matrix of this case gives

$$T_e = \begin{pmatrix} 1 - \frac{Cs}{2} & -\frac{Cs}{2} \\ \frac{Cs}{2} & 1 + \frac{Cs}{2} \end{pmatrix}$$

so that

$$\begin{aligned} f_e(s) &= 1 \\ g_e(s) &= 1 + \frac{Cs}{2} \\ h_e(s) &= -\frac{Cs}{2} \\ \varepsilon &= 1 \end{aligned}$$

and applying the general extraction formulas of Section **D.2**, we find that

$$f_r(s) = \frac{f(s)}{f_e(s)} = f(s)$$

$$g_r(s) = \frac{\varepsilon g_e(-s)g(s) - \varepsilon h_e(-s)h(s)}{f_e^2(s)} = \frac{\left(1 - \frac{Cs}{2}\right)g(s) - \left(\frac{Cs}{2}\right)h(s)}{1} = g(s) - \frac{Cs}{2}(h(s) + g(s))$$

$$h_r(s) = \frac{g_e(s)h(s) - h_e(s)g(s)}{f_e^2(s)} = \frac{\left(1 + \frac{Cs}{2}\right)h(s) + \frac{Cs}{2}g(s)}{1} = h(s) + \frac{Cs}{2}(h(s) + g(s))$$

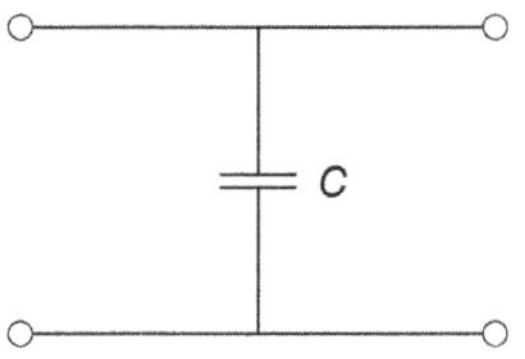

Figure D.6 Parallel capacitor.

In summary

$$f_r(s) = f(s)$$
$$g_r(s) = g(s) - \frac{Cs}{2}(h(s) + g(s))$$
$$h_r(s) = h(s) + \frac{Cs}{2}(h(s) + g(s))$$

Assuming that $g(s)$ and $h(s)$ are polynomials of order n, $g_r(s)$ will have for $(n+1)$th and nth polynomial coefficients

$$g_{rn+1} = -\frac{C}{2}(h_n + g_n) \quad \text{and} \quad g_{rn} = g_n - \frac{C}{2}(h_{n-1} + g_{n-1})$$

In order to reduce the order of the polynomials in the remaining transfer matrix, one must have these two coefficients equal to zero. In the first case, one must have a precondition that $h_n = -g_n$ and for the second case, we can take the capacitor to be

$$C = \frac{2g_n}{g_{n-1} + h_{n-1}}$$

D.3.3 Extracting a Parallel Inductor

We are interested in defining how to extract the parallel inductor shown in Figure D.7.
The transfer matrix of this case gives

$$T_e = \frac{1}{s}\begin{pmatrix} s - \frac{1}{2L} & -\frac{1}{2L} \\ \frac{1}{2L} & s + \frac{1}{2L} \end{pmatrix}$$

so that

$$f_e(s) = s$$
$$g_e(s) = s + \frac{1}{2L}$$
$$h_e(s) = -\frac{1}{2L}$$
$$\varepsilon = -1$$

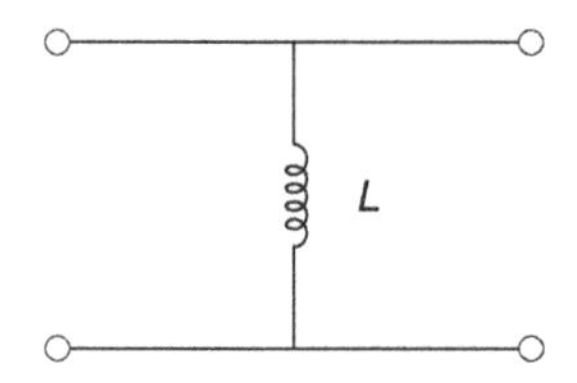

Figure D.7 Parallel inductor.

and applying the general extraction formulas of Section **D.2**, we find that

$$f_r(s) = \frac{f(s)}{f_e(s)} = \frac{f(s)}{s}$$

$$g_r(s) = \frac{\varepsilon g_e(-s)g(s) - \varepsilon h_e(-s)h(s)}{f_e^2(s)} = \frac{-\left(-s+\frac{1}{2L}\right)g(s) + \left(-\frac{1}{2L}\right)h(s)}{s^2} = \frac{sg(s) - \frac{1}{2L}(h(s)+g(s))}{s^2}$$

$$h_r(s) = \frac{g_e(s)h(s) - h_e(s)g(s)}{f_e^2(s)} = \frac{\left(s+\frac{1}{2L}\right)h(s) + \frac{1}{2L}g(s)}{s^2} = \frac{sh(s) + \frac{1}{2L}(h(s)+g(s))}{s^2}$$

In summary

$$f_r(s) = \frac{f(s)}{s}$$

$$g_r(s) = \frac{sg(s) - \frac{1}{2L}(h(s)+g(s))}{s^2}$$

$$h_r(s) = \frac{sh(s) + \frac{1}{2L}(h(s)+g(s))}{s^2}$$

In order for $g_r(s)$ to be a polynomial, the numerator $g_{r'}(s) = sg(s) - \frac{1}{2L}(h(s)+g(s))$ should be divisible by s^2. Therefore, the first and second polynomial coefficients of $g_{r'}(s)$ should be zero:

$$g_{r'0} = -\frac{1}{2L}(h_0 + g_0) = 0 \quad \text{and} \quad g_{r'1} = g_0 - \frac{1}{2L}(h_1 + g_1) = 0$$

For the first coefficient, one must have a precondition such that $h_0 = -g_0$, and for the second coefficient, we can take the inductor to be

$$L = \frac{g_1 + h_1}{2g_0}$$

D.3.4 Extracting a Series Capacitor

We are interested in defining how to extract the series capacitor shown in Figure D.8.

The transfer matrix of this case gives

$$T_e = \frac{1}{s}\begin{pmatrix} s - \frac{1}{2C} & \frac{1}{2C} \\ -\frac{1}{2C} & s + \frac{1}{2C} \end{pmatrix}$$

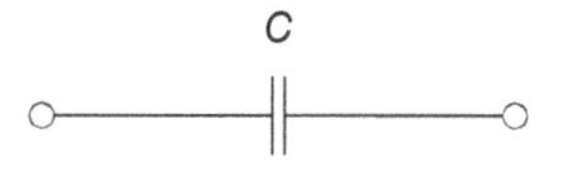

Figure D.8 Series capacitor.

so that

$$\begin{aligned} f_e(s) &= s \\ g_e(s) &= s + \frac{1}{2C} \\ h_e(s) &= \frac{1}{2C} \\ \varepsilon &= -1 \end{aligned}$$

and applying the general extraction formulas of Section **D.2**, we find that

$$f_r(s) = \frac{f(s)}{f_e(s)} = \frac{f(s)}{s}$$

$$g_r(s) = \frac{\varepsilon g_e(-s)g(s) - \varepsilon h_e(-s)h(s)}{f_e^2(s)} = \frac{-\left(-s+\frac{1}{2C}\right)g(s) + \left(\frac{1}{2C}\right)h(s)}{s^2} = \frac{sg(s) + \frac{1}{2C}(h(s)-g(s))}{s^2}$$

$$h_r(s) = \frac{g_e(s)h(s) - h_e(s)g(s)}{f_e^2(s)} = \frac{\left(s+\frac{1}{2C}\right)h(s) - \frac{1}{2C}g(s)}{s^2} = \frac{sh(s) + \frac{1}{2C}(h(s)-g(s))}{s^2}$$

In summary

$$\begin{aligned} f_r(s) &= \frac{f(s)}{s} \\ g_r(s) &= \frac{sg(s) + \frac{1}{2C}(h(s)-g(s))}{s^2} \\ h_r(s) &= \frac{sh(s) + \frac{1}{2C}(h(s)-g(s))}{s^2} \end{aligned}$$

In order for $g_r(s)$ to be a polynomial, the numerator $g_{r'}(s) = sg(s) + \frac{1}{2C}(h(s)-g(s))$ should be divisible by s^2. Therefore, the first and second polynomial coefficients of $g_{r'}(s)$ should be zero:

$$g_{r'0} = \frac{1}{2C}(h_0 - g_0) = 0 \text{ and } g_{r'1} = g_0 + \frac{1}{2C}(h_1 - g_1) = 0$$

For the first coefficient, one must have a precondition such that $h_0 = g_0$, and for the second coefficient, we can take the capacitor to be

$$C = \frac{g_1 - h_1}{2g_0}$$

D.3.5 Extracting a Series LC Cell

We are interested in defining how to extract the series LC cell shown in Figure D.9.

The transfer matrix of this case gives

$$T_e = \frac{1}{s^2+\omega_0^2}\begin{pmatrix} s^2 - \frac{s}{2C} + \omega_0^2 & \frac{s}{2C} \\ -\frac{s}{2C} & s^2 + \frac{s}{2C} + \omega_0^2 \end{pmatrix} \quad \text{with } LC = \frac{1}{\omega_0^2}$$

so that

$$\begin{aligned} f_e(s) &= s^2 + \omega_0^2 \\ g_e(s) &= s^2 + \frac{s}{2C} + \omega_0^2 \\ h_e(s) &= \frac{s}{2C} \\ \varepsilon &= 1 \end{aligned}$$

and applying the general extraction formulas of Section **D.2**, we find that

$$f_r(s) = \frac{f(s)}{f_e(s)} = \frac{f(s)}{s^2+\omega_0^2}$$

$$g_r(s) = \frac{\varepsilon g_e(-s)g(s) - \varepsilon h_e(-s)h(s)}{f_e^2(s)} = \frac{\left(s^2 - \frac{s}{2C} + \omega_0^2\right)g(s) + \frac{s}{2C}h(s)}{\left(s^2+\omega_0^2\right)^2} = \frac{(s^2+\omega_0^2)g(s) + \frac{s}{2C}(h(s)-g(s))}{\left(s^2+\omega_0^2\right)^2}$$

$$h_r(s) = \frac{g_e(s)h(s) - h_e(s)g(s)}{f_e^2(s)} = \frac{\left(s^2 + \frac{s}{2C} + \omega_0^2\right)h(s) - \frac{s}{2C}g(s)}{\left(s^2+\omega_0^2\right)^2} = \frac{(s^2+\omega_0^2)h(s) + \frac{s}{2C}(h(s)-g(s))}{\left(s^2+\omega_0^2\right)^2}$$

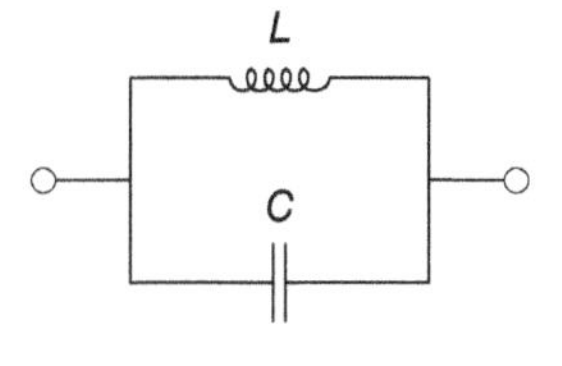

Figure D.9 Series LC cell.

In summary

$$f_r(s) = \frac{f(s)}{s^2 + \omega_0^2}$$

$$g_r(s) = \frac{\left(s^2 + \omega_0^2\right) g(s) + \frac{s}{2C}(h(s) - g(s))}{\left(s^2 + \omega_0^2\right)^2}$$

$$h_r(s) = \frac{\left(s^2 + \omega_0^2\right) h(s) + \frac{s}{2C}(h(s) - g(s))}{\left(s^2 + \omega_0^2\right)^2}$$

Note that

$$g_r(s) = \frac{\left(s^2 + \omega_0^2\right)\left(g(s) + \frac{s}{2C}\frac{(h(s) - g(s))}{\left(s^2 + \omega_0^2\right)}\right)}{\left(s^2 + \omega_0^2\right)^2} = \frac{\left(g(s) + \frac{s}{2C}\frac{h(s) - g(s)}{s^2 + \omega_0^2}\right)}{\left(s^2 + \omega_0^2\right)}$$

In order for $g_r(s)$ to become a polynomial, we need to satisfy two simultaneous conditions:

1. $h(s) - g(s)$ is divisible by the factor $s^2 + \omega_0^2$
2. $g(s) + \frac{s}{2C}\frac{h(s) - g(s)}{s^2 + \omega_0^2}$ is divisible by the factor $s^2 + \omega_0^2$

The first condition is a precondition needed for extracting a series LC cell, and for the second condition, we can make it divisible by the factor $s^2 + \omega_0^2$ by choosing the capacitor value such that it makes it have a root at $s = j\omega_0$:

$$\left[g(s) + \frac{s}{2C}\frac{h(s) - g(s)}{s^2 + \omega_0^2}\right]_{s = j\omega_0} = 0$$

Note that since poles come in complex conjugate pairs for real-valued polynomial coefficients, this condition also makes $s = -j\omega_0$ a root, leading to the term $s^2 + \omega_0^2$. The capacitor to use is then given by

$$C = \left[\frac{s}{s^2 + \omega_0^2}\frac{g(s) - h(s)}{2g(s)}\right]_{s = j\omega_0}$$

and

$$L = \frac{1}{C\omega_0^2}$$

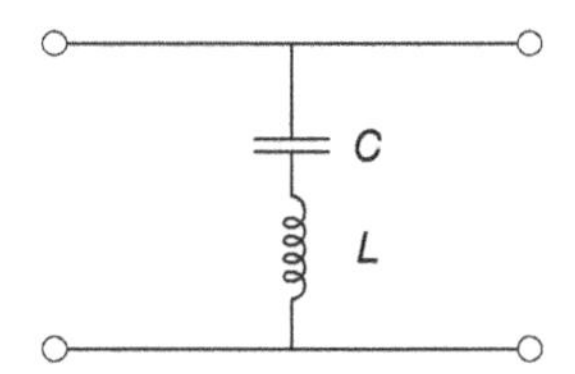

Figure D.10 Parallel *LC* cell.

D.3.6 Extracting a Parallel LC *Cell*

We are interested in defining how to extract the parallel *LC* cell shown in Figure D.10.

The transfer matrix of this case gives

$$T_e = \frac{1}{s^2+\omega_0^2}\begin{pmatrix} s^2 - \dfrac{s}{2L} + \omega_0^2 & -\dfrac{s}{2L} \\ \dfrac{s}{2L} & s^2 + \dfrac{s}{2L} + \omega_0^2 \end{pmatrix} \quad LC = \frac{1}{\omega_0^2}$$

so that

$$\begin{aligned} f_e(s) &= s^2 + \omega_0^2 \\ g_e(s) &= s^2 + \frac{s}{2L} + \omega_0^2 \\ h_e(s) &= -\frac{s}{2L} \\ \varepsilon &= 1 \end{aligned}$$

and applying the general extraction formulas of Section **D.2**, we find that

$$f_r(s) = \frac{f(s)}{f_e(s)} = \frac{f(s)}{s^2+\omega_0^2}$$

$$g_r(s) = \frac{\varepsilon g_e(-s)g(s) - \varepsilon h_e(-s)h(s)}{f_e^2(s)} = \frac{\left(s^2 - \frac{s}{2L} + \omega_0^2\right)g(s) - \frac{s}{2L}h(s)}{\left(s^2+\omega_0^2\right)^2} = \frac{(s^2+\omega_0^2)g(s) - \frac{s}{2L}(h(s)+g(s))}{\left(s^2+\omega_0^2\right)^2}$$

$$h_r(s) = \frac{g_e(s)h(s) - h_e(s)g(s)}{f_e^2(s)} = \frac{\left(s^2 + \frac{s}{2L} + \omega_0^2\right)h(s) + \frac{s}{2L}g(s)}{\left(s^2+\omega_0^2\right)^2} = \frac{(s^2+\omega_0^2)h(s) + \frac{s}{2L}(h(s)+g(s))}{\left(s^2+\omega_0^2\right)^2}$$

In summary

$$f_r(s) = \frac{f(s)}{s^2+\omega_0^2}$$

$$g_r(s) = \frac{(s^2+\omega_0^2)g(s) - \frac{s}{2L}(h(s)+g(s))}{\left(s^2+\omega_0^2\right)^2}$$

$$h_r(s) = \frac{(s^2+\omega_0^2)h(s) + \frac{s}{2L}(h(s)+g(s))}{(s^2+\omega_0^2)^2}$$

Note that

$$g_r(s) = \frac{(s^2+\omega_0^2)\left(g(s) - \frac{s}{2L}\frac{(h(s)+g(s))}{(s^2+\omega_0^2)}\right)}{(s^2+\omega_0^2)^2} = \frac{\left(g(s) - \frac{s}{2L}\frac{h(s)+g(s)}{s^2+\omega_0^2}\right)}{(s^2+\omega_0^2)}$$

In order for $g_r(s)$ to become a polynomial, we need to satisfy two simultaneous conditions:

1. $h(s)+g(s)$ is divisible by the factor $s^2+\omega_0^2$
2. $g(s) - \frac{s}{2L}\frac{h(s)+g(s)}{s^2+\omega_0^2}$ is divisible by the factor $s^2+\omega_0^2$

The first condition is a precondition needed for extracting a parallel *LC* cell, and for the second condition, we can make it divisible by the factor $s^2+\omega_0^2$ by choosing the inductor value such that it makes it have a root at $s=j\omega_0$:

$$\left[g(s) - \frac{s}{2L}\frac{h(s)+g(s)}{s^2+\omega_0^2}\right]_{s=j\omega_0} = 0$$

Note that since poles come in complex conjugate pairs for real-valued polynomial coefficients, this condition also makes $s=-j\omega_0$ a root, leading to the term $s^2+\omega_0{}^2$.

The inductor to use is then given by

$$L = \left[\frac{s}{s^2+\omega_0^2}\frac{g(s)+h(s)}{2g(s)}\right]_{s=j\omega_0}$$

and

$$C = \frac{1}{L\omega_0^2}$$

Index

Microwave Amplifier and Active Circuit Design Using the Real Frequency Technique, First Edition.
Pierre Jarry and Jacques N. Beneat.